Direct Adaptive Control Algorithms:
Theory and Applications

Communications and Control Engineering Series

Editors: B.W. Dickinson · A. Fettweis · J.L. Massey · J.W. Modestino
E.D. Sontag · M. Thoma

CCES published titles include:

Sampled-Data Control Systems
J. Ackermann

Interactive System Identification
T. Bohlin

The Riccatti Equation
S. Bittanti, A.J. Laub and J.C. Willems (Eds.)

Nonlinear Control Systems
A. Isidori

Analysis and Design of Stream Ciphers
R.A. Rueppel

Sliding Modes in Control Optimization
V.I. Utkin

Fundamentals of Robotics
M. Vukobratović

Parametrizations in Control, Estimation and Filtering Problems: Accuracy Aspects
M. Gevers and G. Li

Parallel Algorithms for Optimal Control of Large Scale Linear Systems
Zoran Gajić and Xuemin Shen

Loop Transfer Recovery: Analysis and Design
Ali Saberi, Ben M. Chen and Peddapullaiah Sannuti

Direct Adaptive Control Algorithms: Theory and Applications
Howard Kaufman, Izhak Bar-Kana and Kenneth Sobel

Howard Kaufman Izhak Bar-Kana Kenneth Sobel

Direct Adaptive Control Algorithms:
Theory and Applications

With 154 Figures

Case studies contributed by David S. Bayard and Gregory W. Neat, Pasadena, CA

Springer-Verlag
New York Berlin Heidelberg London Paris
Tokyo Hong Kong Barcelona Budapest

Howard Kaufman
Electrical, Computer, and Systems
Engineering Department
Rensselaer Polytechnic Institute
Troy, NY 12180-3590 USA

Izhak Bar-Kana
Department of Electrical and
Computer Engineering
Drexel University
Philadelphia, PA 19104 USA

Kenneth Sobel
Department of Electrical Engineering
The City College of New York
New York, NY 10031 USA

Library of Congress Cataloging-in-Publication Data
Kaufman, Howard, 1940–
Direct adaptive control algorithms : theory and applications
/ Howard Kaufman, Izhak Bar-Kana, Kenneth Sobel.
p. cm. – (Communications and control engineering series)
Includes bibliographical references and index.
ISBN 0-387-94155-X
1. Adaptive control systems. I. Bar-Kana, Izhak. II. Sobel, Kenneth. III. Title. IV. Series.
TJ217.K38 1994
629.8′36–dc20 93-29789

Printed on acid-free paper.

Production managed by James Harbison; manufacturing supervised by Vincent Scelta.
Camera-ready copy prepared using the authors' TeX files.
Printed and bound by R.R. Donnelley & Sons, Harrisonburg, VA, USA.
Printed in the United States of America.

9 8 7 6 5 4 3 2 1

ISBN 0-387-94155-X Springer-Verlag New York Berlin Heidelberg
ISBN 3-540-94155-X Springer-Verlag Berlin Heidelberg New York

To

Eve, David, Jeffrey, Deborah (HK)

Pnina, Yaniv, Rennan (IBK)

my niece, Jessica, and my friend, E.Y. Shapiro (KS)

and to the memory of

Hilda Stein Kaufman
Rahel and Haim Bercovici
Seymour Sobel
and to Marilyn Sobel who passed away during
the preparation of this book.

Preface

This book is a self contained compendium of simple adaptive control techniques, that have been developed and applied by the authors for over 10 years. The proposed algorithms are easily implementable and do not require explicit process parameter identification. Further, they have been successfully applied to a wide variety of engineering problems including flight control, flexible structure control, blood pressure control, and robotics. In general, these algorithms are suitable for a wide class of multiple input-multiple output control systems containing significant uncertainty as well as disturbances.

The text has been written so that anyone with a basic linear multivariable systems background will be able to develop and apply the adaptive algorithms to their particular problems. Thus, besides developing the theoretical details of the algorithm, the text gives considerable emphasis to the design recipes and to representative applications in flight control, flexible structure control, robotics, and drug infusion control.

The text has been prepared in a manner such that it will be possible for readers to actually use the algorithm even before they thoroughly understand the complete theory. This is the result of a separate emphasis on the algorithm structure, its implementation, and the recommended procedures for tuning. This should appeal to a practicing design engineer. It is also suitable either as a reference or as a text for a graduate course in adaptive control systems.

The key features of the text are

1. Algorithm theory and development

2. Algorithm implementation

3. Algorithm applications

The text is organized to include these items as noted below. Chapter 1 defines the processes, including uncertainties, that will be controlled. Specifications of the desired response in terms of reference model following are discussed, and the need for adaptation is motivated.

The basic theory of model following and adaptive model following is developed in Chapter 2. The passivity constraints for that guarantee asymptotic model following are discussed with supporting additional theory in the Appendix. Procedures for alleviating these restrictions are presented in Chapter 3 along with some illustrative tutorial examples.

Modifications to account for the effects of a wide class of input signal disturbances and noise are discussed in Chapter 4. Some results that account for time variations and certain types of nonlinearitiies are then presented in Chapter 5.

Following the above theoretical development, Chapter 6 summarizes the developed algorithms and discusses their applications. Actual case studies are used for illustration in Chapter 7. These include examples in robotics, flight control, large flexible structure control, and drug infusion.

The reader who is interested primarily in implementation and applications should read:

Chapter 1

Chapter 2, Sections 2.1, 2.2, 2.4.1, 2.4.2, 2.6

Chapter 3, Sections 3.1, 3.2, 3.3.1, 3.3.3, 3.3.4, 3.3.5, 3.4

Chapter 4, Section 4.1, 4.2

Chapter 6

Chapter 7

Acknowledgments

Compilation of this book involved the assistance of many individuals other than the authors. First to be thanked is our super typist Betty Lawson, who braved the New York State Thruway in her daily travels from the scenic Mohawk Valley to the hills of Troy.

Credit for the artwork goes to Rensselaer undergraduate assistants, Laurel Dobert and Julie Caleb. Most of the simulation presentations result from the efforts of our former graduate students. Thanks go to Jeff Sankey and Dave Swift, from Rensselaer and Wangling Yu from City College of New York.

Significant case studies were contributed by Gregory Neat and David Bayard of JPL. In particular, Section 7.2 presents Dave's application of the adaptive algorithms to JPL's Large Spacecraft Control Laboratory Ground Experiment Facility, while Section 7.3 discusses Greg's use of the algorithms for drug delivery control.

We also appreciate the many valuable discussions concerning the text material with Wei Su, GEO Centers Inc., with Amish Mehta and Jeff Sankey at Rensselaer, and with Moshe Kam, Allon Guez, and Ilan Rusnak at Drexel.

Finally we would like to acknowledge that much of the material in this book is based upon research performed by the authors over the past several years. Portions of this research have been sponsored by National Science Foundation Grants ENG-7707446, ECS-9111565, ECS 8016173, ECS-8016225, EET-8620246, and BCS9005678; NASA grants NAGR 33-018-183, NAG-1-171, and NAGW-1333; and the City University of New York PSC-CUNY Research Award Program.

Howard Kaufman
Rensselaer Polytechnic Institute

Izhak Bar-Kana
Drexel University

Kenneth Sobel
City College of New York
1993

Contents

List of Figures

List of Tables

1

Introduction

1.1 Definition of the Problem

This book is a self contained compendium of easily implementable adaptive control algorithms that have been developed and applied by the authors for over 10 years. These algorithms do not require explicit process parameter identification and have been successfully applied to a wide variety of engineering problems including flexible structure control, blood pressure control, and robotics. In general, these algorithms are suitable for a wide class of multiple input-multiple output control systems containing significant uncertainty as well as disturbances.

The text has been written so that persons with a basic linear multivariable systems background will be able to develop and apply the adaptive algorithms to their particular problems. Thus, besides developing the theoretical details of the algorithm, the text gives considerable emphasis to design recipes and to representative applications in flight control, flexible structure control, robotics, and drug infusion control.

The text has been prepared in a manner such that it will be possible for readers to actually use the algorithms even though they do not thoroughly understand the complete theory. This is the result of a separate emphasis on the algorithm structure, its implementation, and the recommended procedures for tuning. This should appeal to a practicing design engineer. It is also suitable either as a reference or as a text for a graduate course in adaptive control systems.

The key features of the text are

- Algorithm theory and development
- Algorithm implementation
- Algorithm applications

Although many definitions of adaptive control have been suggested in the literature [1], [2], [3], [4], [5], a "gold standard" adaptive controller does not appear to exist. However, there does appear to be a set of common considerations in the development of any adaptive controller. These include:

- The selection of a performance measure

- The evaluation of the actual online process behavior with respect to the desired or ideal behavior
- The parameter adjustment procedure used to drive the process towards the desired behavior.

Thus, while fixed conventional and/or robust controllers are designed using *a priori* information, adaptive controllers use both *a priori* and *a posteriori* information.

This book presents a particular adaptive controller that is very simple to understand and implement. With regard to the above considerations, both the performance measure and the evaluation procedure are based upon the concept of following. That is, the desired behavior is defined by the output of a simulated reference model apriori designed to account for standard specifications such as overshoot, settling and rise times, and steady state behavior.

In the development that follows, the system behavior will be evaluated by comparing the process output with the output of a reference model. This instantaneously and easily computed error or difference between the plant and model is in turn used in a set of easily evaluated formulas for adjusting the control gains. This adjustment or adaptation rule successively reduces an index composed of a positive weighted combination of the squared error signals and a positive weighted combination of the differences between the instantaneous gains and their ideal values.

The primary adaptive control problem to be treated in this text is the control of a continuous multivariable linear system subject to uncertainty in the parameters. Some consideration will, however, be given in Chapter 3 to nonlinear and time varying systems. More precisely, the continuous linear model reference control problem is solved for the linear process equations

$$\dot{x}_p(t) = A_p x_p(t) + B_p u_p(t), \tag{1.1}$$

$$y_p(t) = C_p x_p(t), \tag{1.2}$$

where $x_p(t)$ is the $n \times 1$ plant state vector, $u_p(t)$ is the $m \times 1$ control vector, $y_p(t)$ is the $q \times 1$ plant output vector, and A_p and B_p are matrices with the appropriate dimensions. The range of the plant parameters is assumed to be bounded as defined by:

$$\begin{array}{llll} \underline{a}_{ij} \leq a_p(i,j) \leq \bar{a}_{ij}, & i = 1, & \ldots, & n, \\ & j = 1, & \ldots, & n; \\ \underline{b}_{ij} \leq b_p(i,j) \leq b_{ij}, & i = 1, & \ldots, & n, \\ & j = 1, & \ldots, & m; \end{array} \tag{1.3}$$

where $a_p(i,j)$ is the i, jth element of A_p and $b_p(i,j)$ is the i, jth element of B_p. The objective is to find, without explicit knowledge of A_p and B_p, the control $u_p(t)$ such that the plant output vector $y_p(t)$ approximates "reasonably well" the output of the reference model:

$$\dot{x}_m(t) = A_m x_m(t) + B_m u_m(t), \tag{1.4}$$

$$y_m(t) = C_m x_m(t), \tag{1.5}$$

where $x_m(t)$ is the $n_m \times 1$ model state vector, $u_m(t)$ is the $m \times 1$ model input or command, $y_m(t)$ is the $q \times 1$ model output vector, and A_m and B_m are matrices with the appropriate dimensions. The model is assumed to be bounded–input/bounded–state stable. It is important to note that the dimension of the model state may be less than the dimension of the plant state. However, since $y_p(t)$ is to track $y_m(t)$, the number of model outputs must be equal to the number of plant outputs.

Although there are several procedures for designing fixed robust controllers for such a system [6], these can often be unsatisfactory because:

- The amount of uncertainty is too large for compensation by a fixed controller.
- The response at some operating points may have to be overly conservative in order to satisfy specifications at other operating points.
- The controlled process itself varies significantly during operation.

Examples of systems that might benefit from adaptive control include:

- Control of robots in the presence of sudden load changes
- Control of large flexible structures with a large degree of uncertainty
- Control of aircraft over a flight envelope that encompasses significant variation in speed and altitude
- Control of drug delivery for patients whose transfer characteristics are unknown to begin with, and which furthermore vary significantly with time
- Control of electrical motors subject to sudden and uncertain load changes

Of course, if a fixed robust controller is satisfactory for a specific system, it naturally should be used. An alternate approach is to first compensate the process with a fixed robust compensator and then use an adaptive controller for computing the command signal vector to the compensated system. This will in many cases assist in satisfying some of the constraints that, if satisfied, guarantee asymptotic tracking of the reference model output vector. Furthermore, the compensated system presents fewer demands than does the uncompensated system. This can result in smaller control signals and less effort in tuning the adaptive system parameters.

1.2 Prologue to Simple Adaptive Control

This book will try to convince the reader, with practical applications in mind, that the described adaptive techniques are not only easily implementable, but also applicable to real control problems that could not be solved otherwise. Also, it will show that the theory presented is "closed and bounded," and that adaptive techniques can be used to improve the robustness of control systems.

Some warning is required: at no stage is it intended to claim that the control problem is simple, or that adaptive control methods are meant to obviate the need for diligent learning and accumulation of knowledge, the patient modeling of plants, or to replace established control theoretic principles.

This being said, the readers are encouraged to test the described adaptive techniques, either with the numerous and various examples in this book or with their own examples. Only in this way can one get some feeling and understanding of the authors' own surprise and enthusiasm with the performance that can be obtained using minimal prior knowledge about the controlled plant.

It is important to note that many design methods have been developed over the years! Classical control design methods have been using the input–output relationships given by the transfer–function representation of the plant in the frequency domain. Since experience has made it clear that the models used were only crude representations of the real plants, measures of caution were intuitively added to guarantee safe operation of these controllers in connection with real world plants. In particular, feedback has been shown to maintain stability and robustness with (some degree of) uncertainty of the plant models and parameters [7], [8], in so many instances that some designers may call any feedback control system "adaptive."

Modern control techniques, developed in particular because of the difficulty of using classical techniques with multivariable plants, used the state space representation of plants and gave new insight about the internal state variable behavior of the plant [9], [10], [11]. Under ideal conditions, with full state feedback available, it was shown that any desired behavior could have been imposed upon the plant. State space representations also use Lyapunov stability analysis methods, for analysis of nonstationary and nonlinear systems [12], [13], [14], [15], [16]. The problems start when realistic designs try to imitate the performance of the ideal full state feedback system. This is because in general the measurable quantities consist of only a reduced number of variables, rather than all state components. When full state feedback is not available, observers can be used to reconstruct the estimates of the hidden state variables. Observer based controllers were shown to enable any desired pole placing only if the order of the real plant is known. When such controllers try to maintain the order of the closed loop plant equal to the order of the original plant, (in spite of the additional observer dynamics), the result is often noncontrollable or nonobservable closed loop systems due to the pole zero cancellations inherent to most pole placing methods. Control of nonminimum–phase plants seems to be difficult in particular because of the possible need to cancel unstable zeros.

There are many classical and modern control design methods with which good controllers can be designed if the plant parameters are fixed and known. Even if because of uncertainties, the designed controller is not necessarily optimal while interacting with the real world plant, there is some admissible domain where *fixed* control gains maintain some level of satisfactory performance and guarantee stability of the control system if the controlled plant can also be considered to be *linear time invariant*.

However, modern controllers do not necessarily use constant gains, and they do not necessarily deal with linear time invariant plants. Gain scheduling is commonly used when the parameters are expected to change in various operational environments. Intelligent control methods not only use variable parameter controllers, but they even use different controller structures for different situations. Many adaptive control techniques [17], [18] are nonlinear adaptive analogues of pole placement methods in some form. In adaptive control, a dynamic algorithm might be used to estimate the parameters of the controlled plant, and the controller design is then based on the identified parameters (indirect control). Alternatively, adaptive algorithms can be used to directly compute the gains of the controller (direct control).

Adaptive controllers are reemerging as a timely and important class of controller design, as reflected in the recent surge of publications and development efforts in the industry. Adaptive controllers possess the essential ability to cope with the unavoidable challenges imposed by internal uncertainties, as well as by external environmental uncertainties.

This book, in particular, intends to show that one can improve both performance and robustness of control systems by the use of a simple adaptive control algorithm [19]–[20]. This algorithm is called "simple" because it does not use identifiers or observers in the control loop. The procedure involves the use of any available prior knowledge about the stabilizability properties of the plant to be controlled, and then the use of an adaptation algorithm to get the right gain at the right time, in a manner such that stability is maintained. It will be shown that the adaptive controller improves the performance of the system in normal situations and maintains stability when the plant parameters and operational conditions change in a way that could lead to divergence with non–adaptive controllers.

The Simple Adaptive Control (SAC) procedure has been shown to be applicable at present to stationary and nonstationary linear systems [20]–[21], and to the class of nonlinear systems linear in control including robotic manipulators [22]–[23], and many other complicated systems [24].

To understand the "obsession" of adaptive control developers with the problem of stability, assume that some plant is totally known and that one

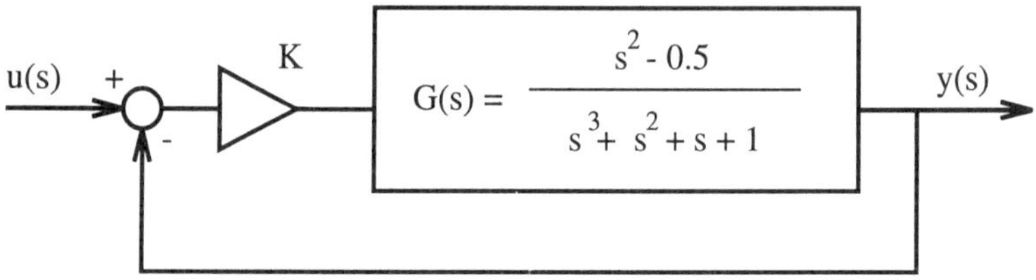

FIGURE 1.1. A Simple Example.

performs a preliminary control design based on fixed linear controllers. Assume that in order to improve the controller's performance, the designer wishes to replace the fixed controller with a nonstationary one that has variable gains. Since it is intuitive to expect trouble if the nonstationary gains violate stability bounds, one would slowly change the gains to determine the stability limits and then respect these limits while using the variable gains. Although this straightforward approach is sometimes used in real world design, it is worth noting that stability is not necessarily maintained even if the "admissible" bounds of stability for fixed gains are respected by the nonstationary gains [25]. A simple example may be useful for understanding this phenomenon. Dewey and Jury [26] used for illustration the system shown in Figure 1.1 with open loop transfer function:

$$G(s) = \frac{s^2 - 0.5}{s^3 + s^2 + s + 1}. \tag{1.6}$$

Although it is easy to show that this system is asymptotically stable for any *fixed* gain K such that $0 < K < 2$, they managed to obtain a limit cycle using a nonlinear gain, even though the gain values did not violate the bounds. The same example is used here to show that using nonstationary gains may even lead to total divergence.

Figure 1.2 shows the response for a time–varying gain $K(t) = (1 + 0.89 sin\ t)$. Note that the system response converges.

However, in Figure 1.3, the response corresponding to the time–varying gain $K(t) = (1 + 0.95 sin\ t)$ diverges despite the fact that the nonstationary gain remains within the "admissible" bounds that would guarantee stability with fixed gains. Actually, the system would diverge with any sinusoidal component of the gain that is larger than 0.9.

Therefore, it is very important that control methods based on nonstationary, nonlinear, adaptive or intelligent control guarantee, stability of the controlled system. On the other hand, in *fixed* controllers, instability can

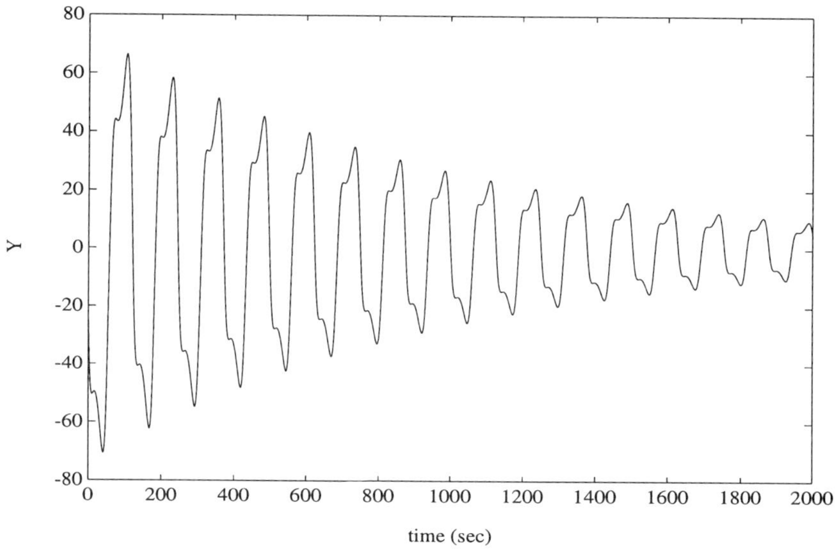

FIGURE 1.2. Response for Time–Varying Gain $K = (1.0 + 0.89 sin\ t)$

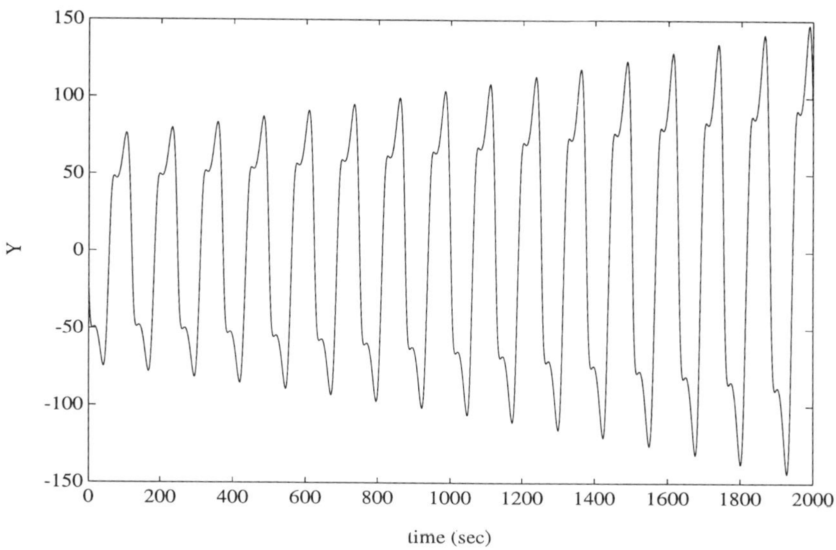

FIGURE 1.3. Response for Time–Varying Gain $K = (1. + 0.95 sin\ t)$

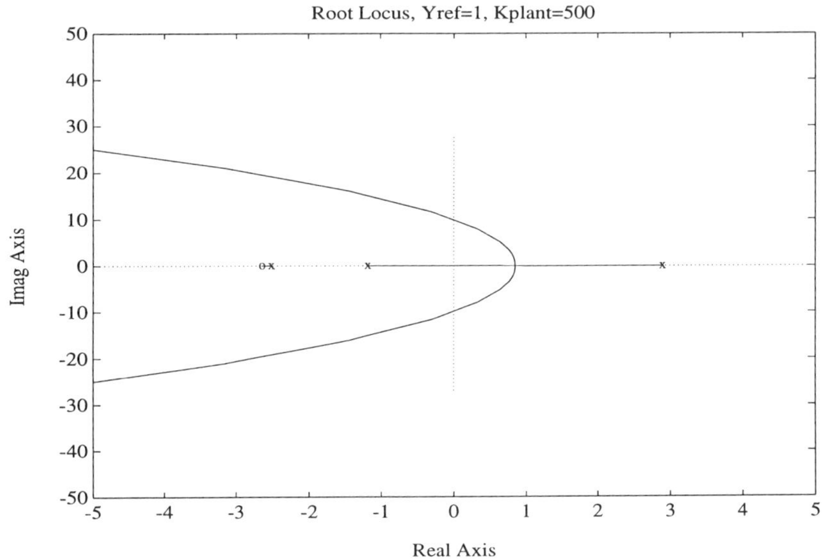

FIGURE 1.4. Root–Locus Diagram Showing High–Gain Stability of the Augmented Plant.

be similarly induced by time varying plant parameters. Thus, in spite of the "danger" related to nonstationary gains, it is desirable to show that if appropriately used, adaptive control can provide solutions in very difficult situations that could not be offered otherwise due to lack of knowledge or due to (unknown) changes in systems parameters.

As stated above, fixed controllers may not be able to provide the desired control when confronted with insufficient prior knowledge or with (unknown) changes in the system parameters and/or the desired task. However, to illustrate the usefulness of adaptive control, consider a simplified linearized approximation of a non–linear generator–voltage control system [27] augmented with a feedforward compensator that will be described in Chapter 3. For now it is sufficient to consider the resulting augmented transfer function:

$$G_a(s) = \frac{.025K_{\text{plant}}(s^2 + 58.217s + 146.607)}{(s+1.17)(s-2.9)(s+2.5)}. \tag{1.7}$$

As shown in Figure 1.4 the nominal closed loop plant is stable for gain values greater than some minimal values.

Variations of plant parameters from the assumed nominal values affect the performance and stability of the plant. For convenience of the presentation in this example, the plant uncertainty is represented by an unknown and arbitrarily time varying gain, K_{plant}.

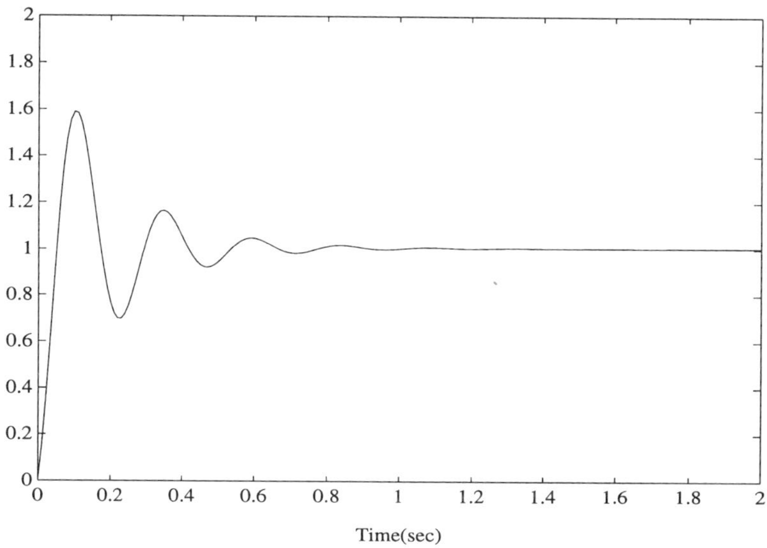

FIGURE 1.5. Stable Response for, $Y_{\text{ref}} = 1.0$, $K_{\text{plant}} = 500$.

If the closed loop control system is required to maintain the output y_p close to the desired value of the reference signal y_{ref} and if the parameters are known, the designer can select from the various possible values of stabilizing gains depending on the desired performance. Of course, the system can become unstable if the gain decreases towards a lower value. A stable situation is illustrated in Figure 1.5, with $y_{\text{ref}} = 1$ and $K_{\text{plant}} = 500$.

If the parameter K_{plant} decreases as shown in Figure 1.6, the quality of regulation is correspondingly affected. As shown in Figure 1.7, the output diverges, when the gain K_{plant} becomes too small.

To add an adaptive correction, consider the output tracking error

$$e_y(t) \;=\; y_{\text{ref}}(t) \;-\; y(t), \tag{1.8}$$

and generate the following adaptive gain [20] in the configuration shown in Figure 1.8:

$$K_{\text{total}} = K(t) + K_{\text{adapt}}(t) \;=\; 500 + \int_0^t \left[\gamma e_y^2(t) \;-\; \sigma K_{\text{adapt}}(t)\right] dt \tag{1.9}$$

where γ and σ are adaptation coefficients and $K(t)$ is the variation shown in Figure 1.6. The adaptive gain $K_{\text{adapt}}(t)$ can then be represented as the output of a first order pole fed by the error square

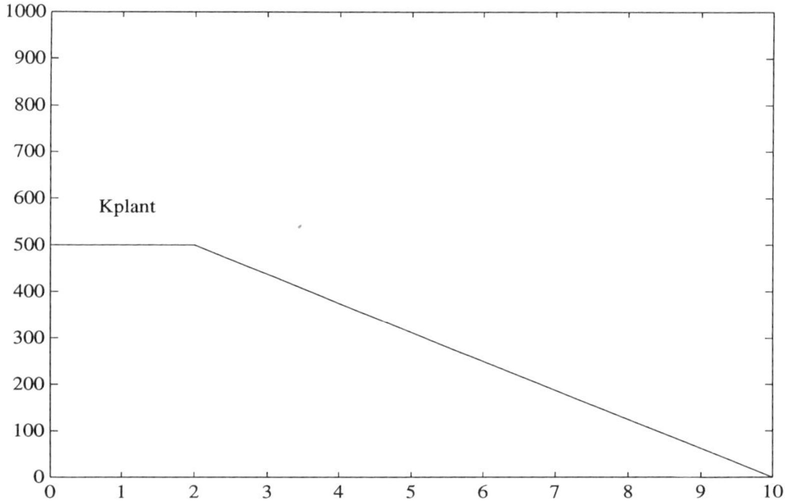

FIGURE 1.6. Gain Variation

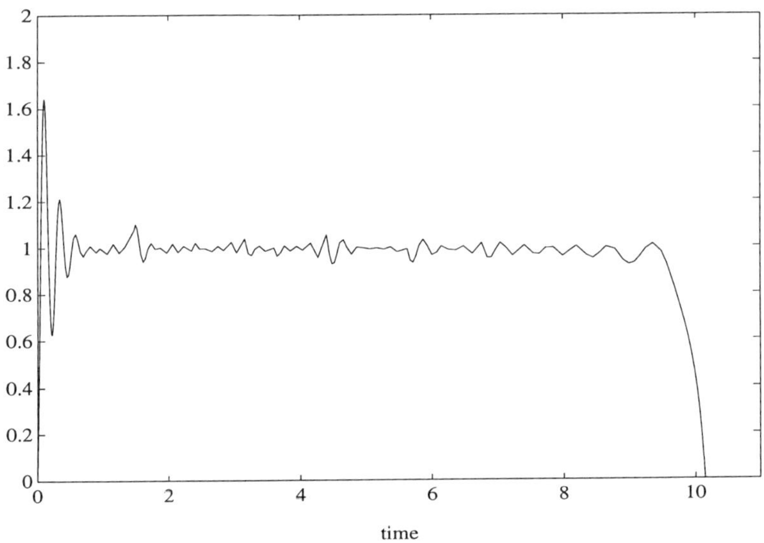

FIGURE 1.7. Unstable Response for K_{plant} of Figure 1.6, $Y_{\text{ref}} = 1.0$

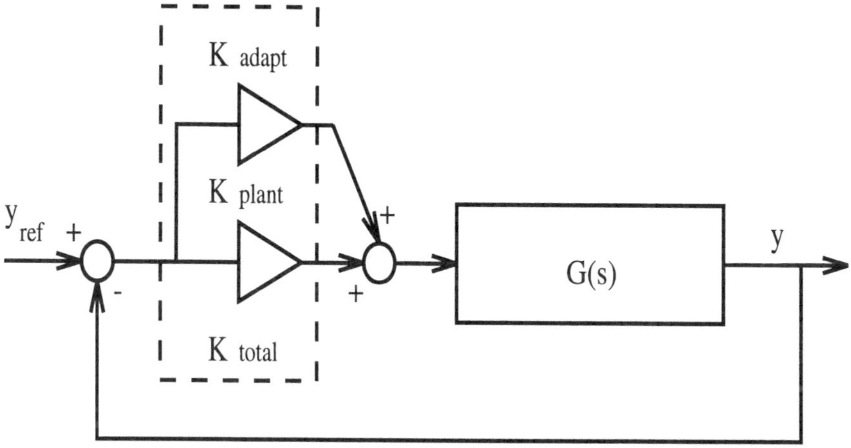

FIGURE 1.8. Adaptive Control System.

$$K_{\text{adapt}}(s) = \frac{\gamma}{\sigma} \frac{1}{1 + \frac{s}{\sigma}} L\left[e_y^2(t)\right] \tag{1.10}$$

where $L\left[e_y^2(t)\right]$ is the Laplace transform of $e_y^2(t)$. Detailed explanations of this and other adaptive control procedures are the object of the subsequent chapters of this book. For the present time, it is sufficient to understand that the adaptive gain (1.9) was selected such that if the output error $e_y(t)$ tends to increase, the adaptive gain which is proportional to the square of the error increases even faster. Due to the term $\sigma K_{\text{adapt}}(t)$, the adaptive gain may then decrease, and even ultimately vanish.

The last (unstable) case was run again, this time with the above adaptive gain, with coefficients $\gamma = 10000$ and $\sigma = 0.1$ (Figure 1.9). It can be seen that stable behavior is maintained, in spite of the large (and unknown) variation of the uncertain parameter K_{plant}. Figure 1.10 shows how the adaptive gain changes and fights the tendency of the errors to increase, thus not only maintaining a stable behavior, but also the desired performance. Notice also that only the regulation error is measured, and that no indication whatsoever on the specific failure is used or needed. Similar improvements were obtained with many arbitrary variations of the parameter K_{plant} [28].

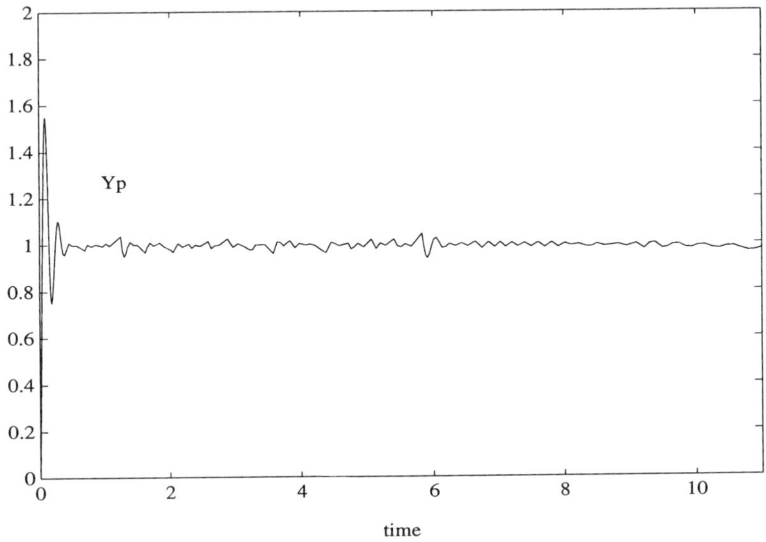

FIGURE 1.9. Stable Response from Adaptive Controller, K_{plant} Defined in Figure 1.6, $\gamma = 10000.$, $\sigma = 0.1$.

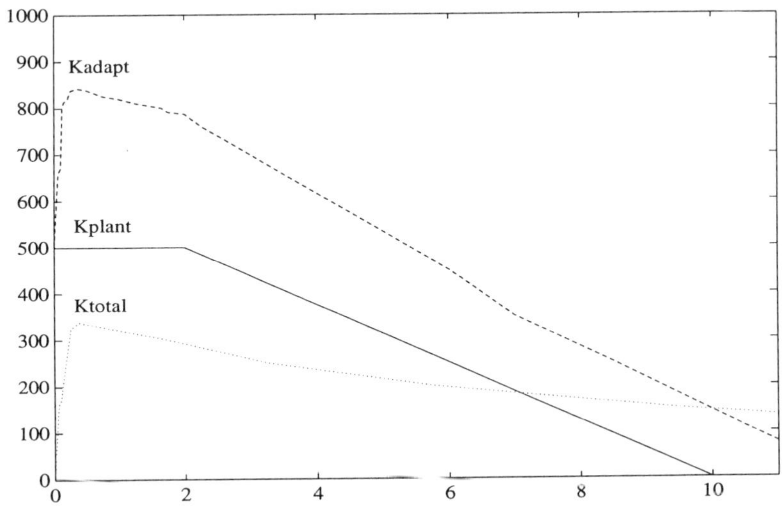

FIGURE 1.10. Adaptive Gains for System of Figure 1.8 $\gamma = 10000.$, $\sigma = 0.1$.

1.3 Background on Adaptive Control Algorithms

Closed loop adaptive control methods may be divided into the two broad categories of direct or implicit control and indirect or explicit control. Indirect control methods utilize separate parameter identification and control schemes. Direct methods merge the identification and control functions into one scheme. In such direct methods, the control gains are computed directly without an explicit identification of the system parameters. With fewer computations to perform, one advantage of direct over indirect adaptive control may be speed.

Model reference adaptive methods might be classified as evolving from three different approaches. First is the full state access method described by Landau [29] which assumes that the state variables are measurable. Second is the input–output method which originates from Monopoli's augmented error signal concept [30]. In this approach, adaptive observers are incorporated into the controller to overcome the inability to access the entire state vector. Third is the simple adaptive control approach originated by the authors of this book [31]. This approach is an output feedback method which requires neither full state feedback nor adaptive observers. Other important properties of this class of algorithms are as follows:

a. they are applicable to non–minimum phase systems,

b. the order of the plant (physical system) may be much larger than the order of the reference model, and

c. this approach considers plants with multiple inputs and outputs.

Direct model reference adaptive control was first proposed in [32] by using a performance index minimization approach which later became known as the MIT design rule. This was later extended to an accelerated gradient method [33], but stability could not be guaranteed with either procedure. The stability of a linear system combined with an adaptive controller is often in question because of the highly nonlinear nature of the closed loop system.

A significant contribution to the theory of direct model reference adaptive control of single-input single-output (SISO) systems was the augmented error signal concept of Monopoli [34]. This technique eliminated the need for either state feedback or derivatives of the output by incorporating adaptive observers into the control law. Monopoli's contribution encouraged much research into this class of algorithms which will be referred to as the input-output approach. Other contributions to this approach include the work of Morse [35], Feuer and Morse [36], and Narendra et.al. [37], [38], [39]. The contribution of Narendra, Lin, and Valavani [39] settled the question of stability for the input-output approach.

Some of the most recent work in the input-output approach to MRAC is stability in the presence of disturbances and unmodeled dynamics. This area has become known as the robust adaptive control problem. The observation that earlier stability proofs were not valid in the presence of disturbances and unmodeled dynamics is discussed by Rohrs et.al. [40]. Astrom [41], in response to Rohrs' questions, emphasized the need for excitation in the proper frequency range to perform estimation. However, it is generally accepted that robust stability is a necessary characteristic of any adaptive control algorithm. Robust MRAC laws for SISO systems have been proposed which utilize a deadzone [42], [43], a fixed σ–modification [44], or a switching σ -modification [45], [46]. However, the sufficient conditions for stability require that the relative degree of each element of the modeled part of the plant transfer matrix be known, and that either the left or right interacter matrix be known and diagonal. Furthermore, the choice of the reference model is restricted based upon a relationship involving an upper bound on the observability index of a quantity related to the modeled part of the plant transfer matrix. Nevertheless, the extension of the input-output approach to multi-input multi-output plants is an interesting research area.

During the same time period that Monopoli was working on the augmented error signal concept, Landau [29], Lindorff and Carroll [47], and others were proposing a full state access approach to direct MRAC. This approach uses a control structure which is a linear combination of feedforward of the model states and inputs and feedback of the error between plant and model states. Stability was ensured by using either Lyapunov's stability theory or Popov's hyperstability theory. Although this approach applied to multi–input multi– output plants, the satisfaction of Erzberger's [48] perfect model following conditions was required.

The simple approach to direct MRAC of multi–input multi–output plants was first proposed by Sobel, Kaufman, and Mabius [19] in 1979. This approach uses a control structure which is a linear combination of feedforward of the model states and inputs and feedback of the error between plant and model outputs. This class of algorithms requires neither full state access nor satisfaction of the perfect model following conditions. Asymptotic stability is ensured provided that the plant is almost strictly positive real (ASPR). That is, for a plant represented by the triple (A, B, C), or equivalently by the state equation system:

$$\dot{x} = Ax + Bu \tag{1.11}$$

$$y = Cx \tag{1.12}$$

there exists a feedback gain K_e (not needed for implementation) such that the system defined by $(A - BK_eC, B, C)$ is strictly positive real (as defined in Section 2.4).

The appealing characteristics of this simple adaptive control algorithm over indirect and other direct model reference adaptive methods include:

- lack of dependence on plant parameter estimates,
- applicability to MIMO plants,
- sufficiency conditions which are independent of plant dimension,
- control calculation which does not require adaptive observers or full state feedback,
- ease of implementation, and
- successful experimental validation.

The ease of its implementation and its robustness properties make this adaptive control approach attractive to the practitioner. Among its realistic applications, one can find large flexible space structures [49], [20], [50], [51], [52], [53], [54], [55], [56], [57]; robotic manipulators [23],[58], [59]; ship steering control [60]; DC motors [61], [62], [63]; boilers [64]; drug infusion [65]; aircraft [20], [66], [67], [68]; missiles [20]; and nonlinear servomechanisms with time varying uncertainties [24].

This adaptive control approach is described in detail by Sobel, Kaufman, and Mabius [31] and Sobel and Kaufman [66]. They also propose another algorithm which enlarges the class of allowable plants. Unfortunately, the control law requires knowledge of the unknown ideal plant input. However, boundedness of the output error was ensured when an approximation to the ideal plant input was utilized in the adaptive control law.

Bar-Kana and Kaufman [50], removed the restriction on the algorithm in [19], [66], [31] that the model input be a step command. They also incorporated [49], [69], [20] Ioannou's fixed sigma–algorithm [44] into the basic algorithm. This results in a robust MRAC law for ASPR multi–input multi–output plants. Later, Bayard, Ih, and Wang [52] extended the MRAC algorithm by introducing "branch" noise filters into the adaptive controller which suppress measurement noise while ensuring global stability of the adaptive algorithm. Such branch filters allow filtering of the output error without introducing phase lag into the adaptive loop.

The ASPR algorithm was extended in [69], [70] to the class of non–ASPR plants for which there exists a known dynamic output stabilizing feedback with transfer matrix $H(s)$. In this case, it is shown [71] that an augmented system consisting of the plant in parallel with $H^{-1}(s)$ is ASPR. This approach does not guarantee perfect output following in general because the augmented plant includes a contribution from the supplementary feedforward. However, in [69], [70] it is argued that if the supplementary feedforward can be made small, the true output error can also be made small.

Thus to date, the major limitation of this approach was the inability to guarantee asymptotic output tracking. To this effect, Kaufman et. al. [72] suggested a new modification that incorporates part of the supplementary feedforward into the reference model in a manner such that asymptotic tracking of the augmented plant and model outputs implies asymptotic tracking of the original plant and model outputs. Such asymptotic tracking has indeed been observed in various illustrative applications including the control of robotic dynamics [73].

An alternative approach for asymptotic output tracking has been proposed by Su and Sobel [74] using adaptive gains for the plant parallel supplementary dynamics in addition to adaptive gains for the controller. Su and Sobel [74] show how to choose the design parameters to guarantee asymptotic output tracking when the model input is constant for $t \geq t_1$.

The modifications of Kaufman et. al. [72] and Su and Sobel [74] appear to alleviate a seemingly very restrictive positive real constraint and thus greatly expand the class of processes that can now be adaptively controlled with zero output error.

1.4 Objectives and Overview

1.4.1 Objectives

In designing adaptive control systems, consideration should be given to indirect or explicit adaptive systems in which online parameter identification is performed and to direct or implicit systems which do not utilize explicit parameter identification.

Although explicit adaptive controllers lend themselves to a wide class of adjustable control algorithms, their implementation requirements are generally more demanding. Assumptions are needed with regard to the system order and to the modeling of the uncertainty to be identified. Thus, explicit or indirect adaptive control can be very difficult to implement for high dimensional systems, (e.g., flexible structures), multiple input multiple output systems, and systems in which the parameters are varying relatively fast.

Although various direct adjustment algorithms have been proposed, model reference adaptive control is very appealing for the following reasons:

- It provides (continuously or at each sample) a measure of system performance.
- It is not difficult to design a reference model that incorporates the desired response specifications.
- It leads itself to elegant stability analysis.

Thus, taking into account the large body of theory and successful applications of the simple direct model reference adaptive control algorithm discussed in the previous sections, this text, with the designer in mind, covers the relevant theory and design procedures. Thus, this text might be considered as a treatise on an especially useful form of direct model reference adaptive control.

1.4.2 Relation with Other Texts

It is important to note that this text, unlike other adaptive control books, is unique in that it concentrates on developing the theory and design tasks for a specific direct model reference adaptive controller. This controller, as will be shown, is easily designed and applied to a wide variety of systems. More general discussions of alternate approaches are contained in references [1], [2], [3], [4], [5], [75].

1.4.3 Overview of Text

The development of the basic adaptive algorithm is self contained in Chapter 2, which includes theoretical developments, algorithm summaries, and illustrative examples. Further extensions that alleviate the constraints defined in Chapter 2 and thus widen the scope of applicability are presented in Chapter 3. Robustness with respect to noise, disturbance, and nonlinearities, is treated in Chapter 4. Design procedures, including recipes and rules of thumb that have been inferred from years of experience, are detailed in Chapter 5. Finally detailed case studies are presented in Chapter 6, and conclusions are discussed in Chapter 7.

2

Basic Theory of Simple Adaptive Control

The basic direct model reference adaptive controllers are developed in this chapter. To this effect, Sections 2.1 and 2.2 discuss basic model following concepts. Of importance is the development in Section 2.2 of the output model following control algorithms which subsequently serve as the basis for the adaptation procedures.

In order to fully comprehend the stability analysis of the adaptive control algorithms, it is useful to have a basic understanding of Lyapunov function based stability analysis and some background on positive real transfer functions. These concepts are summarized with adequate references in Section 2.3. This section may of course, be omitted by any reader who already has an understanding of these concepts.

Finally the basic direct model reference adaptive controller for output tracking is presented in Section 2.4 for step commands and then generalized in Section 2.5 for a wider class of commands.

2.1 Model Following

Model reference control is based upon matching the response of a system or "plant" to that of a reference model or "model". Plant design specifications are often incorporated within the model such that a step input to the model would cause its outputs to respond with the specified rise time, overshoot, and settling time. The reference inputs are fed into the model, which responds in accordance with the design specifications that have been built into it. If the control system is properly designed, the inputs to the plant (which are generated from the model inputs, the model states, and the error between plant and model outputs) drive the outputs of the plant to equal the outputs of the model. A block diagram of a typical model reference control system is shown in Figure 2.1.

When the designer has limited knowledge of the plant parameters, it may be desirable to utilize adaptive control, to adjust the control law on-line in order to reduce the effects of the unknown parameters. This adaptive feature allows satisfactory performance with only limited knowledge of the plant parameters. As discussed in the preceding Chapter, model reference adaptive control systems can be divided into the following two classes:

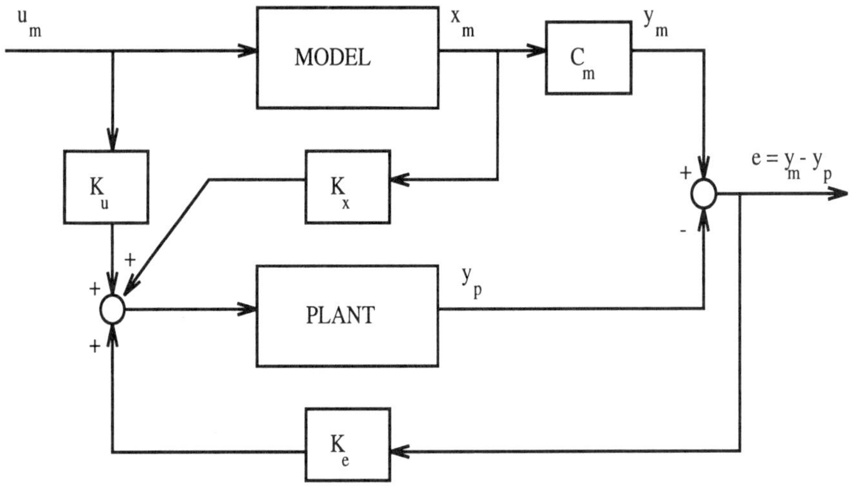

FIGURE 2.1. Model Reference Control System.

(1) indirect or explicit adaptive controllers in which on–line estimates of the plant parameters are used for control law adjustment and (2) direct or implicit adaptive controllers in which no effort is made to identify the plant parameters; that is, the control law is directly adjusted to minimize the error between plant and model states or outputs. In this text, only direct model reference adaptive controllers will be considered.

In the development of model following controllers, it is important to note that many of the earlier developments were concerned with the concept of perfect model following (PMF). A PMF controller is designed for a system of the form:

PLANT:

$$\dot{x}_p(t) = A_p x_p(t) + B_p u_p(t) \tag{2.1}$$

MODEL:

$$\dot{x}_m(t) = A_m x_m(t) + B_m u_m(t). \tag{2.2}$$

A controller of the form:

$$u_p(t) = K_e(x_m - x_p) + K_x x_m(t) + K_m u_m(t) \tag{2.3}$$

is to be found such that

1. If the plant state vector x_p is equal to the model state vector x_m at some time, (i.e., $x_p(t_1) = x_m(t_1)$), then $x_p(t) = x_m(t),\ t > t_1$.

2. If the plant state x_p, is not equal to the model state x_m it should asymptotically approach the model state.

Thus, a necessary condition for PMF control is that the reference model must have the same number of states as the plant being controlled. Furthermore, design of a PMF controller is quite demanding on the plant model structure. For example, Erzberger [48] has shown that a PMF controller exists if gains K_x and K_u can be found such that

$$(A_m - A_p) = B_p K_x \tag{2.4}$$

$$B_p\, K_u = B_m. \tag{2.5}$$

These equations will, in general, have a solution if the number of controls is equal to the number of states, or if the plant and model structure conform such that (2.4) and (2.5) have solutions. For example if there are four states and two controllers, then (2.4) and (2.5) will in general be solvable if the first two rows of A_m and A_p are identical (e.g., if they are in phase variable form) and if the first two rows of both B_p and B_m are null, and at least two rows of B_p constitute a nonsingular matrix $\tilde{B}_p$. Then (2.4) and (2.5) become:

$$\begin{pmatrix} 0 & 0 & 0 & 0 \\ 0 & 0 & 0 & 0 \\ X & X & X & X \\ X & X & X & X \end{pmatrix} = \begin{pmatrix} 0 & 0 \\ 0 & 0 \\ \multicolumn{2}{c}{\tilde{B}_p} \end{pmatrix} K_x \tag{2.6}$$

$$\begin{pmatrix} 0 & 0 \\ 0 & 0 \\ \multicolumn{2}{c}{\tilde{B}_p} \end{pmatrix} K_u = \begin{pmatrix} 0 & 0 \\ 0 & 0 \\ \multicolumn{2}{c}{\tilde{B}_m} \end{pmatrix}. \tag{2.7}$$

Then

$$K_x = \tilde{B}_p^{-1} \times (\text{last 2 rows of } [A_m - A_p]) \tag{2.8}$$

$$K_u = \tilde{B}_p^{-1}\, \tilde{B}_m. \tag{2.9}$$

The conditions that must be satisfied for (2.4) and (2.5) to have solutions are known as the perfect model following conditions.

In general, however, a PMF controller is not easily implemented because:

- It is usually not possible to measure all the plant states.
- The PMF conditions are often not satisfied.

To alleviate these problems, output model following (rather than full state following) controllers have been developed. These are designed such that the process output vector (which is usually of a much lower dimension than the state vector) tracks the model output vector with all states remaining bounded. Such an output model following controller is described in the next section and then used subsequently as the basis for development of a simple adaptive controller.

2.2 Output Model Following

The continuous linear model reference control problem is solved for the linear process equations

$$\dot{x}_p(t) = A_p x_p(t) + B_p u_p(t), \tag{2.10}$$

$$y_p(t) = C_p x_p(t), \tag{2.11}$$

where $x_p(t)$ is the $n \times 1$ plant state vector, $u_p(t)$ is the $m \times 1$ control vector, $y_p(t)$ is the $q \times 1$ plant output vector, and A_p and B_p are matrices with the appropriate dimensions. The range of the plant parameters is assumed to be bounded as defined by:

$$\begin{aligned} \underline{a}_{ij} \le a_p(i,j) \le \bar{a}_{ij}, \quad & i = 1, \ldots, n, \\ & j = 1, \ldots, n; \\ \underline{b}_{ij} \le b_p(i,j) \le \bar{b}_{ij}, \quad & i = 1, \ldots, n, \\ & j = 1, \ldots, m; \end{aligned} \tag{2.12}$$

where $a_p(i,j)$ is the i,jth element of A_p and $b_p(i,j)$ is the i,jth element of B_p. It is also assumed that (i) all possible pairs A_p, B_p are controllable and output stabilizable, (ii) all possible pairs A_p, C_p are observable, and (iii) B_p has maximum rank. The objective is to find, without explicit knowledge of A_p and B_p, the control $u_p(t)$ such that the plant output vector $y_p(t)$ approximates "reasonably well" the output of the reference model:

$$\dot{x}_m(t) = A_m x_m(t) + B_m u_m(t), \tag{2.13}$$

$$y_m(t) = C_m x_m(t), \tag{2.14}$$

where $x_m(t)$ is the $n_m \times 1$ model state vector, $u_m(t)$ is the $m \times 1$ model input or command, $y_m(t)$ is the $q \times 1$ model output vector, and A_m and B_m are matrices with the appropriate dimensions. The model is assumed to be bounded–input/bounded–state stable. It is important to note that the dimension of the model state may be less than the dimension of the plant state. However, since $y_p(t)$ is to track $y_m(t)$, the number of model outputs is of course equal to the number of plant outputs.

2.2.1 Command Generator Tracker Description

To facilitate adaptive control algorithm development, it is useful to incorporate the command generator tracker concept (CGT) developed by O'Brien and Broussard [76]. The CGT is a model reference control law for linear time–invariant systems with *known* coefficients. In this section, the basic elements of the command generator tracker will be reviewed.

When perfect output tracking occurs (i.e., when $y_p = y_m$ for $t \geq 0$), the corresponding state and control trajectories are defined to be the ideal state and ideal control trajectories, respectively. These ideal trajectories will be denoted by $x_p^*(t)$ and $u_p^*(t)$. By definition, the ideal plant is such that it satisfies the same dynamics as the real plant. In addition, the output of the ideal plant is defined to be identically equal to the model output. Mathematically then:

$$\dot{x}_p^* = A_p x_p^* + B_p u_p^* \text{ for all } t \geq 0 \tag{2.15}$$

and

$$y_p^* = y_m = C_p x_p^* = C_m x_m. \tag{2.16}$$

Hence, when perfect tracking occurs, the real plant trajectories become the ideal plant trajectories, and the real plant output becomes the ideal plant output, which is defined to be the model output.

It will be assumed that the ideal trajectories are linear functions of the model state and model input. Mathematically:

$$\begin{bmatrix} x_p^*(t) \\ u_p^*(t) \end{bmatrix} = \begin{bmatrix} S_{11} & S_{12} \\ S_{21} & S_{22} \end{bmatrix} \begin{bmatrix} x_m(t) \\ u_m \end{bmatrix}. \tag{2.17}$$

In (2.13), u_m is presently restricted to be a constant input; otherwise, derivatives of the model input may be required. This is a temporary assumption that will be relaxed in subsequent sections. Combining the ideal plant state equation (2.15) with the ideal plant output equation yields

$$\begin{bmatrix} \dot{x}_p^* \\ y_p^* \end{bmatrix} = \begin{bmatrix} A_p & B_p \\ C_p & 0 \end{bmatrix} \begin{bmatrix} x_p^* \\ u_p^* \end{bmatrix}. \tag{2.18}$$

Substituting (2.17) into (2.18), gives:

$$\begin{bmatrix} \dot{x}_p^* \\ y_p^* \end{bmatrix} = \begin{bmatrix} A_p & B_p \\ C_p & 0 \end{bmatrix} \begin{bmatrix} S_{11} & S_{12} \\ S_{21} & S_{22} \end{bmatrix} \begin{bmatrix} x_m \\ u_m \end{bmatrix}. \tag{2.19}$$

Differentiating the first equation in (2.17) gives

$$\dot{x}_p^* = S_{11}\dot{x}_m + S_{12}\dot{u}_m \tag{2.20}$$

or since u_m is a constant input:

$$\dot{x}_p^* = S_{11}\dot{x}_m. \tag{2.21}$$

Now, substituting the equation for the model dynamics into (2.21) gives:

$$\dot{x}_p^* = S_{11}A_m x_m + S_{11}B_m u_m. \tag{2.22}$$

Concatenating (2.22) with (2.16) yields:

$$\begin{bmatrix} \dot{x}_p^* \\ y_p^* \end{bmatrix} = \begin{bmatrix} S_{11}A_m & S_{11}B_m \\ C_m & 0 \end{bmatrix} \begin{bmatrix} x_m \\ u_m \end{bmatrix}. \tag{2.23}$$

Equating the right–hand sides of (2.19) and (2.23) results in:

$$\begin{aligned} &\begin{bmatrix} S_{11}A_m & S_{11}B_m \\ C_m & 0 \end{bmatrix} \begin{bmatrix} x_m \\ u_m \end{bmatrix} = \\ &\begin{bmatrix} A_p & B_p \\ C_p & 0 \end{bmatrix} \begin{bmatrix} S_{11} & S_{12} \\ S_{21} & S_{22} \end{bmatrix} \begin{bmatrix} x_m \\ u_m \end{bmatrix}. \end{aligned} \tag{2.24}$$

Since x_m and u_m are arbitrary, (2.24) is satisfied if:

$$\begin{bmatrix} S_{11}A_m & S_{11}B_m \\ C_m & 0 \end{bmatrix} = \begin{bmatrix} A_p & B_p \\ C_p & 0 \end{bmatrix} \begin{bmatrix} S_{11} & S_{12} \\ S_{21} & S_{22} \end{bmatrix}. \tag{2.25}$$

The above matrix equation represents a set of linear simultaneous equations to be solved for the elements in the S_{ij} matrices. Specifically there are $(n + q) \times (n_m + m)$ equations with $(n\,n_m + n\,n_m + m\,n_m + m^2)$ unknowns. When m, the number of controls, is greater than or equal to q, the number of outputs to be controlled, there are at least as many unknowns as equations. For this reason, the CGT solution almost always exists. In the rare case when a singular situation occurs, it is usually possible to modify the selected reference model parameters so that a solution will indeed exist. A more elegant, but not always solvable approach, is to define

$$\begin{bmatrix} \Omega_{11} & \Omega_{12} \\ \Omega_{21} & \Omega_{22} \end{bmatrix} = \begin{bmatrix} A_p & B_p \\ C_p & 0 \end{bmatrix}^{-1}. \tag{2.26}$$

Then the partitional matrix equations to be solved are

$$S_{11} = \Omega_{11}S_{11}A_m + \Omega_{12}C_m, \tag{2.27}$$

$$S_{12} = \Omega_{11}S_{11}B_m, \tag{2.28}$$

$$S_{21} = \Omega_{21}S_{11}A_m + \Omega_{22}C_m, \tag{2.29}$$

$$S_{22} = \Omega_{21}S_{11}B_m. \tag{2.30}$$

The existence of the inverse in (2.26) requires that the number of controls m equal the number of outputs q. If $m > q$, a pseudoinverse may be required, while the case $m < q$ does not usually have a solution.

Note that (2.27) is a Lyapunov equation, the solution of which exists if the product of the ith eigenvalue of Ω_{11} and the inverse of the jth eigenvalue of A_m does not equal unity for all i, j.

When y_p differs from y_m at $t = 0$, asymptotic tracking is achievable provided stabilizing output feedback is included in the control law. To see this, write the error equation as:

$$\dot{e}_x = \dot{x}_p^* - \dot{x}_p = A_p x_p^* + B_p u_p^* - A_p x_p - B_p u_p \tag{2.31}$$

$$= A_p e_x + B_p(u_p^* - u_p). \tag{2.32}$$

If the control law is

$$u_p = u_p^* + K(y_m - y_p) = u_p^* + KC_p e_x, \tag{2.33}$$

then the error equation becomes

$$\dot{e}_x = (A_p - B_p K C_p)e_x \tag{2.34}$$

and the error will approach zero provided that K is a stabilizing output feedback gain.

The development of the adaptive algorithm will begin by defining a new error

$$e_x = x_p^*(t) - x_p(t). \tag{2.35}$$

A controller is to be found which guarantees that $e_x \to 0$ as $t \to \infty$. It is important to note that since $x_p(t) = x_p^*(t)$, $C_p x_p(t) = C_p x_p^*(t)$. From (2.16), $C_p x_p^*(t) = C_m x_m(t)$, and therefore, $C_p x_p = C_m x_m$, which is the desired result.

It is important to observe that since $x_p \to x_p^*$, and x_p^* is a linear combination of model states and commands, x_p will be ultimately bounded.

The perfect model following conditions, (2.4) and (2.5), are together a special case of the command generator tracker when the state vector is available, and it is assumed that $x_p^*(t) = x_m(t)$. Since $x_p^*(t) = S_{11}x_m(t) + S_{12}u_m$, the PMF conditions imply that $S_{11} = I$ and $S_{12} = 0$. The ideal plant input is $u_p^*(t) = S_{21}x_m(t) + S_{22}u_m$, where S_{21} and S_{22} satisfy

$$A_m - A_p = B_p S_{21}, \tag{2.36}$$

$$B_p S_{22} = B_m. \tag{2.37}$$

These equations have a solution when the column vectors of the difference matrix $A_m - A_p$ and of the matrix B_m are linearly dependent on the column vectors of the matrix B_p.

If the matrices S_{21} and S_{22} which satisfy (2.36) and (2.37) exist, then a valid PMF controller becomes

$$u_p(t) = u_p^*(t) + K(x_m - x_p), \tag{2.38}$$

where K is a stabilizing feedback gain.

It should be noted that even though the CGT–based analysis is valid only when u_m is a step command, any command signal which can be described as the solution of a differential equation forced by a step input (or zero) can be used. This is true provided that the time–varying portion of the command signal is augmented to the model state vector and not to the model output vector.

For example, suppose it is desired to follow a continuous model forced by the scalar command $r_m(t) = 1 + sin\ t$. Then it should be noted that the following augmented system represents the original model forced by r_m:

$$\begin{bmatrix} \dot{x}_m(t) \\ \dot{z}_1(t) \\ \dot{z}_2(t) \end{bmatrix} = \begin{bmatrix} A_m & B_m & 0 \\ 0 \dots 0 & 0 & 1 \\ 0 \dots 0 & -1 & 0 \end{bmatrix} \begin{bmatrix} x_m(t) \\ z_1(t) \\ z_2(t) \end{bmatrix} + \begin{bmatrix} B_m \\ 0 \\ 0 \end{bmatrix} u_m, \tag{2.39}$$

$$z_1(0) = 0, \quad z_2(0) = 1. \tag{2.40}$$

B_m would be the control distribution vector multiplying $z_1(t) + u_m$, u_m would be the unit step function, and $z_1(t)$, which equals $sin\ t$, would be the output of the (z_1, z_2) system of differential equations.

Summary

The CGT output model reference controller is:

$$u_p(t) = S_{21} x_m + S_{22} u_m + K(y_m - y_p) \tag{2.41}$$

where S_{21}, S_{22} are solutions to (2.27)–(2.30), and

$$\textbf{REAL}\ [\text{eigenvalues}\ (A_p - B_p K C_p)] < 0. \tag{2.42}$$

2.2.2 Modifications for the Tracking of a Larger Class of Input Commands

In general, the adaptive control system should track various input commands that are not step inputs, and also arbitrary external inputs that are not generated by the command generator described in Section 2.2.1. Recall from (2.10) and (2.11) that the controlled plant is represented by

$$\dot{x}_p(t) = A_p x_p(t) + B_p u_p(t) \tag{2.43}$$

$$y_p(t) = C_p x_p(t) \tag{2.44}$$

where $x_p(t)$ is the $n \times 1$ plant state vector, $u_p(t)$ is the $m \times 1$ control vector, y_p is the $m \times 1$ output vector, and A_p, B_p, C_p are matrices with the appropriate dimensions.

The plant output y_p is required to track the output y_m of an asymptotically stable reference model which is described by:

$$\dot{x}_m(t) = A_m x_m(t) + B_m u_m(t) \tag{2.45}$$

$$y_m(t) = C_m x_m(t), \tag{2.46}$$

where x_m is the $n_m \times 1$ model state vector, u_m is the $m \times 1$ model command, and y_m is the $m \times 1$ model output vector. The reference model is designed to meet some desired performance properties and has the same number of outputs as the plant, but is otherwise independent of the controlled plant. Further, it will be permissible to have

$$n = \dim(x_p) >> \dim\,(x_m) = n_m. \tag{2.47}$$

Define the output tracking error as:

$$e_y(t) = y_m(t) - y_p(t) \tag{2.48}$$

and consider controllers of the form

$$u_p(t) = \tilde{K}_e e_y(t) + \tilde{K}_x x_m(t) + \tilde{K}_u u_m(t) \tag{2.49}$$

where $\tilde{K}_e$ is the unknown stabilizing output feedback matrix, and $\tilde{K}_x$ and $\tilde{K}_u$ are some unknown "ideal" control gains to be defined subsequently. For example, a system of order (say) 20 may be required (as in the example of Chapter 6) to track the output of a model of order four, or a large plant may be required to reproduce the input-output behavior of a well–designed second order model.

Since the reference model may be very low order compared with the controlled plant, it cannot be assumed that the state of the plant follows the state of the model. Therefore, it is only specified that the plant *outputs* track asymptotically the outputs of the model. Yet, it is not intuitively clear that this is always possible, and at first, it was thought that only constant (set–point) input commands could be tracked in the low order case.

In order to get a measure for the complexity of the acceptable commands that allow perfect tracking, represent the input command $u_m(t)$ as the output signal of an unknown command generating system of the form

$$\dot{v}_m(t) = A_v v_m(t) \tag{2.50}$$

$$u_m(t) = C_v v_m(t) \tag{2.51}$$

where $v_m(t)$ is the command state vector. Of course, this representation is only needed for the subsequent analysis; the matrices A_v and C_v are unknown, and only measurements of the input $u_m(t)$ are permitted. The commands $u_m(t)$ are thus represented as generalized Fourier terms of the form $\sum_{i\ j} t^j\ e^{\alpha_i t}\ cos\ \beta_i t$.

When the reference model defined in (2.45) is supplied with the input defined by (2.50)–(2.51), its response can be written as the sum of the general and the particular solution:

$$x_m(t) = E v_m(t) + e^{A_m t}\delta_0. \tag{2.52}$$

Substitution of (2.52) into (2.45) shows that the constant matrix E satisfies the relation [49]

$$A_m E - E A_v + B_m C_v = 0 \tag{2.53}$$

and that

$$\delta_0 = x_m(0) - E v_m(0). \tag{2.54}$$

Note that the stable differential equation (2.45), with input command defined by (2.50)–(2.51) always has a solution. It should be noted that if A_m and A_v both have the same dimensions, then a unique solution to (2.53) can be guaranteed only if A_m and A_v have no common eigenvalues [77]. Examples at the end of this section will better illustrate these comments.

If the plant is perfectly tracking the model, i.e., when

$$y_p(t) = y_m(t) \tag{2.55}$$

the control (2.49) becomes the "ideal" control $u_p^*(t)$ defined as

$$u_p^*(t) = \tilde{K}_x x_m(t) + \tilde{K}_u u_m(t) \tag{2.56}$$

and the plant state will be moving along some "ideal trajectories" $x_p^*(t)$ of the form

$$x_p^*(t) = X_{11} x_m(t) + X_{12} u_m(t) \tag{2.57}$$

that satisfy $y_p^*(t) = C_p x_p^*(t) = y_m(t)$.

In general, it is not known if perfect following is possible, or whether or not the ideal values $x_p^*(t)$ and $u_p^*(t)$ satisfy both plant equations (2.43)–(2.44), simultaneously and for all time. Differentiation of (2.57) with the use of (2.51), and comparison with (2.43)–(2.44) shows that if the following relations are satisfied:

$$\begin{aligned}&\left[\left(A_pX_{11}-X_{11}A_m+B_p\tilde{K}_x\right)E+A_pX_{12}C_v-X_{12}A_v\right.\\&\left.-X_{11}B_mC_v+B_p\tilde{K}_uC_v\right]\ v_m(t)=0\end{aligned} \tag{2.58}$$

$$[(C_pX_{11}-C_m)E+C_pX_{12}]\,v_m(t)=0, \tag{2.59}$$

then the ideal curves satisfy the desired conditions asymptotically i.e.:

$$\dot{x}_p^*(t)=A_px_p^*(t)+B_pu_p^*(t)-\left(A_pX_{11}-X_{11}A_m+B_p\tilde{K}_x\right)e^{A_mt}\delta_0 \tag{2.60}$$

$$y_p^*(t)=y_m(t)+(C_pX_{11}-C_m)\,e^{A_mt}\delta_0. \tag{2.61}$$

For notational convenience, define

$$X=X_{11}E+X_{12}C_v \tag{2.62}$$

$$K=\tilde{K}_xE+\tilde{K}_uC_v. \tag{2.63}$$

Substituting (2.62) and (2.63) into (2.58) and (2.59) gives:

$$\begin{bmatrix}A_p & B_p\\ C_p & 0\end{bmatrix}\begin{bmatrix}X\\ K\end{bmatrix}=\begin{bmatrix}XA_v\\ C_mE\end{bmatrix}. \tag{2.64}$$

Equation (2.64) has a *unique* solution if the plant does not have transmission zeros at the origin ($s=0$) and if no transmission zero of the plant is equal to an eigenvalue of the command matrix A_v [76], [78], [79]. These are only sufficient conditions, and they are also very mild. However, this is not enough yet, because explicit solutions of $\tilde{K}_x$ and $\tilde{K}_u$ in (2.63) are needed for implementation of the control (2.56). These solutions exist, in general, only if the number of equations is not larger than the number of variables, or equivalently only if

$$\dim(v_m)\leq\dim(u_m)+\dim(x_m). \tag{2.65}$$

Relation (2.65) restricts the complexity of the input commands that can be managed when *perfect* output tracking is sought. In the subsequent sections, it is shown that conditions (2.62)–(2.65) can be eliminated if bounded error rather than asymptotically perfect tracking is permitted. Furthermore, note that perfect tracking with fixed gains and arbitrary inputs is practically impossible because the ideal gains depend on the parameters of the input commands and may differ for different inputs. However, since the solutions exist, it is the role of the adaptive controllers to fit the right gain to the right command (and also to the right operational conditions).

Summary

The controller for tracking inputs defined by (2.50) and (2.51) is of the form:

$$u_p(t) = \tilde{K}_x x_m + \tilde{K}_u u_m + \tilde{K}_e(y_m - y_p) \tag{2.66}$$

where

$$\begin{gathered} \tilde{K}_x \text{ and } \tilde{K}_u \text{ are solutions of (2.63) and} \\ \tilde{K}_e \text{ stabilizes } (A_p - B_p\tilde{K}_e C_p). \end{gathered} \tag{2.67}$$

2.2.3.1 Two Simple Examples

Two examples have been selected to illustrate the output tracking properties discussed in this section.

Example 1: Let the plant have the following representation

$$\dot{x}_p(t) = x_p(t) + u_p(t) \tag{2.68}$$

$$y_p(t) = x_p(t). \tag{2.69}$$

The stable reference model is

$$\dot{x}_m(t) = -x_m(t) + u_m(t) \tag{2.70}$$

$$y_m(t) = x_m(t). \tag{2.71}$$

The input command is a sinusoidal wave generated by

$$\dot{v}_m(t) = \begin{bmatrix} 0 & 2 \\ -2 & 0 \end{bmatrix} v_m(t) \tag{2.72}$$

$$u_m(t) = \begin{bmatrix} 0 & 1 \end{bmatrix} v_m(t). \tag{2.73}$$

Substituting the numerical values into (2.53) gives

$$-1\,[e_1 \;\; e_2] - [e_1 \;\; e_2] \begin{bmatrix} 0 & 2 \\ -2 & 0 \end{bmatrix} + [0\ 1] = 0 \tag{2.74}$$

$$E = [e_1 \quad e_2] = \begin{bmatrix} \frac{2}{5} & \frac{1}{5} \end{bmatrix}. \tag{2.75}$$

Define

$$X = [X_1 \ X_2]; \quad K = [K_1 \quad K_2] \tag{2.76}$$

and get from (2.64)

$$\begin{bmatrix} 1 & 1 \\ 1 & 0 \end{bmatrix} \begin{bmatrix} X_1 & X_2 \\ K_1 & K_2 \end{bmatrix} = \begin{bmatrix} [X_1 \ X_2] & \times & \begin{bmatrix} 0 & 2 \\ -2 & 0 \end{bmatrix} \\ \frac{2}{5} & \frac{1}{5} & \end{bmatrix}. \tag{2.77}$$

This has solutions

$$X = \begin{bmatrix} \frac{2}{5} & \frac{1}{5} \end{bmatrix} \tag{2.78}$$

$$K = \begin{bmatrix} -\frac{4}{5} & \frac{3}{5} \end{bmatrix}. \tag{2.79}$$

Equations (2.62)–(2.63) then become

$$X_{11} \begin{bmatrix} \frac{2}{5} & \frac{1}{5} \end{bmatrix} + [0 \ X_{12}] = \begin{bmatrix} \frac{2}{5} & \frac{1}{5} \end{bmatrix} \tag{2.80}$$

$$\tilde{K}_x \begin{bmatrix} \frac{2}{5} & \frac{1}{5} \end{bmatrix} + \tilde{K}_u [0 \quad 1] = \begin{bmatrix} -\frac{4}{5} & \frac{3}{5} \end{bmatrix}. \tag{2.81}$$

One possible solution is:

$$X_{11} = 1 \quad X_{12} = [0 \ 0] \tag{2.82}$$

$$\tilde{K}_x = -2 \ \tilde{K}_u = 1 \tag{2.83}$$

and the ideal control is then

$$u_p^*(t) = -2x_m(t) + u_m(t). \tag{2.84}$$

Example 2: Since the differential equation (2.45) has a solution, the algebraic equation (2.53) also has a solution. However, the following example shows that one must be careful with the representation of the (fictitious) input generator system (2.50)–(2.51), in order to avoid any apparent contradictions in more complex situations, when the eigenvalues of the input command are identical to the eigenvalues of the model. For the same plant and model as in example 1, and for an exponential input command, assume that the input command is to be e^{-t}.

This could be generated using:

$$\dot{v}_m(t) = -v_m(t) \tag{2.85}$$

$$u_m(t) = v_m(t). \tag{2.86}$$

However, in this case, $A_m = A_v$ and (2.53) will not have a solution. An alternate way to interpret this result is to note that the particular solution $x_m(t)$ of (2.70) with input e^{-t} is a weighted combination of the independent functions e^{-t} and te^{-t}. Since this particular solution must be equivalent to $Ev_m(t)$, it is necessary that $v_m(t)$ contain both these functions. To avoid this problem, replace the representation of the fictitious command generator using

$$\dot{v}_m(t) = \begin{bmatrix} -1 & 1 \\ 0 & -1 \end{bmatrix} v_m(t). \tag{2.87}$$

Thus

$$c_v = [0 \ 1]. \tag{2.88}$$

Application of (2.53) gives:

$$-1 [e_1 \ \ e_2] - [-e_1 \ \ e_2] \begin{bmatrix} -1 & 1 \\ 0 & -1 \end{bmatrix} + [0 \ \ 1] = 0 \tag{2.89}$$

$$[-e_1 \ \ -e_2] - [-e_1 \ \ e_1 - e_2] + [0 \ \ 1\,] = 0 \tag{2.90}$$

which gives

$$E = [e_1 \ \ e_2] = [1 \ \ 0]. \tag{2.91}$$

Equation (2.64) then becomes

$$\begin{bmatrix} 1 & 1 \\ 1 & 0 \end{bmatrix} \begin{bmatrix} X_1 & X_2 \\ K_1 & K_2 \end{bmatrix} = \begin{bmatrix} [X_1 \ \ X_2] & \times \begin{bmatrix} -1 & 1 \\ 0 & -1 \end{bmatrix} \\ +1 & 0 \end{bmatrix} \tag{2.92}$$

which has the solutions

$$K = [-2 \ \ 1] \tag{2.93}$$

$$X = [1 \ \ 0]. \tag{2.94}$$

The ideal control is thus

$$u_p^*(t) = -2x_m(t) + u_m(t). \tag{2.95}$$

2.2.3 The General Tracking Problem

The previous sections have developed the conditions required for *perfect* tracking in the presence of certain type commands. It will be shown in the subsequent chapters that under appropriate conditions, the adaptive controller automatically finds the right gains that are needed to ultimately reach the perfect tracking conditions.

However, one cannot assume that perfect tracking can always be achieved, nor is it desirable to build a control system that would perfectly follow any occasional high frequency signal that may enter the input connections. Furthermore, a main property of the adaptive control techniques presented in this book, is robust stability and performance in the presence of a wide class of input signals and input or output disturbances. In this text, the concept of robust stability means that the adaptive control system remains stable with respect to boundedness of all states, errors, and adaptive gains, for a very wide class of input commands, and input and output disturbances. Robust performance is measured by the quality of tracking, and it will be shown that the adaptive controller can maintain small tracking errors in nonideal environments.

For this general situation, perfect tracking is only an ideal target, that the actual plant might eventually reach at those times when the input signal can be followed and when no disturbances are present. In more difficult situations, perfect tracking cannot be achieved, and the designer is confined to tracking with errors. It will be shown that the adaptive control systems proposed here remain stable and that the tracking errors remain bounded, in spite of the disturbances.

In a manner similar to Section 2.2.2, general classes of the input command $u_m(t)$ can be represented as output signals of an unknown and fictitious command generating system of the form (2.50)–(2.51), only this time no particular assumption on the dimension of the unknown matrices A_v and C_v, and thus on the complexity of $u_m(t)$, is used.

As satisfactory tracking remains the task that the control system has to accomplish in the best possible manner, define for reference those ideal target trajectories, x_p^*, such that, if the plant could reach and move along them, its output would perfectly track the output of the model, or in other words

$$y_p^*(t) = C_p x_p^*(t) = C_m x_m(t) = y_m(t). \tag{2.96}$$

As in (2.56), the ideal trajectories are defined as

$$x_p^*(t) = X_{11} x_m(t) + X_{12} u_m(t). \tag{2.97}$$

Substituting $x_p^*(t)$ from (2.97) into (2.96) gives as the condition of existence of such ideal target trajectories:

$$C_p X_{11} x_m(t) + C_p X_{12} u_m(t) = C_p X_{11} x_m(t) + C_p X_{12} C_v v_m(t) \\ = C_m x_m(t) \tag{2.98}$$

or

$$C_p X_{11} = C_m \tag{2.99}$$

$$C_p X_{12} C_v = 0. \tag{2.100}$$

Solutions for X_{11} and X_{12} from (2.99) and (2.100) exist, in general, because this system consists of more variables than equations. This result implies the existence of some bounded trajectories in the $x_p(t)$ space that the plant needs to attain for perfect tracking. The question is whether or not the ideal control configuration defined as

$$u_p^*(t) = \tilde{K}_x x_m(t) + \tilde{K}_u u_m(t) \tag{2.101}$$

can make the plant move along those ideal "target trajectories." In other words, one may ask whether there are trajectories of the plant, that when driven by the ideal control (2.101) become "ideal trajectories" or alternatively, whether the ideal trajectories satisfy the plant differential equation.

Differentiating $x_p^*(t)$ in (2.97) gives

$$\dot{x}_p^*(t) = X_{11}\dot{x}_m(t) + X_{12}\dot{u}_m(t) \\ = X_{11} A_m x_m(t) + X_{11} B_m C_v v_m(t) + X_{12} C_v A_v v_m(t). \tag{2.102}$$

To satisfy the plant equation in the ideal case, the ideal trajectories would have to satisfy the plant equation, i.e.,

$$\dot{x}_p^*(t) = A_p x_p^*(t) + B_p u_p^*(t). \tag{2.103}$$

Then using the ideal control configuration (2.101) and also substituting (2.97) into (2.103) gives

$$\dot{x}_p^*(t) = A_p X_{11} x_m(t) + A_p X_{12} u_m(t) + B_p \tilde{K}_x x_m(t) + B_p \tilde{K}_u u_m(t). \tag{2.104}$$

While the ideal trajectory does satisfy its differential equation (2.102), it is not clear under what general conditions it can satisfy either of the plant equations, (2.103) or (2.104). To check how close the plant trajectories can be to the ideal trajectory, add the right side of (2.103) to (2.102) and then subtract the right side of (2.104), to get

$$\begin{aligned}\dot{x}_p^*(t) &= X_{11}A_m x_m(t) + X_{11}B_m C_v v_m(t) + X_{12}C_v A_v v_m(t) \\ &+ A_p x_p^*(t) + B_p u_p^*(t) \\ &- A_p X_{11} x_m(t) - A_p X_{12} u_m(t) - B_p \tilde{K}_x x_m(t) - B_p \tilde{K}_u u_m(t).\end{aligned} \tag{2.105}$$

This can be rewritten as:

$$\dot{x}_p^*(t) = A_p x_p^*(t) + B_p u_p^*(t) - E(t) \tag{2.106}$$

where

$$\begin{aligned}E(t) &= (A_p X_{11} - X_{11}A_m + B_p\tilde{K}_x)x_m \\ &+ (A_p X_{12} C_v + B_p \tilde{K}_u C_v - X_{11} B_m C_v - X_{12} C_v A_v) v_m.\end{aligned} \tag{2.107}$$

Perfect tracking is possible whenever it is possible to eliminate this 'bias' term $E(t)$, or whenever:

$$\begin{aligned}E(t) &= (A_p X_{11} - X_{11}A_m + B_p\tilde{K}_x)x_m \\ &+ (A_p X_{12} C_v + B_p \tilde{K}_u C_v - X_{11} B_m C_v - X_{12} C_v A_v) v_m = 0.\end{aligned} \tag{2.108}$$

Since (2.108) must hold for any u_m and x_m, $\tilde{K}_x$ and $\tilde{K}_u$ must satisfy:

$$A_p X_{11} - X_{11}A_m + B_p\tilde{K}_x = 0 \tag{2.109}$$

$$A_p X_{12} C_v + B_p \tilde{K}_u C_v - X_{11} B_m C_v - X_{12} C_v A_v = 0. \tag{2.110}$$

However, these conditions contain many more equations than variables, and thus do not, in general, have a solution, as expected. Indeed, one cannot guarantee that perfect tracking is attainable for any arbitrary input command. Actually, it may not be even desirable to seek perfect tracking for any possible input command, such as, for example, any high frequency disturbance signals that may occasionally appear at the plant input. Therefore, the ideal trajectories are only "targets" that the plant tries to reach or at least be close to, in order to have bounded tracking errors (that hopefully are as small as possible).

To further extend the preliminary analysis to the most general case, recall that in the development the input $u_m(t)$ was assumed to be generated by the differential equations (2.50)–(2.51) and thus was differentiable. For the sake of generality, any particular assumptions about the input signal $u_m(t)$, except for its boundedness will now be removed. Since $u_m(t)$ is not even required to be differentiable, and also for convenience of future proofs of robust stability, it is useful to seek target trajectories of the form

$$x_p^*(t) = X x_m(t) \tag{2.111}$$

because in this manner $x_p^*(t)$ remains differentiable. Substituting (2.111) into (2.96) shows that such bounded ideal target trajectories exist if the equation

$$C_p X = C_m \tag{2.112}$$

has a solution for the matrix X. Notice that (2.112) has a solution for the matrix X if

$$\operatorname{rank}\left[C_p \quad C_m\right] = \operatorname{rank}\left[C_p\right] \tag{2.113}$$

which is satisfied, in general, because the system has more variables than equations, especially if the number of plant states greatly exceeds the number of model states.

Differentiating (2.111) gives:

$$\dot{x}_p^*(t) = X\dot{x}_m(t) = XA_m x_m(t) + XB_m u_m(t). \tag{2.114}$$

Perfect tracking with the ideal control of (2.101) is considered again with respect to whether the ideal trajectories can satisfy the plant differential equation

$$\dot{x}_p^*(t) = A_p x_p^*(t) + B_p u_p^*(t). \tag{2.115}$$

Substituting (2.101) and (2.111) into (2.115) gives

$$\dot{x}_p^*(t) = A_p X x_m(t) + B_p \tilde{K}_x x_m(t) + B_p \tilde{K}_u u_m(t). \tag{2.116}$$

Adding the right side of (2.115) to (2.114) and subtracting the right side of (2.116) gives:

$$\begin{aligned} \dot{x}_p^*(t) &= XA_m x_m(t) + XB_m u_m(t) \\ &+A_p x_p^*(t) + B_p u_p^*(t) \\ &-A_p X x_m(t) - B_p \tilde{K}_x x_m(t) - B_p \tilde{K}_u u_m(t). \end{aligned} \tag{2.117}$$

This can be written as:

$$\dot{x}_p^*(t) = A_p x_p^*(t) + B_p u_p^*(t) - E(t) \tag{2.118}$$

where

$$E(t) = (A_pX - XA_m + B_p\tilde{K}_x)x_m + (B_p\tilde{K}_u - XB_m)u_m. \quad (2.119)$$

This implies that the ideal trajectories can be plant trajectories, and that perfect tracking is possible if

$$E(t) = (A_pX - XA_m + B_p\tilde{K}_x)x_m + (B_p\tilde{K}_u - XB_m)u_m = 0. \quad (2.120)$$

Since (2.120) must hold for any u_m and x_m;

$$A_pX - XA_m + B_p\tilde{K}_x = 0 \quad (2.121)$$

$$B_p\tilde{K}_u - XB_m = 0. \quad (2.122)$$

Conditions (2.121), (2.122), that must be satisfied by the matrices $\tilde{K}_x$, and $\tilde{K}_u$, in general, contain more equations than variables. The ideal trajectories therefore remain as reference target trajectories that the plant attempts to reach.

The linear analysis of this section will be used in the next chapters, where adaptive algorithms are used to compute the control gains. In the most general tracking cases, the "ideal" controller gains $\tilde{K}_x$ and $\tilde{K}_u$, will be thought of as those solutions that minimize the 'bias' terms (2.107) or (2.119).

2.3 Stability and Positivity Concepts

Because the adaptive controllers developed in this text, must be analyzed using stability and positivity concepts, this section serves as an introduction to these subjects. While Section 2.3.1 elaborates on the conceptual differences between fixed and varying control gains, Sections 2.3.2 and 2.3.3 develop the tools needed for the subsequent analysis.

2.3.1 Introduction: Stability with Fixed Controllers Versus Nonlinear Adaptive Controllers

When it is not expected that fixed gains can guarantee stability and performance over large ranges of operational conditions, the control designer may want to use nonstationary gains, that would change when operating conditions change. When the plant is unknown, the designer may want to use adaptive control that hopefully, would fit the appropriate values of the control gains to the appropriate situation.

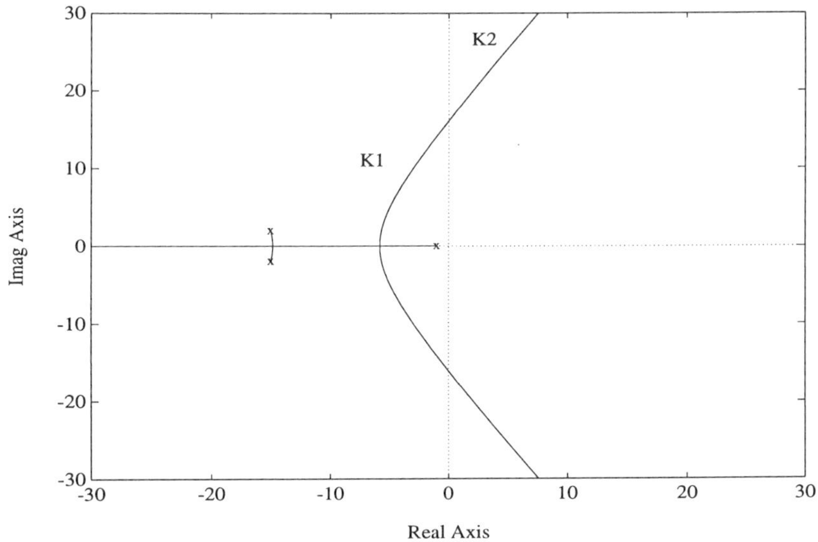

FIGURE 2.2. Root–Locus for the Rohrs' Example.

This book is all about using (simple) adaptive control (SAC) in realistic difficult situations and thus maintaining robustness of stability and performance in changing environments. However, even before the various adaptive algorithms and their rigorous proofs of stability are presented in detail, it is worthwhile to show how various elements of the SAC methodology prevent those peculiar things that may happen whenever nonstationary rather than stationary controllers are used (with no apparent reason, except for the fact that the gains become nonstationary).

It is useful to consider Rohrs' example that was used to test various adaptive controllers in a seminal paper [80] in adaptive control. The object of the test [80] is a "nominal" first order system with a pair of complex unmodeled poles described by

$$G_p(s) = \frac{2}{s+1} \frac{229}{s^2 + 30s + 229}. \tag{2.123}$$

Under the assumption that the true order of the plant is unknown, first order adaptive controllers were used for simulations. As is well–known today, all the adaptive control algorithms considered were shown to lead to instability under these conditions. Apparently, the reduced (first) order of the nonlinear adaptive controller that did not account for the "unmodeled dynamics" was to blame.

However, the Root–locus plot Figure 2.2, that could be used in a *linear* control design of such a linear time–invariant plant, may provide some very useful *qualitative* intuitive observations about the stability of the control system. The open loop plant (2.123) used here is stable, and the plot shows that the corresponding closed loop system remains stable for any *constant* gain, say, K_1, which is of course a controller of order zero. It is clear, however, that the plant becomes unstable if the gain, say K_2, is larger than some admissible limit, $K_{\max}$.

Then what happens when nonstationary gains are used? It is very attractive to apply this root–locus based intuition with the nonlinear adaptive controllers and use this upper admissible bound with the adaptive gains in order to prevent the adaptive system from becoming unstable. As this book will show, these upper bounds can indeed be used within the framework of the *particular* adaptive controllers proposed here. Notice though, that unfortunately, when the control gains are nonstationary or nonlinear, as adaptive gains are, this intuition can not be used, *in general*.

The stationary versus nonstationary control problem was first questioned by Lurie [25]: He considered a linear plant known to be stable for any *fixed* value of the gain K, between the minimal value $K_{\min}$ and the maximal value $K_{\max}$. Of interest was the stability if the fixed gain were replaced by an arbitrary nonlinear and/or nonstationary gain $K(x,t)$ within the same bounds.

Unfortunately, the famous Aizerman conjecture, namely, that the bounds of stability established for fixed gains would also guarantee stability with nonlinear gains, was shown to be false, *in general* [25], [81], [82] and, as counterexamples show, the nonstationary gains may drive the control system to total divergence. Thus, the control designer who may wish to use nonlinear (adaptive) controllers instead of fixed controllers in order to reach some improved performance, may find out that mere stability of adaptive system can not be guaranteed, not to mention performance, even under the same conditions that do guarantee stability with fixed controllers.

This situation will not happen with the simple adaptive control (SAC) methodology of this book. First, it is shown here that although root–locus arguments may not be fit for stability analysis of *general* nonlinear or nonstationary controllers, the intuition given by these arguments remains useful when designing certain classes of nonlinear controllers. For example, it can be shown that the stability bounds established for *fixed* controllers are sufficient to guarantee robust stability of the nonlinear adaptive controllers proposed in this book.

Assume for illustration that the (possibly unstable) SISO system

$$\dot{x}_p(t) = A_p x_p(t) + B_p u_p(t) \tag{2.124}$$

$$y_p(t) = C_p x_p(t) \tag{2.125}$$

becomes asymptotically stable if a fixed output feedback controller of the form $u_p(t) = -K y_p(t)$, where the fixed gain K satisfies $K_{\min} \leq K \leq K_{\max}$, is used. Then the following theorem proven in [81] is of interest:

Theorem: The adaptive system obtained from (2.124)–(2.125) with the nonlinear adaptive controller

$$u_p(t) = -K(t)y_p(t) \tag{2.126}$$

$$\dot{K}(t) = \begin{cases} \gamma \times y_p^2(t) & K < K_{\max},\ \gamma > 0 \\ 0 & K \geq K_{\max} \end{cases} \tag{2.127}$$

is asymptotically stable with bounded gains for any initial gain $K_0 = K(t=0) < K_{\max}$.

If the initial gain is small and the output has some nonzero value, the adaptive gain starts increasing. When the gain is sufficiently large, it has a stabilizing effect, and the state is driven towards the equilibrium $x(t) = 0$. Finally, the state reaches $x(t) = 0$, and the adaptive gain reaches some fixed value in the admissible stabilizing region. If the adaptive gain tends to increase too much, the upper bound imposed by (2.127) is sufficient to guarantee stability of the closed–loop system. The rigorous proof of this simple theorem is rather elaborate and can be found in reference [81].

The upper bound for the admissible control gains is merely a measure of caution, rather than real knowledge. Since it is known that real plants do not necessarily tolerate very high gains, one imposes some upper limits on their eventual range. For example, assume that the plant is stable for any fixed gain K that satisfies $-50 \leq K \leq 400$, while any "good" performance can only be obtained around the (unknown) gain $K \approx 30$. Therefore, the value 400 can not be considered real knowledge and is of no use in the linear control design. In the adaptive case, it only prevents those difficult situations when the adaptive gain might reach high values. The adaptive controller is then called to fit the right gains to the right situation in order to maintain the desired performance (small tracking errors) with various and changing tasks and operational environments.

2.3.2 Basic Stability Concepts

When dealing with adaptive controllers (as well as with any nonlinear and/or nonstationary system), one can not always use those frequency domain analysis techniques that have become so common and intuitive in control engineering practice. Although analyses based on local linearization may add some intuition on the internal structure of the systems, in general one cannot use the location of (nonstationary) poles as arguments to guarantee stability of the system or to determine performance of the control. Fortunately, Lyapunov stability analysis techniques and their extensions by LaSalle [83] to nonlinear nonautonomous systems constitute very useful means that suit the proofs of robust stability of adaptive control techniques of this book rather effectively.

To illustrate these concepts, consider the system:

$$\dot{x}(t) = f(x, t) \tag{2.128}$$

together with the following definitions:

Definition: A continuous function $W(x)$ is said to be *a locally positive definite (locally positive semidefinite)* function if $W(0) = 0$, and $W(x) > 0$ $(W(x) \geq 0)$ in some neighborhood around the origin.

Definition: A continuous function $W(x)$ is said to be a *positive definite (positive semidefinite)* function if $W(0) = 0$, and $W(x) > 0$ in R^n ($W(x) \geq 0$ in R^n) and there exists some $r > 0$ such that

$$\inf_{\| x \| > r} \quad (W(x) > 0) \quad .$$

Note that the definition for positive definite functions requires that $W(x)$ be strictly positive, and that it differentiates between functions such as x^2, which *is* positive definite, and $x^2/(1+x^4)$, which is *not* positive definite.

Definition: A continuous function $V(x,t)$ is said to be a (locally) *positive definite* function if there exists a (locally) positive definite function $W(x)$, such that $V(x,t) \geq W(x)$ for all x in R^n.

Definition: A continuous function $W(x,t)$ is said to be a *negative definite (negative semidefinite)* function if $-W(x,t)$ is *positive definite (positive semidefinite)*.

To prove stability of systems of the form (2.128) using Lyapunov techniques, one attempts to associate with (2.128) a "Lyapunov function" namely, a positive definite function of x, defined as $V(x)$, or $V(x,t)$ if it is also time–variable, that increases if $\| x \|$ increases, and decreases if $\| x \|$ decreases. If it can be shown that such a function decreases "along the trajectories" of (2.128), then the only possible conclusion would be that $V(x,t)$ can only decrease whenever $x \not\to 0$, and finally reaches the value $V(x,t) = 0$. Because $V(x,t)$ is positive definite, this result implies that ultimately $x \equiv 0$. Thus, because $V(x,t)$ cannot increase, $x(t)$ is guaranteed to be bounded, and because ultimately $V(x,t) = 0, x(t)$ is ultimately attracted to the equilibrium point $x = 0$. To test the variation of $V(x,t)$ one differentiates it "along the trajectories" of (2.128) and checks whether the derivative

$$\dot{V}(x,t) = \frac{\partial V(x,t)}{\partial t} + \frac{\partial V(x,t)}{\partial x}\dot{x}(t) = \frac{\partial V(x,t)}{\partial t} + \frac{\partial V(x,t)}{\partial x}f(x,t)$$

is negative definite.

Consider for example, the stability of linear time invariant systems of the form

$$\dot{x}(t) = Ax(t). \tag{2.129}$$

A Lyapunov function that suits the analysis of such system is the positive definite quadratic form

$$V(x) = x^T(t)Px(t) \tag{2.130}$$

where P is some positive definite symmetric matrix. The time derivative of $V(x)$ is

$$\dot{V}(x) = x^T(t)P\dot{x}(t) + \dot{x}^T(t)Px(t). \tag{2.131}$$

To find the derivative of the Lyapunov function "along the trajectories" of (2.129) one substitutes $\dot{x}(t)$ from (2.129) into (2.131) to obtain

$$\dot{V}(x) = x^T(t)PAx(t) + x^T(t)A^TPx(t) = x^T(t)\left[PA + A^TP\right]x(t). \tag{2.132}$$

Now define for convenience

$$PA + A^TP = -Q. \tag{2.133}$$

Since (2.132) must be negative definite, one gets the following theorem of stability for linear time invariant systems:

Theorem: The linear time invariant system (2.129) is *asymptotically* stable if and only if for any positive definite matrix Q there exits a positive definite matrix P that satisfies the algebraic Lyapunov equation (2.133) [12], [13]. To avoid any misinterpretation, this theorem means that if the system is asymptotically stable, then for any arbitrary selection of $Q > 0$, a positive definite solution for P exists. Yet, any single arbitrary selection is sufficient for the stability test.

It should be observed that if positive definite matrices P and Q that satisfy (2.133) exist, then $V(x)$ in (2.130) is positive definite and its derivative (2.132) is negative definite. If the derivative is not negative definite but only semidefinite, it still guarantees that $V(x)$ cannot increase, and thus $x(t)$ remains bounded. However, $x(t)$ may not go to zero. Thus consider the next theorem of stability:

Theorem: The linear time invariant system (2.129) is stable if there exists a positive definite matrix P and a positive semidefinite matrix Q that satisfy the algebraic Lyapunov equation (2.133) [12], [13].

The next step involves stability of linear time–varying systems of the form

$$\dot{x}(t) = A(t)x(t). \tag{2.134}$$

One can still use a Lyapunov function of the form (2.130), but one cannot expect that a constant matrix P can be used for weighting. Therefore the following Lyapunov equation is used

$$V(x,t) = x^T(t)P(t)x(t) \tag{2.135}$$

where $P(t)$ is some uniformly positive definite symmetric matrix. The time derivative of $V(x,t)$ is

$$\dot{V}(x,t) = x^T(t)\dot{P}(t)x(t) + x^T(t)P(t)\dot{x}(t) + \dot{x}^T(t)P(t)x(t). \tag{2.136}$$

To find the derivative of the Lyapunov function "along the trajectories" of (2.134) substitute $\dot{x}(t)$ from (2.134) into (2.136) to obtain

$$\dot{V}(x,t) = x^T(t)\dot{P}(t)x(t)+x^T(t)P(t)A(t)x(t)+x^T(t)A^T(t)P(t)x(t) \tag{2.137}$$

$$\dot{V}(x,t) = x^T(t)\left[\dot{P}(t) + P(t)A(t) + A^T(t)P(t)\right]x(t). \tag{2.138}$$

Now define for convenience

$$\dot{P}(t) + P(t)A(t) + A^T(t)P(t) = -Q(t). \tag{2.139}$$

From (2.138)–(2.139) the following theorem of stability is obtained:

Theorem: The linear time varying system (2.134) is uniformly *asymptotically* stable if and only if there exist two positive definite matrices, $P(t)$ and $Q(t)$, that satisfy the differential Lyapunov equation (2.139). The system is stable if the equation can only be satisfied for a positive semidefinite matrix $Q(t)$ [12], [13].

A nonlinear autonomous differential equation (when time is not an explicit variable)

$$\dot{x}(t) = f(x) \tag{2.140}$$

may have more than one equilibrium point, and one must differentiate between *local* stability around an equilibrium point and *global* stability that includes the entire state–space. Without loss of generality, it is assumed that the origin is an equilibrium point. The following theorem of stability can be formulated:

Theorem: The equilibrium point $x = 0$ of the autonomous nonlinear system (2.140) is (locally) globally and uniformly *asymptotically* stable if there exists a (locally) positive definite function $V(x)$ whose derivative "along the trajectories" of (2.140) is negative definite. The system is stable if the derivative is negative semidefinite [12], [13].

For the nonlinear nonautonomous differential equation (i.e., when time is an explicit variable)

$$\dot{x}(t) = f(x,t) \tag{2.141}$$

one formulates the following theorem of stability:

Theorem: The nonlinear system (2.141) is (locally) globally uniformly *asymptotically* stable if there exists a (locally) positive definite function

$V(x,t)$ whose derivative "along the trajectories" of (2.141) is (locally) negative definite. The system is stable if the derivative is negative semidefinite [12], [13].

In nontrivial systems, even if the system is asymptotically stable, it may be difficult to find appropriate positive definite functions with negative definite derivatives. LaSalle has extended the Lyapunov techniques with his "invariance principle" for those cases where either $V(x,t)$ is not positive definite, or the derivative is only negative semidefinite. A modified version of the invariance principle, that is needed when the definiteness of either $V(x,t)$ or $\dot{V}(x,t)$ is affected by transient terms is now considered [49]. Basically, this is an extension to LaSalle's Invariance Principle for nonautonomous nonlinear differential equations for the case when, due to transient terms, $\dot{V}(x,t)$ is not necessarily negative definite or even semi–definite.

To illustrate these concepts let:

$$\dot{x}(t) = f(x,t) \tag{2.142}$$

be a general nonlinear nonautonomous differential equation and assume that at any time either

Assumption 1: $f(x,t)$ is bounded for any bounded value of x, or

Assumption 2: $\mid \int_a^b f(x,t)dt \mid < \ \mu(b-a)$

where the function $\mu(t)$ is a 'modulus of continuity' for the integral [83] and implies that $\mu(b-a)$ is bounded for any finite interval $(b-a)$ (where a and b are the limits of the integral). **Assumptions 1** or **2** above imply that either the velocities are bounded or, that the system can not pass an infinite distance in a finite amount of time. Notice that only one of the conditions above is required, and **Assumption 2**, allows for tests of stability when $f(x,t)$ is not necessarily bounded, such as, for example, in the case when $f(x,t) = g(x) + h(x)u(t)$ and $u(t)$ contains a finite number of impulse functions.

Let $V(x,t)$ be a differentiable function bounded from below. Notice that $V(x,t)$ is not required to be positive definite. Assume that the derivative $\dot{V}(x,t)$, along the trajectories of (2.142) is

$$\dot{V}(x,t) = W_1(x,t) + W_2(x,t) \tag{2.143}$$

such that

$$W_1(x,t) \ \leq \ \tilde{W}_1(x) \leq 0 \tag{2.144}$$

where $\tilde{W}_1(x)$ is a continuous function of x and where $W_2(x,t)$ is a continuous function of x and piecewise continuous in t satisfying

$$\lim_{t\to\infty} W_2(x,t) = 0 \tag{2.145}$$

Thus $\dot{V}(x,t) \rightarrow W_1(x,t)$ as $t \rightarrow \infty$, and $W_1(x,t)$ is defined as the "limiting derivative of the Lyapunov function."

Theorem: Based upon the satisfaction of **Assumption 1** or **2**, and (2.143)–(2.145), all bounded solutions of (2.128) approach asymptotically the set

$$\Omega \triangleq \{x \mid \tilde{W}_1(x) \equiv 0\} \tag{2.146}$$

Proof: In order to analyze stability of trajectories it is useful to define limit points (or accumulation points). Define an infinite sequence of discrete times $t_1,\ t_2,\ t_3,\ \ldots\ t_i, \ldots$ and the corresponding points on the trajectory $x(t_1), x(t_2), x(t_3), \ldots, x(t_i), \ldots$. A limit point is defined as a point about which any arbitrarily small neighborhood contains an infinite number of points $x(t_i)$ of the trajectory. An example of a limit point is a stable equilibrium point that is asymptotically approached by the trajectory as time goes to infinity. Thus, for any neighborhood around the equilibrium point there exists some finite time t_0 such that all points on the corresponding trajectory belong to this neighborhood for all times larger than t_0. However, the class of limit points is more general then the class of equilibrium points, and the following description is needed.

It is easy to see that any bounded trajectory contains a limit point. If the trajectory is bounded, it is contained in some finite hypercube. The bounded trajectory contains an infinite number of points $x(t_i)$. If one divides the hypercube in two parts, then at least one part must contain an infinite number of points. Dividing again this part, one must conclude that a smaller subdivision contains an infinite number of points. The division can be continued, until one concludes that some arbitrarily small subdivision, contains an infinite number of points, and therefore the trajectory must have, at least, a limit point.

Let $x(t)$ be a bounded trajectory of (2.128) and let p be a limit point of this trajectory. Then, for this finite point p, there exists some finite constant c, such that

$$V(p,t) < \infty, \qquad \lim_{t\to\infty} V(p,t) = c < \infty \tag{2.147}$$

It is desired to show that p must belong to Ω, such that any trajectory ultimately belongs to Ω. To show this by a contradiction argument, assume that p does *not* belong to Ω. Then

$$\tilde{W}_1(p) < 0 \tag{2.148}$$

Let $S_\varepsilon(p)$ be a neighborhood of p such that, if $x \in S_\varepsilon(p)$, then $\tilde{W}_1(x) < -d_1 < 0$ for some $d_1 > 0$ sufficiently small, i.e.,

$$S_\varepsilon(p) \stackrel{\Delta}{=} \{x \epsilon \Omega \colon \parallel x - p \parallel < \varepsilon\} \tag{2.149}$$

Because $W_2(x,t)$ eventually vanishes, there exists a finite time t_1, such that

$$\dot{V}(x,t) \le \tilde{W}_1(x) + W_2(x,t) < -d_2 < 0 \tag{2.150}$$

for all $t \ge t_1$ and all $x \epsilon\ S_\varepsilon(p)$ and for some d_2 where $0 < d_2 \le d_1$.

Assume first that $x(t)$ approaches the limit point p asymptotically. In other words, after some finite time, $x(t)$ enters the neighborhood of p and remains in this neighborhood thereafter. There exists a time $t_2 \ge t_1$ such that

$$x(t)\ \epsilon\ S_\varepsilon(p) \quad \text{for all} \quad t \ge t_2 \tag{2.151}$$

In this case

$$V(x,t) = V(x,t_2) + \int_{t_2}^{t} \dot{V}(x,t)dt \tag{2.152}$$

and thus

$$V(x,t) \to -\infty \quad \text{as} \quad t \to \infty \tag{2.153}$$

which contradicts the assumptions.

Before one can conclude that p belongs to Ω, one must eliminate the eventuality of another possible situation. The trajectory $x(t)$ may approach the limit point p in a different way, namely, $x(t)$ may enter $S_\varepsilon(p)$ and leave it immediately, an infinite number of times. This way, the total amount of time that $x(t)$ is within $S_\varepsilon(p)$, could still remain finite, and in this case the integral (2.152) could remain finite. However, by **Assumptions 1** or **2**, the trajectories cannot pass an infinite distance in a finite amount of time.

One must therefore conclude that (2.148) is impossible and that $\tilde{W}_i(p) = 0$, so that:

$$p \in \Omega \tag{2.154}$$

and, since p stands for any limit point of (2.128), all bounded trajectories approach asymptotically the set Ω defined by (2.146).

This is a very general result, and allows for analysis of complex nonlinear systems even when one cannot find the "best" Lyapunov functions. In particular cases, one must analyze the meaning of the set Ω, because it could turn out to be a collection of equilibrium points, limit cycles, etc.. In some cases, using other information, it may be possible to show that all trajectories are bounded, and that Ω is actually reduced to a unique equilibrium point.

In particular, let $V(x)$ be a positive definite quadratic function of x, with the form

$$V(x) = x^T P x. \tag{2.155}$$

Then, one can easily prove the following theorems [83], [12], [13], [14], [15], [16]:

Theorem: Assume that there exists a compact (closed and bounded) set S_b around the origin such that $\dot{V}(x,t) < 0$ for any $x(t)$ outside S_b. Then $x(t)$ is bounded and ultimately bounded by $x(t) \in S_b$.

Theorem: Assume that $\dot{V}(x,t) < 0$ for any $x(t) \neq 0$ (or for any $x(t) \neq 0$ in some neighborhood around the origin) and $\dot{V}(x,t) = 0$ for $x(t) = 0$. Then the equilibrium point $x(t) = 0$ is globally (or locally) asymptotically stable.

2.3.3 Positive Real Analysis

Motivation

To motivate the subsequent concepts, consider the single input single output system:

$$\begin{aligned} \dot{x}_p(t) &= A_p x_p(t) + B_p u_p(t) \\ y_p(t) &= C_p x_p(t) \end{aligned} \tag{2.156}$$

where x_p is the $(n \times 1)$ state vector, and u_p and y_p are the scalar control and output signals respectively.

Stability will be considered for various possible controls based upon the Lyapunov function:

$$V(x_p) = x_p^T P x_p \tag{2.157}$$

where P is a positive definite symmetric matrix.

The derivative of $V(x_p)$ is then:

$$\begin{aligned} \dot{V}(x_p) &= \dot{x}_p^T P x_p + x_p^T P \dot{x}_p = x_p^T P A x_p + x_p^T A_p^T P x_p \\ &= x_p^T (PA + A_p^T P) x_p = -x_p^T Q x_p. \end{aligned} \tag{2.158}$$

Then as shown in Section 2.3.2, the linear time invariant system is asymptotically stable if and only if there exist positive definite matrices P and Q, such that the following Lyapunov equation is satisfied:

$$PA_p + A_p^T P = -Q. \tag{2.159}$$

Consider the plant (2.156) again, and use the Lyapunov equation to determine if the plant can be stabilized by the constant output feedback controller

$$u_p(t) = -K_e y_p(t) = -K_e C_p x_p(t). \tag{2.160}$$

The closed loop system is then

$$\begin{aligned} \dot{x}_p(t) = A_p x_p(t) + B_p u_p(t) = A_p x_p(t) - B_p K_e C_p x_p(t) = \\ (A_p - B_p K_e C_p) x_p(t). \end{aligned} \tag{2.161}$$

Then, the closed loop system is stable, or the original plant is stabilizable, if there exists some positive definite matrices P and Q, such that the following Lyapunov equation is satisfied:

$$P(A_p - B_p K_e C_p) + (A_p - B_p K_e C_p)^T P = -Q. \tag{2.162}$$

Now, assume that the plant *can* be stabilized by some unknown output feedback gain K_e. To illustrate the use of adaptive control and the resulting stability analysis, consider a controller of the form:

$$\begin{aligned} u_p(t) &= -K(t) y_p(t); \\ \dot{K}(t) &= \gamma y_p^T(t), \ \gamma > 0. \end{aligned} \tag{2.163}$$

The closed loop system is then

$$\dot{x}_p(t) = A_p x_p(t) - B_p K(t) C_p x_p(t) = [A_p - B_p K(t) C_p] \, x_p(t). \tag{2.164}$$

Adding and subtracting $B_p K_e C_p x_p(t)$ gives:

$$\dot{x}_p(t) = [A_p - B_p K_e C_p] \, x_p(t) - B_p \, [K(t) - K_e] \, C_p x_p(t) \tag{2.165}$$

where K_e is a stabilizing gain which is known to exist but is unknown in value. To check the stability of the closed–loop adaptive system, one must also take into account the adaptive gain. Therefore, one must select a quadratic Lyapunov function that includes all dynamic values of the system, i.e. both $x_p(t)$ and $[K(t) - K_e]$ (as the state is expected to reach zero, and the adaptive gain $K(t)$ is expected to reach the gain K_e). Consider:

$$V = x_p^T P x_p + tr\{[K(t) - K_e] \, [K(t) - K_e]^T\}/\gamma \tag{2.166}$$

(where the trace operator, denoted as 'tr' above, is simply the sum of the diagonal entries). The derivative of the Lyapunov function is

$$\dot{V}(x) = \dot{x}_p^T P x_p + x_p^T P \dot{x}_p + tr\{[K(t) - K_e]\, \dot{K}^T(t)\}/\gamma +$$
$$tr\{\dot{K}(t)\, [K(t) - K_e]^T\}/\gamma$$
$$= x_p^T P\, [A_p - B_p K_e C_p]\, x_p + x_p^T\, [A_p - B_p K_e C_p]^T\, P x_p \tag{2.167}$$
$$-x_p^T P B_p\, [K(t) - K_e]\, x_p - x_p^T\, [K(t) - K_e]^T\, B_p^T P x_p$$
$$+tr\{[K(t) - K_e]\, \dot{K}^T(t)\}/\gamma + tr\{\dot{K}(t)\, [K(t) - K_e]^T\}/\gamma.$$

Substituting $\dot{K}(t) = \gamma y^2(t)$ gives:

$$\dot{V}(x) = x_p^T P\, [A_p - B_p K_e C_p]\, x_p + x_p^T\, [A_p - B_p K_e C_p]^T\, P x_p$$
$$-x_p^T P B_p\, [K(t) - K_e]\, C_p x_p - x_p^T C_p^T\, [K(t) - K_e]^T\, B_p^T P x_p$$
$$+tr\{[K(t) - K_e]\, \gamma y^2(t)\}/\gamma + tr\{\gamma y^2(t)\, [K(t) - K_e]^T\}/\gamma$$
$$= x_p^T \{P\, [A_p - B_p K_e C_p] + [A_p - B_p K_e C_p]^T\, P\} x_p$$
$$-x_p^T P B_p\, [K(t) - K_e]\, C_p x_p - x_p^T C_p^T\, [K(t) - K_e]^T\, B_p^T P x_p$$
$$+y_p^T\, [K(t) - K_e]\, y_p + y_p^T\, [K(t) - K_e]\, y_p \tag{2.168}$$
$$= -x_p^T Q x_p - x_p^T P B_p\, [K(t) - K_e]\, C_p x_p$$
$$-x_p^T C_p^T\, [K(t) - K_e]^T\, B_p^T P x_p$$
$$+x_p^T C_p^T\, [K(t) - K_e]\, C_p x_p + x_p^T C_p^T\, [K(t) - K_e]\, C_p x_p$$
$$= -x_p^T Q x_p - x_p^T\, \left[P B_p - C_p^T\right]\, [K(t) - K_e]\, C_p x_p$$
$$-x_p^T C_p^T\, [K(t) - K_e]^T\, \left[B_p^T P - C_p\right]\, x_p.$$

The first term in (2.168) is indeed negative definite, but the second and third terms are not necessarily sign definite. Yet, if the matrix P that satisfies (2.162) happens to also satisfy the relation

$$P B_p = C_p^T \tag{2.169}$$

the second and third terms of $\dot{V}$ vanish, and one gets from (2.168)

$$\dot{V}(x_p) = -x_p^T Q x_p. \tag{2.170}$$

This guarantees that all states and gains are bounded, and also that the states eventually reach the region $\Omega \stackrel{\Delta}{=} \{x_p \mid \dot{V}(x_p) \equiv 0 \}$, which from (2.170) is equivalent to $x_p = 0$. Thus, for such a system, stability with the adaptive gain is guaranteed. Adding together the assumptions used for the proof of stability with nonlinear adaptive gains, one can claim that the adaptive system is stable if there exist two positive definite symmetric matrices, P and Q, such that the fictitious closed–loop system using the unknown gain K_e satisfies simultaneously the following conditions:

$$P\left[A_p - B_p K_e C_p\right] + \left[A_p - B_p K_e C_p\right]^T P = -Q < 0 \tag{2.171}$$

$$P B_p = C_p^T. \tag{2.172}$$

It will be subsequently shown that these equations (2.171)–(2.172) are related to the positive realness property of the corresponding input–output transfer function of the system. In particular, if these relations are satisfied for a positive semidefinite matrix Q, the system is called "positive real" (PR). If Q is positive definite, the system is denoted as "strictly positive real" (SPR). Notice that it is not the original plant, but some stabilized closed–loop plant that is required to be SPR. However, because in this particular case only a constant feedback gain separates the original controlled plant from strict positive realness, the plant is denoted as "almost strictly positive real" (ASPR). The subsequent adaptive algorithms and the proofs of stability will be applied first to plants that satisfy such ASPR conditions. Afterwards, the techniques will be generalized to increase the scope of applicability.

It is important to note that ASPR transfer functions remain stable for any high gain positive definite feedback, be it fixed or nonlinear and nonstationary. For example, assume that the feedback control signal is

$$u_p(t) = -\left[K_e + \Delta K(x,t)\right] y_p(t) \tag{2.173}$$

where $\Delta K(x,t)$ is any arbitrary positive definite function. The closed loop system is then

$$\dot{x}_p(t) = \left[A_p - B_p(K_e + \Delta K(x,t))C_p\right] x_p(t). \tag{2.174}$$

Using again the Lyapunov function $V(x_p) = x_p^T P x_p$, and differentiating gives:

$$\dot{V}(x_p) = \dot{x}_p^T P x_p + x_p^T P \dot{x}_p =$$

$$x_p^T \{P\left[A_p - B_p K_e C_p\right] + \left[A_p - B_p K_e C_p\right]^T P\} x_p \tag{2.175}$$

$$-x_p^T P B \Delta K(t) C x_p - x_p^T C^T \Delta K(t) B^T P x_p.$$

Substituting (2.174) results in:

$$\begin{aligned} \dot{V}(x_p) &= \dot{x}_p^T P x_p \;+\; x_p^T P \dot{x}_p \\ &= x_p^T \{P\left[A_p - B_p K_e C_p\right] + \left[A_p - B_p K_e C_p\right]^T P\} x_p \\ &\quad -2 x_p^T C^T \Delta K(t) B^T C x_p \\ &= -x_p^T Q x_p - 2 y_p^T \Delta K(t) y_p \end{aligned} \tag{2.176}$$

which is negative definite for $\Delta K \;>\; 0$, and therefore (2.174) remains asymptotically stable with any gain that is higher than some minimal value. This property does not imply that use of a high gain is recommended. On the contrary, adaptive controllers try to maintain the adaptive gains as small as possible. However, this concept will be needed for robustness of adaptive control in nonideal environments when due to internal and external disturbances, the adaptive gains generated by the tracking errors may possibly become very large.

Positive Real Concepts

This section contains a few useful definitions and lemmas regarding the positive–realness property in linear *multivariable* time–invariant systems. A few definitions are presented first in the time–domain. Afterwards, the relationship between time–domain and frequency domain will be discussed.

The transfer function $G(s) = D + C(sI - A)^{-1}B$ is called "strictly proper" if $D = 0$ and "proper" if $D \neq 0$.

Definition 1: The m-input m-output proper *minimal* (controllable and observable) realization

$$\dot{x}_p(t) = A_p x_p(t) + B_p u_p(t) \tag{2.177}$$

$$y_p(t) = C_p x_p(t) + D_p u_p(t) \tag{2.178}$$

where $A_p \epsilon R^{n \times n}$, $B_p \epsilon R^{n \times m}$, $C_p \epsilon R^{m \times n}$, $D_p \epsilon R^{m \times m}$, $D_p \;>\; 0$ and B_p and C_p are of maximal rank is called "positive real" (PR) if the following time–domain relations are satisfied [17], [84]:

$$P A_p + A_p^T P - P = -Q - L^T L \;\leq\; 0 \tag{2.179}$$

$$PB_p = C_p^T - L^T W \tag{2.180}$$

$$D_p + D_p^T = W^T W \tag{2.181}$$

where $L \epsilon R^{m \times n}$, and $W \epsilon R^{m \times m}$, and where $P \epsilon R^{n \times n}$ is positive *definite* and $Q \epsilon R^{n \times n}$ is positive *semidefinite.*

Definition 2: A strictly proper minimal realization, with $D_p = 0$ is called "positive real" (PR) if the following time–domain relations are satisfied [17], [84]:

$$PA_p + A_p^T P = -Q \leq 0 \tag{2.182}$$

$$PB_p = C_p^T. \tag{2.183}$$

Though useful in many applications, positive–realness is not sufficient for proving robust stability of various control systems, in particular when nonstationary, nonlinear or adaptive controllers are involved, because (2.179) or (2.182) may imply only stability rather than asymptotic stability. Stronger "strict positive–realness" conditions are usually needed, as the designer wants Q to be positive definite. The desired definitions are presented below.

Definition 3: The m-input m-output proper minimal realization (2.177)–(2.178) is called "strictly positive real" (SPR) if the following time–domain relations are satisfied [85]:

$$PA_p + A_p^T P = -Q - L^T L < 0 \tag{2.184}$$

$$PB_p = C_p^T - L^T W \tag{2.185}$$

$$D_p + D_p^T = W^T W. \tag{2.186}$$

Solving (2.185) for L and then substituting this and (2.186) into (2.184) gives [86]:

$$PA_p + A_p^T P + \left[PB_p - C_p^T\right]\left[D_p + D_p^T\right]^{-1}\left[B_p^T P - C_p\right] = -Q < 0 \tag{2.187}$$

that will be used in the subsequent proofs of stability.

Definition 4: A strictly proper minimal realization, with $D_p = 0$, is called "strictly positive real" (SPR) if the following time–domain relations are satisfied [85]:

$$PA_p + A_p^T P = -Q < 0 \tag{2.188}$$

$$PB_p = C_p^T. \tag{2.189}$$

Notice that (2.184) and (2.188) imply asymptotic stability.

The name "positive real" originates in frequency domain representations. Then when is a transfer function positive real? Assume that a single input single output system has the transfer function $g(s)$. By definition, $g(s)$ is positive real if it is real for any real value of s, and if its real part is nonnegative on the closed right plane, or

$$Re\{g(s)\,|_{Re(s)\geq 0}\} \geq 0. \tag{2.190}$$

It can be shown that the positive real $g(s)$ is stable and minimum–phase, and of relative degree not exceeding unity. In other words, strictly proper transfer functions of physical (strictly proper) systems have n poles and $(n-1)$ zeros. (Note, for completion, that proper but not strictly proper positive real transfer functions have the same number of poles and zeros, and nonproper positive real transfer function have $n-1$ poles and n zeros.) For example, one can easily check that

$$g_1(s) = \frac{1}{s+1} \tag{2.191}$$

and

$$g_2(s) = \frac{s+1.5}{s^2+3s+2} \tag{2.192}$$

are positive real transfer functions. One can plot the Nyquist (polar) plot of $g(j\omega)$ and check that the phase $\phi\{g(j\omega)\}$ satisfies $-90° \leq \phi\{g(j\omega)\} \leq 90°$, which guarantees (2.190).

It can be shown that the time domain definitions of positive realness are equivalent to the following frequency–domain definition

Definition 5: The $m \times m$ transfer function matrix $G_p(s)$ is called "positive real" (PR) if [17], [84]:

5a. All elements of $G_p(s)$ are analytic in $Re(s) > 0$ (i.e., they do not have poles in $Re\,[s] > 0$).

5b. $G_p(s)$ is real for real s.

5c. $G_p(s) + G_p^*(s) \geq 0$ for $Re\,[s] \geq 0$.

It can be shown [17], [84] that positive real systems are stable (but not necessarily asymptotically stable) and minimum–phase (but not necessarily strictly minimum–phase).

In proper but not strictly proper transfer functions plants (with $D_p > 0$), it can be shown that **Definition 3** is equivalent to the following frequency domain definition for "strict" positive realness [17], [84]:

Definition 6: The $m \times m$ proper but not strictly proper transfer function matrix $G_p(s) = D_p + C_p(sI - A_p)^{-1}B_p$ with $D_p > 0$, is called "strictly positive real" (SPR) if:

6a. All elements of $G_p(s)$ are analytic in $Re(s) \geq 0$ (i.e., they do not have poles in $Re\,[s] \geq 0$).

6b. $G_p(s)$ is real for real s.

6c. $G_p(s) + G_p^*(s) > 0$ for $Re\,[s] \geq 0$.

The frequency domain SPR conditions for strictly proper systems are more complex. One would expect that "strict" positivity would indeed imply that $G_p(s) + G_p^*(s) > 0$ for $Re\,[s] \geq 0$ as in 6c above. However, strictly proper transfer functions cannot satisfy condition 6c. because they vanish for $|\, s \,| \rightarrow \infty$ for $Re\,[s] \geq 0$ (as one can easily see from (2.191) and (2.192) for the SISO case). Prior tentative definitions assumed that it might be sufficient to require that SPR systems satisfy $G_p(s)+G_p^*(s) > 0$ for $Re[s] \geq 0$ and s finite. Counterexamples [87] show that this tentative definition does not necessarily imply (2.188) with $Q > 0$.

Finally, Wen [85] shows that strict positivity (**Definition 4**) is equivalent to

Definition 7: The $m \times m$ strictly proper transfer function matrix $G_p(s) = C_p(sI - A_p)^{-1}B_p$ is called "strictly positive real" (SPR) if:

7a. All elements of $G_p(s)$ are analytic in $Re(s) \geq 0$.

7b. $G_p(s)$ is real for real s.

7c. $G_p(s) + G_p^*(s) \geq 0$ for $Re\,[s] \geq 0$, $G_p(s) + G_p^*(s) > 0$ for finite s, and $\lim_{\omega \rightarrow \infty} \{\omega^2 \left[G_p(j\omega) + G_p^*(j\omega)\right]\} > 0$.

Although very useful, the SPR properties are not very common among real world systems. This book will therefore show how the basic stabilizability properties of plants can be used in order to design augmentation dynamics that enable satisfaction of the desired SPR conditions. To this end, define "almost strictly positive real" (ASPR) systems as follows.

Definition 8: [71], [58], [70]: Let $G_a(s)$ be a $m \times m$ transfer matrix. Assume that there exists a positive definite constant gain matrix, K_e,not necessarily known, such that the closed–loop transfer function

$$G_s(s) = [I + G_a(s)K_e]^{-1}\, G_a(s) \tag{2.193}$$

is SPR. Because only a constant output feedback is needed to satisfy the strict positive realness, $G_a(s)$ is called "almost strictly positive real".

It can be shown [71], [88], [89], that any transfer function matrix with n poles and $n - m$ minimum–phase (transmission) zeros is ASPR. (For more information on poles, zeros, and relative degree in multivariable systems, see Appendix 2C). This claim is summarized in the following lemma.

Lemma 1: Let $G(s)$ be a strictly minimum–phase $m \times m$ transfer matrix of relative degree m. (n poles and n-m zeros). Let $G(s)$ have the minimal realization $\{A,\ B,\ C\}$ where $CB\ >\ 0$ (positive definite). Then, $G(s)$ is ASPR.

In other words, there exists a (unknown) positive definite gain K_e such that the fictitious closed–loop system satisfies the strict–positive–realness conditions (2.171)-(2.172).

Lemma 2: Any proper but not strictly proper (i.e., $\det D_p\ \neq\ 0$) and strictly minimum–phase $m \times m$ transfer matrix $G(s) = D_p + C_p(sI - A_p)^{-1}B_p$ is ASPR [71].

It is worth mentioning again that multivariable ASPR systems, which are strictly minimum–phase systems of relative degree zero or m, maintain stability with high (positive definite) gains. This property is important when nonstationary or nonlinear (adaptive) control is used, because the robustness of the control system is maintained in various operational conditions, even if the time–varying gains become too large.

2.4 Adaptive Control Based on CGT

An algorithm is described for direct model reference adaptive control of MIMO systems which do not satisfy the perfect model following conditions. Asymptotic stability is proven provided that certain inequality constraints independent of the model are satisfied for all admissible plant parameter values. The algorithm only requires that the plant outputs and reference model states are available for measurement. Furthermore, the order of the plant is allowed to be much larger than the order of the reference model. The basic algorithms presented in this section assume that the controllable plants are "almost strictly positive real" (ASPR). This ASPR restriction will be relaxed in Chapter 3.

A summary of the algorithm may be found in Table 2.1.

2.4.1 Controller Structure

The CGT control for plants with known parameters was shown in Section 2.2 to be

$$u_p(t) \;=\; S_{21}x_m(t) \;+\; S_{22}u_m \;+\; K\left[y_m(t) \;-\; y_p(t)\right]. \tag{2.194}$$

The adaptive control law is chosen to have a similar form to the CGT control law and is described by

$$u_p(t) \;=\; K_x(t)x_m(t) \;+\; K_u(t)u_m \;+\; K_e(t)[y_m(t) \;-\; y_p(t)]. \tag{2.195}$$

Substituting the definition from (2.16) that $y_m = y_p^*$ into (2.195), yields,

$$u_p(t) = K_x(t)x_m(t) + K_u(t)u_m + K_e(t)C_p e_x(t), \tag{2.196}$$

where the gains $K_x(t), K_u(t)$, and $K_e(t)$ are adaptive.

To simplify later computations, the adaptive gains are concatenated into the $m \times n_r$ matrix $K_r(t)$, which is defined as

$$K_r(t) = [K_e(t),\ K_x(t),\ K_u(t)]\,. \tag{2.197}$$

Correspondingly, the states are put into respective locations in the $n_r \times 1$ vector $r(t)$, which is defined as

$$r(t) = \begin{bmatrix} C_p e_x(t) \\ x_m(t) \\ u_m \end{bmatrix} = \begin{bmatrix} y_m(t) - y_p(t) \\ x_m(t) \\ u_m \end{bmatrix}, \tag{2.198}$$

then,

$$u_p(t) = K_r(t)r(t). \tag{2.199}$$

The concatenated gain $K_r(t)$ is defined here as the sum of a proportional gain $K_p(t)$ and an integral gain $K_I(t)$, each of which is adapted as follows:

$$K_r(t) = K_p(t) + K_I(t), \tag{2.200}$$

$$K_p(t) = v(t)r^T(t)\bar{T}, \tag{2.201}$$

$$\dot{K}_I(t) = v(t)r^T(t)T, \tag{2.202}$$

$$K_I(0) = K_{I0}, \tag{2.203}$$

$$v(t) = C_p e_x(t) = (y_m - y_p), \tag{2.204}$$

where $T, \bar{T}$ are $n_r \times n_r$ time–invariant weighting matrices, K_{I0} is the initial integral gain, and C_p is the time–invariant $m \times n$ plant output matrix. Selection of the weighting matrices T and $\bar{T}$ and the plant output matrix C_p is limited by the sufficient conditions for stability.

In partitioned form, the above gain equations become:

$$K_e(t) = K_{I_e}(t) + K_{p_e}(t) \tag{2.205}$$

$$K_x(t) = K_{I_x}(t) + K_{p_x}(t) \tag{2.206}$$

$$K_u(t) \;=\; K_{I_u}(t) + K_{p_u}(t) \tag{2.207}$$

where

$$\dot{K}_{I_e}(t) \;=\; v(t)e^T(t)T_e, \quad T_e \;>\; 0 \tag{2.208}$$

$$K_{p_e}(t) \;=\; v(t)e^T(t)\bar{T}_e, \quad \bar{T}_e \;\geq\; 0 \tag{2.209}$$

$$\dot{K}_{I_x}(t) \;=\; v(t)x_m^T(t)T_x \quad T_x \;>\; 0 \tag{2.210}$$

$$K_{p_x}(t) \;=\; v(t)x_m^T(t)\bar{T}_x, \quad \bar{T}_x \;\geq\; 0 \tag{2.211}$$

$$\dot{K}_{I_u}(t) \;=\; v(t)u_m^T(t)T_u, \quad T_u \;>\; 0 \tag{2.212}$$

$$K_{p_u}(t) \;=\; v(t)u_m^T(t)\bar{T}_u \quad \bar{T}_u \;\geq\; 0. \tag{2.213}$$

Note that (2.25)–(2.26) assume that the number of controls, denoted by m, is equal to the number of outputs, denoted by q. If m is greater than q, then another output can be included or a control might be eliminated. An alternative approach might be to linearly combine two controls.

2.4.2 Stability Analysis

The adaptive control algorithm is now analyzed for stability. The equations which govern the closed–loop system are summarized in Table 2.1.

Asymptotic stability of the output error for the adaptive control algorithm described in Table 2.1 will be shown by using a Lyapunov approach which involves (i) finding a Lyapunov candidate V, positive definite in the state variables, and (ii) evaluating the closed–loop stability by analyzing the sign of the derivative $\dot{V}$. This stability result is summarized in the following theorem.

Theorem 2.1: Suppose there exits a real symmetric positive definite matrix P and real matrices Q and $\tilde{K}_e$ such that

$$P(A_p - B_p\tilde{K}_eC_p) + (A_p - B_p\tilde{K}_eC_p)^T P \;<\; 0 \tag{2.214}$$

$$
\begin{aligned}
r(t) &= \begin{bmatrix} y_m(t) - y_p(t) \\ x_m(t) \\ u_m \end{bmatrix} \\
K_r(t) &= [K_e(t),\ K_x(t),\ K_u(t)] \\
u_p(t) &= K_r(t) r(t) \\
K_r(t) &= K_I(t) + K_p(t) \\
\dot{K}_I(t) &= v(t) r^T(t) T, \quad T > 0 \\
K_p(t) &= v(t) r^T(t) \bar{T}, \ \bar{T} \geq 0 \\
v(t) &= C_p e_x(t) = y_m(t) - y_p(t)
\end{aligned}
$$

TABLE 2.1. Algorithm Summary

$$C_p = Q B_p^T P \tag{2.215}$$

where the matrices T and $\bar{T}$ are positive definite symmetric and positive semi–definite symmetric, respectively. Then, all states and gains in the adaptive system as defined in Table 2.1 are bounded, and the output error is asymptotically stable.

Proof: See Appendix 2A.

The constraints given by (2.214)–(2.215) are equivalent to requiring that the plant be almost strictly positive real (ASPR). That is, the equations are equivalent to the existence of a real matrix $\tilde{K}_e$, which may be a function of A_p and B_p, such that the transfer function matrix

$$H(s) = C_p(sI - A_p + B_p \tilde{K}_e C_p)^{-1} B_p \tag{2.216}$$

is strictly positive real. Note that the matrix $\tilde{K}_e$ is not needed for implementation of the adaptive control algorithm.

2.4.3 System Constraints

In order to solve the adaptive control problem with time invariant A_p and B_p, it is sufficient that the constraints given by (2.214)–(2.215), or equivalently by the strict positive realness of $H(s)$ as defined in (2.216), be satisfied for all permissible A_p and B_p. Thus, this section discusses techniques for satisfying the constraints for a bounded set of parameters.

Frequency Domain Approach

From Section 2.3.2, $Z(s)$ is strictly positive real if and only if [17], [85]:

1. all elements of $Z(s)$ are analytic in the closed–right–half plane (i.e.,they do not have poles in $Re(s) \geq 0$),
2. the matrix $Z(j\omega) + Z^T(-j\omega)$ is positive definite Hermitian for all real ω.
3.
$$\lim_{\omega \to \infty} \left[\omega^2 (Z(j\omega) + Z^T(-j\omega))\right] > 0.$$

Based upon the above definition, a modification of a procedure originally proposed by Mabius [90] is presented for validating that $Z(s)$ is strictly positive real for some matrix $\tilde{K}_e$.

Step 1. Choose the matrix product $\tilde{K}_e C_p$ such that the eigenvalues of $(A_p - B_p \tilde{K}_e C_p)$ have negative real parts.

Step 2. Define $Z(s) = C_p \left(sI - A_p + B_p \tilde{K}_e C_p\right)^{-1} B_p$, and define $F(\omega) = Z(j\omega) + Z^T(-j\omega)$.

Step 3. Validate that C_p is such that $F(\omega)$ is positive definite for all ω.

This last step is perhaps best carried out by checking that all m principal minors of $F(\omega)$ are positive. Each such minor can be expanded as a ratio of two polynomials in ω^2. The coefficients of each power of ω^2 (i.e., ω^{2i}) are functions of C_p, A_p, B_p, and $\tilde{K}_e$. In such an expansion the denominator can always be made positive and the numerator can then be written as

$$\sum_{i=0}^{N_m} f_i \left(C_p,\ A_p,\ B_p,\ \tilde{K}_e\right) \omega^{2i}, \tag{2.217}$$

where N_m depends on the number of states and the order of the minor. In order to guarantee that $F(\omega)$ is positive for all ω, it is sufficient that each coefficient f_i in each minor be positive for all permissible A_p, B_p. If not all the coefficients are positive, it is still possible that the principal minors are positive. To this effect, it may be desirable to test the positivity of the principal minors by using the Routh algorithms suggested by Siljak [91].

Furthermore, since each minor must be positive for all possible value of (A_p, B_p, C_p), within the given bounds, it is suggested that recent results [92], [93] in determining robust positive realness for so–called interval plants will be useful. In particular, if maximum and minimum values for each of the minor polynominal coefficients can be established, then the positivity of the polynominal for all permissible parameter variations can be inferred from the positivity properties of only eight associated polynominals.

Alternately from Section 2.3.2 an $m \times m$ transfer function is ASPR if it is minimum–phase and if it has n poles and $n - m$ zeros. Thus define $G(s) = C_p(sI - A_p)^{-1}B_p$ and validate that C_p is such that that $C_pB_p > 0$ and that the zeros, or the roots of

$$\det \begin{bmatrix} sI - A_p & B_p \\ -C_p & 0 \end{bmatrix} = 0 \tag{2.218}$$

are in the open left half plane.

Time Domain Approach

A time domain approach for showing strict positive realness of the transfer matrix

$$Z(s) = J + C_p(sI - A_p + B_p\tilde{K}_eC_p)^{-1}B_p \tag{2.219}$$

is based upon a test for the discrete positive realness of a transformed system [84]. To this effect define the following quantities:

$$A = (I + A_p - B_p\tilde{K}_eC_p)(I - A_p + B_p\tilde{K}_eC_p)^{-1}, \tag{2.220}$$

$$B = \frac{1}{\sqrt{2}}(A + I)B_p, \tag{2.221}$$

$$C^T = \frac{1}{\sqrt{2}}(A^T + I)C_p^T, \tag{2.222}$$

$$J_D = J + C^T(A + I)^{-1}B, \tag{2.223}$$

$$U = J_D + J_D^T. \tag{2.224}$$

Then $Z(s)$ as defined in (2.219) will be positive real if and only if the following recursive difference equation has a negative definite steady state solution [94]:

$$\begin{aligned} \pi(n+1) &= A^T\pi(n)A - \left[A^T\pi(n)B + C\right]\left[U + B^T\pi(n)B\right]^{-1} \\ &\times \left[B^T\pi(n)A + C^T\right], \\ \pi(0) &= 0. \end{aligned} \tag{2.225}$$

2.4.4 An Illustrative Example

This section presents an example to illustrate the application of the adaptive control algorithm. For illustration, consider a plant whose parameters are allowed to vary linearly between two operating points. It will be shown that the adaptive controller is effective for this time varying plant even though the stability results of Section 2.5 are only valid for a time invariant plant.

The second order time varying plant is described at operating point number 1 by

$$\begin{bmatrix} \dot{x}_{p1} \\ \dot{x}_{p2} \end{bmatrix} = \begin{bmatrix} -3.598 & 0 \\ 0.9947 & 0 \end{bmatrix} \begin{bmatrix} x_{p1} \\ x_{p2} \end{bmatrix} + \begin{bmatrix} 14.65 \\ 0 \end{bmatrix} u_p \tag{2.226}$$

and at operating point number 2 by

$$\begin{bmatrix} \dot{x}_{p1} \\ \dot{x}_{p2} \end{bmatrix} = \begin{bmatrix} -10.22 & 0 \\ 0.9997 & 0 \end{bmatrix} \begin{bmatrix} x_{p1} \\ x_{p2} \end{bmatrix} + \begin{bmatrix} 77.86 \\ 0. \end{bmatrix} u_p \tag{2.227}$$

The reference model is described by

$$\begin{bmatrix} \dot{x}_{m1} \\ \dot{x}_{m2} \end{bmatrix} = \begin{bmatrix} -10 & 0 \\ 1 & 0 \end{bmatrix} \begin{bmatrix} x_{m1} \\ x_{m2} \end{bmatrix} + \begin{bmatrix} 20 \\ 0 \end{bmatrix} u_m \tag{2.228}$$

where u_m is a unit step input.

The plant matrices are of the form

$$A_p = \begin{bmatrix} a_{11} & 0 \\ a_{21} & 0 \end{bmatrix}, \quad B_p = \begin{bmatrix} b_1 \\ 0 \end{bmatrix} \tag{2.229}$$

where

$$\begin{aligned} &-10.22 \leq a_{11} \leq -3.598, \; 0.9947 \leq a_{21} \leq 0.9997 \\ &14.65 \leq b_1 \leq 77.86. \end{aligned} \tag{2.230}$$

The selection of an allowable plant output matrix using the positive real procedure, described in Section 2.4.3, will now be detailed. The characteristic equation of the closed loop plant is

$$\lambda^2 + \left(-a_{11} + b_1 \tilde{k}_e c_1\right) \lambda + a_{21} b_1 \tilde{k}_e c_2 = 0. \tag{2.231}$$

If the coefficients of this equation are always greater than zero, then the real part of the eigenvalues of $A_p - B_p \tilde{K}_e C_p$ will be negative as required in step one of the positive real procedure.

Thus, c_1, c_2, and $\tilde{k}_e$ must be chosen such that

$$-a_{11} + b_1\tilde{k}_e c_1 > 0, \tag{2.232}$$

$$a_{21}b_1\tilde{k}_e c_2 > 0. \tag{2.233}$$

Substituting (2.230) into (2.232) and (2.233) gives

$$\tilde{k}_e c_1 > \frac{a_{11}}{b_1} > \frac{-3.598}{77.86} = -0.0462, \tag{2.234}$$

$$\tilde{k}_e c_2 > 0. \tag{2.235}$$

Next, using the definition of $Z(s)$ from (2.219) yields:

$$Z(s) = \frac{b_1(c_1 s + a_{21}c_2)}{s^2 + (-a_{11} + b_1\tilde{k}_e c_1)s + a_{21}b_1\tilde{k}_e c_2} \tag{2.236}$$

and

$$Z(j\omega) = \frac{b_1\left[c_1 j\omega + a_{21}c_2\right]}{-\omega^2 + j\omega(-a_{11} + b_1\tilde{k}_e c_1) + a_{21}b_1\tilde{k}_e c_2}. \tag{2.237}$$

Thus, as a result of using $F(\omega) = Z(j\omega) + Z^T(-j\omega)$,

$$F(\omega) = \frac{2(-a_{21}b_1c_2 - a_{11}b_1c_1 + b_1^2c_1^2\tilde{k}_e)\omega^2 + 2c_2^2b_1^2a_{21}^2\tilde{k}_e}{\left[a_{21}b_1\tilde{k}_e c_2 - \omega^2\right]^2 + \left[(-a_{11} + b_1\tilde{k}_e c_1)\omega\right]^2}. \tag{2.238}$$

If the coefficients of ω^2 and ω^0 in the numerator of (2.238) are positive for all A_p and B_p, then $F(\omega)$ is positive for all ω. Thus, using (2.232)–(2.233),

$$\tilde{k}_e > 0, \tag{2.239}$$

$$c_2 < 3.6c_1 + 14.65c_1^2\tilde{k}_e. \tag{2.240}$$

Furthermore, with these conditions

$$\lim_{\omega\to\infty} \omega^2 F(\omega) > 0.$$

The selection of $C_p = [c_1\ c_2]$ must be such that (2.234), (2.235), (2.239), and (2.240) are satisfied. The selection $C_p = [1\ 0.1]$ can be shown to satisfy these equations.

Another approach based upon showing that $CB > 0$ for a nonminimum phase system follows by noting that the plant matrices are of the form

$$A_p = \begin{bmatrix} a_{11} & 0 \\ a_{21} & 0 \end{bmatrix} \qquad B_p = \begin{bmatrix} b_1 \\ 0 \end{bmatrix}. \tag{2.241}$$

For $C_p = [c_1 \;\; c_2]$, the transfer function of the plant is

$$G_p(s) = \frac{(c_{11}s + a_{21}c_{12})b_1}{s(s - a_{11})}. \tag{2.242}$$

In this case, $CB > 0$ means $c_{11} > 0$ (the sign of the high frequency gain). Further, the system is minimum phase if the root of

$$c_{11}s + a_{21}c_{12} = 0, \tag{2.243}$$

(namely, $s = -a_{21}c_{12}/c_{11}$), is negative. In other words, if $a_{21} > 0$, the plant is ASPR for any $c_{11} > 0$, $c_{12} > 0$.

A simulation was performed for a ten second time interval during which the plant parameters vary linearly from operating point #1 to operating point #2 over the first five seconds and then back to operating point #1 over the next five seconds. This simulation uses MATLAB^{TM} subroutine ODE45 which implements fourth and fifth order Runge-Kutta formulas. The model output matrix C_m is chosen to be the same as the plant output matrix C_p. The weighting matrices T and $\bar{T}$ are both chosen as $0.1I$ where I is the fourth order identity matrix.

It should be observed from Figure 2.3 that the plant output converges to the model output in approximately two seconds. This illustrates the asymptotic stability property of the adaptive control algorithm. The plant and model states x_{p1}, x_{m1} and x_{p2}, x_{m2} are shown in Figures 2.4 and 2.5, respectively. Note that small steady state errors exist between the plant states and the corresponding model states. This is expected in as much as the adaptive control algorithm only guarantees that the state errors will be bounded.

2.5 The Adaptive Algorithm with General Input Commands

2.5.1 Controller Structure

In general, when the adaptive controlled system is expected to track arbitrary external input commands that are not step inputs, the stability analysis is based on the results of Section 2.2.2. Assuming that the "ideal" solutions of Section 2.2.2. exist, then a requirement of the controller would be:

$$x_p(t) \rightarrow x_p^*(t) \text{ as } t \rightarrow \infty \tag{2.244}$$

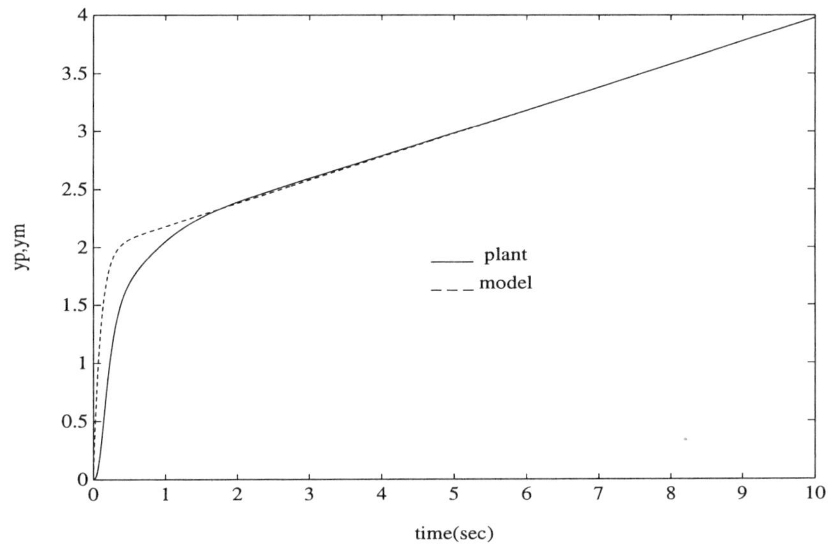

FIGURE 2.3. Plant and Reference Model Outputs

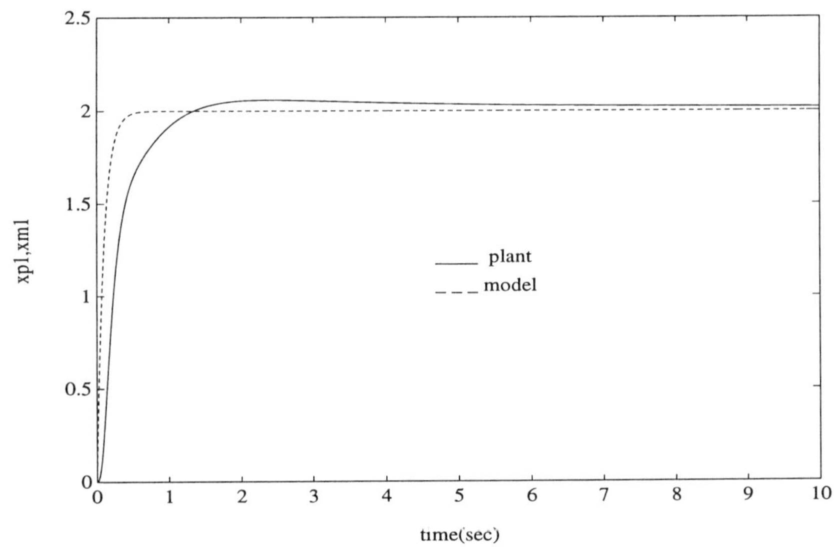

FIGURE 2.4. First Component of Plant and Reference Model State Vectors

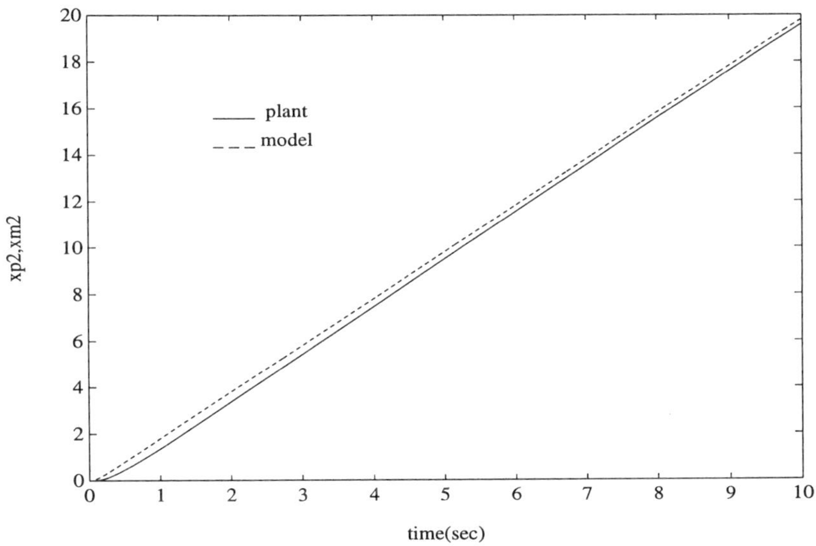

FIGURE 2.5. Second Component of Plant and Reference Model State Vectors

and that

$$y_p(t) \rightarrow y_p^*(t) \rightarrow y_m(t) \text{ as } t \rightarrow \infty \tag{2.245}$$

where it is assumed that the input commands can be represented as outputs of the unknown generating system (2.50)–(2.51).

To this effect define the state error as

$$e_x(t) = x_p^*(t) - x_p(t) \tag{2.246}$$

where $x_p^*(t)$ is the ideal trajectory defined in (2.57), and $x_p(t)$ is the actual plant trajectory. The output tracking error is

$$e_y(t) = y_m(t) - y_p(t). \tag{2.247}$$

Substituting $y_m(t)$ from (2.61) gives

$$e_y(t) = y_p^*(t) - y_p(t) - (C_p X_{11} - C_m)e^{A_m t}\delta_0. \tag{2.248}$$

Since the plant is unknown, the actual control of the plant will be generated by the adaptive algorithm:

$$u_p(t) = K(t)r(t) = K_e(t)e_y(t) + K_x(t)x_m(t) + K_u(t)u_m(t) \tag{2.249}$$

where

$$K(t) \triangleq [K_e(t) \;\; K_x(t) \;\; K_u(t)] \tag{2.250}$$

$$r^T(t) \triangleq \left[e_y^T(t) \;\; x_m^T(t) \;\; u_m^T(t)\right]. \tag{2.251}$$

The adaptive gains are obtained as a combination of the "proportional" and "integral" terms

$$K(t) = K_p(t) + K_I(t) \tag{2.252}$$

$$K_p(t) = e_y(t) r^T \bar{T} \tag{2.253}$$

$$\dot{K}_I(t) = e_y r^T(t) T \tag{2.254}$$

where T and $\bar{T}$ are respectively (selected) positive definite and positive semidefinite adaptation coefficient matrices.

Also denote the concatenation of the unknown ideal gains as:

$$\tilde{K} \triangleq \left[\tilde{K}_e \;\; \tilde{K}_x \;\; \tilde{K}_u\right]. \tag{2.255}$$

Differentiating $e_x(t)$ in (2.246) gives

$$\begin{aligned} \dot{e}_x(t) = A_p e_x(t) - B_p K(t) r(t) - [(A_p X_{11} - X_{11} A_m) x_m(t) - \\ -(A_p X_{12} - X_{12} A_v - X_{11} B_m C_v) v_m(t)] \end{aligned} \tag{2.256}$$

Substituting (2.52)–(2.56) and manipulating the algebraic expressions gives the following differential equation of the state error:

$$\dot{e}_x(t) = \left[A_p - B_p \tilde{K}_e C_p\right] e_x(t) - B_p \left[K(t) - \tilde{K}\right] r(t) - \tilde{F} e^{A_m t} \delta_0 \tag{2.257}$$

where

$$\tilde{F} \triangleq A_p X_{11} - X_{11} A_m + B_p \tilde{K}_x - B_p \tilde{K}_e (C_p X_{11} - C_m). \tag{2.258}$$

2.5.2 Stability Analysis

The following quadratic Lyapunov function is used to prove stability of the adaptive system represented by (2.249) and (2.257):

$$V(t) = e_x^T(t)Pe_x(t) + tr\{\left[K_I(t) - \tilde{K}\right] T^{-1} \left[K_I(t) - \tilde{K}\right]^T\} \qquad (2.259)$$

where "tr" denotes trace.

If the positive realness relations (2.215)–(2.216) are satisfied, then the derivative of V along the trajectories defined by (2.254)–(2.257) is formed by adding (2B.5) and (2B.8) shown in Appendix 2B as follows:

$$\begin{aligned}\dot{V}(t) = &-e_x^T(t)Qe_x(t) - 2e_y^T(t)e_y(t)r^T(t)\bar{T}r(t) - 2e_x^T(t)P\tilde{T}e^{A_m t}\delta_0 \\ &-2r^T(t)\left[K(t) - \tilde{K}\right]^T (C_pX_{11} - C_m)e^{A_m t}\delta_0\end{aligned} \qquad (2.260)$$

Note that $\dot{V}(t)$ is not necessarily negative definite or semi–definite due to the last two terms. However, by applying the Gronwall–Bellman Lemma as cited in Appendix 2B, it can be shown that the Lyapunov function $V(t)$ is bounded; the quadratic form of $V(t)$ then guarantees that the state error $e_x(t)$ and the gains $K_I(t)$ are bounded. In this case, the first two terms in (2.260) vanish as $t \to \infty$. Subsequent applications in Appendix 2B of the Modified Invariance Principle for nonlinear nonautonomous systems gives the following theorem of stability for the adaptive system:

Theorem 2.2: Assume that the controlled plant is ASPR. In other words, assume that there exist two positive definite matrices P, Q, and a gain matrix $\tilde{K}_e$ (all unknown and not needed for implementation) such that the positive realness conditions (2.215)–(2.216) are satisfied. Let T and $\bar{T}$ be (selected) positive definite matrices and assume that (2.62)–(2.64) have solutions for the matrices X_{11}, X_{12}, $\tilde{K}_x$, and $\tilde{K}_u$. Then all states and gains of the adaptive system defined in Table 2.1 are bounded, and the output tracking error vanishes asymptotically.

2.5.3 An Illustrative Example

To illustrate the tracking of more general commands, the illustrative example of Section 2.4.4 was repeated but with $u_m(t) = \sin(5t)$. Since the dimension of the generating system (2.50) is two, the inequality constraint (2.65) is clearly satisfied. Plots of the plant and model outputs and state components are shown in Figures 2.6, 2.7, and 2.8.

2.6 Summary of Adaptive Algorithms

PLANT:

$$\dot{x}_p(t) = A_px_p(t) + B_pu_p(t), \qquad (2.261)$$

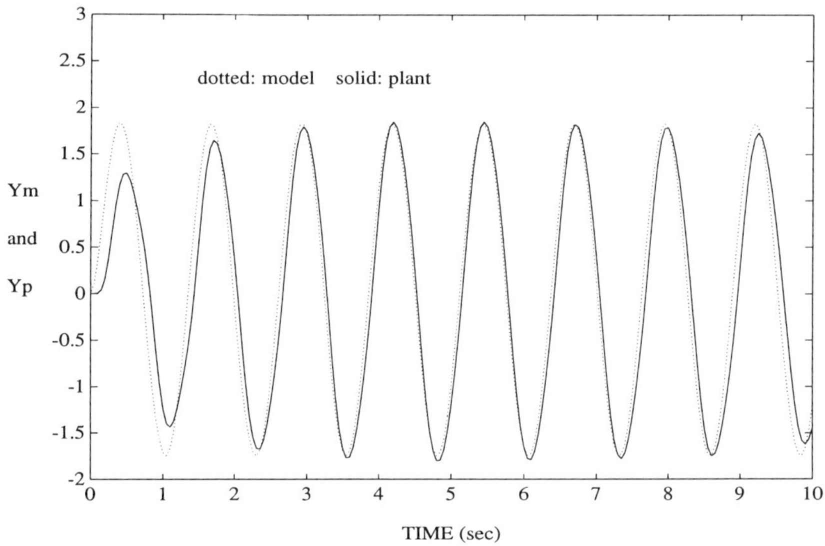

FIGURE 2.6. Plant and Reference Model Outputs for Sine Wave Command

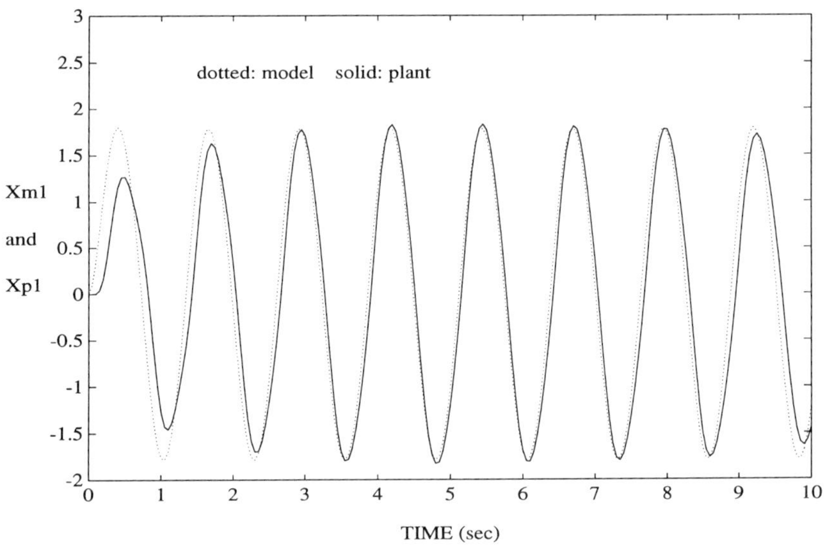

FIGURE 2.7. First Component of Plant and Reference Model State Vectors for Sine Wave Command

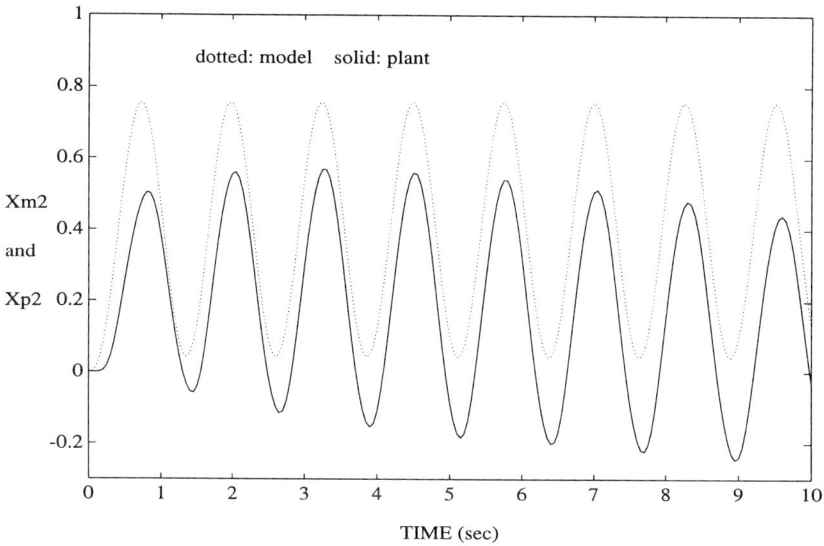

FIGURE 2.8. Second Component of Plant and Reference Model State Vectors for Sine Wave Command

$$y_p(t) = C_p x_p(t) \tag{2.262}$$

MODEL:

$$\dot{x}_m(t) = A_m x_m(t) + B_m u_m(t), \tag{2.263}$$

$$y_m(t) = C_m x_m(t). \tag{2.264}$$

ADAPTIVE CONTROLLER:

$$r(t) = \begin{bmatrix} y_m(t) - y_p(t) \\ x_m(t) \\ u_m \end{bmatrix} \tag{2.265}$$

$$K_r(t) = [K_e(t), K_x(t), K_u(t)] \tag{2.266}$$

$$u_p(t) = K_r(t) r(t) \tag{2.267}$$

$$K_r(t) = K_I(t) + K_p(t) \tag{2.268}$$

$$\dot{K}_I(t) = (y_m(t) - y_p(t)) r^T(t) T, \ T > 0 \tag{2.269}$$

$$K_p(t) = (y_m(t) - y_p(t))r^T(t)\bar{T}, \ \bar{T} \geq 0. \tag{2.270}$$

PROPERTIES:

1) $y_p \rightarrow y_m$ for u_m a step if:

$$H(s) = C_p(sI - A_p + B_p\tilde{K}_eC_p)^{-1}B_p \tag{2.271}$$

is strictly positive real for some matrix $\tilde{K}_e$ (not needed for implementation) and a solution exists to the CGT equation (2.25).

2) $y_p \rightarrow y_m$ for $u_m(t)$ satisfying

$$\dot{v}_m(t) = A_v v_m(t) \tag{2.272}$$

$$u_m(t) = c_v v_m(t) \tag{2.273}$$

where

$$\dim(v_m) \leq \dim(u_m) \text{ and } \dim(x_m) \tag{2.274}$$

if $H(s)$ as defined above is strictly positive real and if (2.62)–(2.64) can be solved.

Appendix 2A

Proof of Theorem 2.1

The first step in the proof is to form a quadratic function which is positive definite in the state variables of the adaptation system, $e_x(t)$ and $K_I(t)$. Before doing so, it is assumed that T^{-1} is a positive definite matrix, which is the first sufficient condition. Then, an appropriate choice of a positive definite Lyapunov function is

$$V = e_x^T P e_x + Tr\left[S\left(K_I - \tilde{K}\right)T^{-1}\left(K_I - \tilde{K}\right)^T S^T\right], \qquad (2A.1)$$

and its time derivative is

$$\dot{V} = e_x^T P \dot{e}_x + \dot{e}_x^T P e_x + 2\,Tr\left[S\left(K_I - \tilde{K}\right)T^{-1}\dot{K}_I^T S^T\right] \qquad (2A.2)$$

where P is an $n \times n$ positive definite symmetric matrix, $\tilde{K}$ is an $m \times n_r$ matrix (unspecified), and S is an $m \times m$ nonsingular matrix.

Since the matrix $\tilde{K}$ appears only in the function V and not in the control algorithm, it is referred to as a dummy gain matrix. It has the same dimensions as $K_r(t)$, which appears in V and can be partitioned as $\tilde{K} = \left[\tilde{K}_e,\ \tilde{K}_x,\ \tilde{K}_u\right]$ so that

$$\tilde{K}r = \tilde{K}_e C_p e_x + \tilde{K}_u u_m + \tilde{K}_x x_m, \qquad (2A.3)$$

where the three gains $\tilde{K}_x$, $\tilde{K}_u$, and $\tilde{K}_e$ are, like $\tilde{K}$, dummy gains.

Next, derive the error equation by using the definition that $e_x = x_p^* - x_p$ to obtain

$$\begin{aligned} \dot{e}_x &= A_p x_p^* + B_p u_p^* - B_p u_p \\ &= A_p\left[x_p^* - x_p\right] + B_p\left[u_p^* - u_p\right] \qquad (2A.4) \\ &= A_p e_x + B_p\left[u_p^* - u_p\right]. \end{aligned}$$

Substituting the definitions for u_p^* from (2.17) and u_p from (2.195), gives:

$$\dot{e}_x = A_p e_x + B_p \left[S_{21} x_m + S_{22} u_m - K_x x_m - K_u u_m - K_e C_p e\right]. \tag{2A.5}$$

$$= A_p e_x + B_p \left[S_{21} x_m + S_{22} u_m - K_I r - C_p e r^T \bar{T} r\right]. \tag{2A.6}$$

Thus, the adaptive system is described by

$$\dot{e}_x = A_p e_x + B_p \left[S_{21} x_m + S_{22} u_m - K_I r - C_p e_x r^T \bar{T} r\right], \tag{2A.7}$$

$$\dot{K}_I = C_p e_x r^T T. \tag{2A.8}$$

Substituting (2A.7) and (2A.8) into (2A.2), gives,

$$\begin{aligned} \dot{V} &= e_x^T P \left[A_p e_x - B_p C_p e r^T \bar{T} r - B_p K_I r + B_p S_{21} x_m + B_p S_{22} u_m\right] \\ &+ \left[A_p e_x - B_p C_p e_x r^T \bar{T} r - B_p K_I r \right. \\ &\left. + B_p S_{21} x_m + B_p S_{22} u_m\right]^T P e \\ &+ 2 e_x^T C_P^T S^T S \left(K_I - \tilde{K}\right) r, \end{aligned} \tag{2A.9}$$

$$\begin{aligned} \dot{V} &= e_x^T \left(P A_p + A_p^T P\right) e_x - e_x^T \left(P B_p C_p\right. \\ &+ \left. C_p^T B_p^T P\right) e_x r^T \bar{T} r - 2 e_x^T B_p K_I r \\ &+ 2 e_x^T P B_p (S_{21} x_m + S_{22} u_m) + 2 e_x^T C_p^T S^T S (K_I - \tilde{K}) r, \end{aligned} \tag{2A.10}$$

$$\begin{aligned} \dot{V} &= e_x^T (P A_p + A_p^T P) e_x - e_x^T (P B_p C_p + C_p^T B_p^T P) e_x r^T \bar{T} r \\ &+ 2 e_x^T \left[C_p^T S^T S - P B_p\right] K_I r - 2 e_x^T C_p^T S^T S \tilde{K} r \\ &+ 2 e_x^T P B_p \left[S_{21} x_m + S_{22} u_m\right]. \end{aligned} \tag{2A.11}$$

The elimination of cross product terms in the derivative establishes the second sufficient condition for stability, which is the following constraint on the output matrix C_p:

$$C_p = QB_p^T P \;\; \forall A_p,\; B_p, \tag{2A.12}$$

where $Q = (S^T S)^{-1}$.

Then, the Lyapunov derivative becomes

$$\begin{aligned}\dot{V} &= e_x^T(PA_p + A_p^T p)e_x - 2e_x^T PB_p(S^T S)^{-1}B_p^T Pe_x r^T \bar{T} r \\ &- 2e_x^T C_p^T S^T S\tilde{K} r \\ &+ 2e_x^T PB_p(S_{21}x_m + S_{22}u_m).\end{aligned} \tag{2A.13}$$

Observe that $\tilde{K}r$ may be expanded as

$$\begin{aligned}\tilde{K}r &= \tilde{K}_e(y_m - y_p) + \tilde{K}_x x_m + \tilde{K}_u u_m \\ &= \tilde{K}_e(C_p x_p^* - C_p x_p) + \tilde{K}_x x_m + \tilde{K}_u u_m \\ &= \tilde{K}_{e_x} C_p e + \tilde{K}_x x_m + \tilde{K}_u u_m.\end{aligned} \tag{2A.14}$$

Substituting (2A.13) into (2A.12) yields,

$$\begin{aligned}\dot{V} &= e_x^T \left[P(A_p - B_p\tilde{K}_e C_p) + (A_p - B_p\tilde{K}_e C_p)^T P\right] e_x \\ &- 2e_x^T PB_p(S^T S)^{-1}B_p^T Pe_x r^T \bar{T} r \\ &+ 2e_x^T PB_p\left[(S_{21} - \tilde{K}_x)x_m + (S_{22} \quad \tilde{K}_u)u_m\right].\end{aligned} \tag{2A.15}$$

Next, with the choice $\tilde{K}_x = S_{21}$ and $\tilde{K}_u = S_{22}$, neither of which is required for implementation, the derivative of the function V becomes

$$\begin{aligned}\dot{V} = e_x^T(t)\left[P\left(A_p - B_p\tilde{K}_e C_p\right) + \left(A_p - B_p\tilde{K}_e C_p\right)^T P\right] e_x(t) \\ - 2e_x^T(t)PB_p(S^T S)^{-1}B_p^T Pe_x(t)r^T(t)\bar{T}r(t).\end{aligned} \tag{2A.16}$$

This derivative consists of two terms. If $\bar{T}$ is positive semidefinite (the third sufficient condition), then the second term is negative semidefinite in $e_x(t)$. Requiring the quadratic term to be negative definite in $e_x(t)$ establishes the fourth sufficient condition for stability. This condition is a constraint on P which can be stated as follows: the matrix P must be chosen such that there exists a $\tilde{K}_e$ such that

$$P\left(A_p - B_p\tilde{K}_eC_p\right) + \left(A_p - B_p\tilde{K}_eC_p\right)^T P \tag{2A.17}$$

is negative definite for all A_p, B_p. Thus, two of the sufficient conditions for stability are constraints on the matrices C_p and P. With these sufficient conditions holding, the derivative of the function V is negative definite in the error $e_x(t)$. Therefore, the plant output will asymptotically approach the model output. Furthermore, since the derivative of the function V is negative semidefinite in the augmented state $[e_x(t),\ K_I(t)]$, the adaptive gains will be bounded.

Appendix 2B

Proof of Theorem 2.2

The derivative of the Lyapunov function (2.259)

Let

$$V(t) = V_1(t) + V_2(t) \tag{2B.1}$$

where

$$V_1(t) = e_x^T(t) P e_x(t) \tag{2B.2}$$

and where

$$V_2(t) = tr\{\left[K_I(t) - \tilde{K}\right] T^{-1} \left[K_I(t) - \tilde{K}\right]^T\}. \tag{2B.3}$$

Then

$$\dot{V}_1(t) = \dot{e}_x^T(t) P e_x(t) + e_x^T(t) P \dot{e}_x(t). \tag{2B.4}$$

Substituting $\dot{e}_x^T(t)$ form (2.257) gives

$$\dot{V}_1(t) = -e_x^T(t)(P\tilde{A}_p + \tilde{A}_p^T P) e_x(t)$$

$$-2e_x^T(t) P B_p \left[K(t) - \tilde{K}\right] r(t) - 2e_x^T(t) P \tilde{F}_e^{A_m t} \delta_0$$

$$\dot{V}_2(t) = 2tr\{\left[K_I(t) - \tilde{K}\right] T^{-1} \dot{K}_I^T(t)\} \tag{2B.5}$$

$$= 2tr\{\left[K_I(t) - \tilde{K}\right] T^{-1} (e_y(t) r^T(t) T)^T\}$$

$$= 2e_y^T(t) \left[K_I(t) - \tilde{K}\right] r(t)$$

Substituting

$$K_I(t) = K(t) - K_p(t) = K(t) - e_y(t) r^T(t) \bar{T} \tag{2B.6}$$

gives

$$\dot{V}_2(t) = 2e_y^T(t) \left[K_I(t) - \tilde{K}\right] r(t) - 2e_y^T(t) e_y(t) r^T(t) \bar{T} r(t). \tag{2B.7}$$

Now substituting $e_y(t)$ from (2.248) gives

$$\begin{aligned}\dot{V}_2(t) &= 2e_x^T(t)C_p^T\left[K_I(t)-\tilde{K}\right]r(t) - 2e_y^T(t)e_y(t)r^T(t)\bar{T}r(t) \\ &\quad -2r^T(t)\left[K(t)-\tilde{K}\right]^T(C_pX_{11}-C_m)e^{A_mt}\delta_0.\end{aligned} \tag{2B.8}$$

Adding now (2B.5) and (2B.8) gives (2.260).
The subsequent analysis utilizes the following lemma:
Gronwall–Bellman Lemma [13]: If

$$\dot{V}(t) \leq V(t)k(t) \tag{2B.9}$$

or

$$V(t) \leq V(0) + \int_0^t V(\tau)k(\tau)d\tau \tag{2B.10}$$

where $V(t) \geq 0$, $k(t) \geq 0$, then

$$V(t) \leq V(0)exp\left[\int_0^t k(\tau d\tau\right]. \tag{2B.11}$$

For the rigorous proof of stability of the adaptive system defined by (2.254) and (2.257), consider the derivative (2.260) of the Lyapunov function defined in (2.259). For a complete proof, define the derivative of the Lyapunov function in the following form

$$\dot{V}(t) = f_1 + f_2 + f_3 + f_4 + f_5 + f_6 + f_7 + f_8 + f_9 + f_{10} \tag{2B.12}$$

where

$$f_1 = -e_x^T(t)Qe_x(t) \tag{2B.13}$$

$$f_2 = -2e_y^Te_y(t)e_y^T(t)\bar{T}_ee_y(t) \tag{2B.14}$$

$$f_3 = -2e_y^T(t)e_y(t)x_m^T(t)\bar{T}_xx_m(t) \tag{2B.15}$$

$$f_4 = -2e_y^T(t)e_y(t)u_m^T(t)\bar{T}_uu_m(t) \tag{2B.16}$$

$$f_5 = -2e_y^T(t)\bar{T}_ee_y(t)e_y^T(t)(C_pX_{11}-C_m)e^{A_mt}\delta_0 \tag{2B.17}$$

$$f_6 = -2x_m^T(t)\bar{T}_xx_m(t)e_y^T(t)(C_pX_{11}-C_m)e^{A_mt}\delta_0 \tag{2B.18}$$

$$f_7 = -2u_m^T(t)\bar{T}_u u_m(t)e_y^T(t)(C_pX_{11} - C_m e^{A_m t}\delta_0 \qquad (2B.19)$$

$$f_8 = -2r^T(t)\left[K_I(t) - \tilde{K}\right]^T (C_pX_{11} - C_m)e^{A_m t}\delta_0 \qquad (2B.20)$$

$$f_9 = -2e_x^T(t)C_p\tilde{K}_e(C_pX_{11} - C_m)e^{A_m t}\delta_0 \qquad (2B.21)$$

$$f_{10} = -2e_x^T(t)P(A_pX_{11} - X_{11}A_m + B_p\tilde{K}_e)e^{A_m t}\delta_0 \qquad (2B.22)$$

where

$$Q \geq 0, \bar{T}_e > 0, \bar{T}_x > 0, \bar{T}_u > 0. \qquad (2B.23)$$

The following relationships between real numbers will be subsequently used to evaluate the different terms of $\dot{V}(t)$:

1) $\forall\ x,\ y \quad \exists\ k_1 > 0, \alpha > 0, \beta > 0,$ s. t. $\quad | xy | \leq k_1 | \alpha x^2 + \beta y^2 |$

2) $\forall\ x,\ y \qquad \exists\ k_2 > 0, \alpha > 0, \beta > 0,$ such that

$x^2 \leq\ k_2(\alpha x^2 + \beta y^2)$

$y^2 \leq\ k_2(\alpha x^2 + \beta y^2)$

3) for x, y large enough $\quad \exists\ k_3 > 0, \alpha > 0, > 0, \beta > 0,$ s. t.

$x \leq k_3(\alpha x^2 + \beta y^2)$

$y \leq k_3(\alpha x^2 + \beta y^2)$

Since

$$V(t) = e_x^T(t)Pe_x(t) + tr\{\left[K_I(t) - \tilde{K}\right]T^{-1}\left[K_I(t) - \tilde{K}\right]^T \qquad (2B.24)$$

then there exists $\alpha_1 > 0,\ \alpha_2 > 0.$ such that

$$V(t) > \alpha_1 \parallel e_x(t) \parallel^2 + \alpha_2 \parallel K_I(t) - \tilde{K} \parallel^2. \qquad (2B.25)$$

The terms of $\dot{V}(t)$ satisfy

$$f_1 \leq 0, \quad f_2 \leq 0, \quad f_3 \leq 0, \quad f_4 \leq 0 \qquad (2B.26)$$

and there exists $a_1 > 0, a_2 > 0$ such that

$$f_5 + f_6 + f_7 \ \leq\ a_1 V(t) \parallel e^{A_m t} \parallel . \parallel e_y(t) \parallel. \qquad (2B.27)$$

$$f_8 + f_9 + f_{10} \ \leq\ a_2 V(t) \parallel e^{A_m t} \parallel \qquad (2B.28)$$

Assume that the output error can increase without bound. Then there exist a time t_1 such that

$$| f_2 |>| f_5 + f_6 + f_7 | \quad \text{for all times } t > t_1. \tag{2B.29}$$

Then

$$V(t) \leq V(t_1) + \int_{t_1}^{t} a_2 V(\tau) \parallel e^{A_m \tau} \parallel d\tau\} \leq M_1 \leq \infty \tag{2B.30}$$

for some $M_1 > 0$.

Because $V(t)$ is bounded, its positive definite quadratic form implies that $e_x(t)$ and $K_i(t)$ are bounded and thus, $e_y(t)$ cannot be unbounded. However, $e_x(t)$ could still be unbounded even if $e_y(t)$ remains bounded. Since $e_y(t)$ was shown to be bounded,

$$f_5 + f_6 + f_7 \leq a_1 V(t) \parallel e^{A_m t} \parallel . \parallel e_y(t) \parallel \leq \alpha_1 V(t) \parallel e^{A_m t} \parallel \tag{2B.31}$$

for some finite $\alpha_1 > 0$.

Therefore,

$$V(t) \leq V_0 + \int_0^t ((f_5 + f_6 + f_7) + (f_8 + f_9 + f_{10}))d\tau \tag{2B.32}$$

$$V(t) \leq V_0 + \int_0^t (\alpha_1 + a_2) V(\tau) \parallel e^{A_m \tau} \parallel d\tau. \tag{2B.33}$$

Again, the Gronwall–Bellman inequality gives

$$V(t) \leq V_0 \exp\{(\alpha_1 + a_2) \int_0^t \parallel e^{A_m \tau} \parallel d\tau < M_2 < \infty \tag{2B.34}$$

for some $V > 0$, and since $V(t)$ is positive definite quadratic, $e_x(t)$ and $K_I(t)$ are bounded.

Observe now that $\dot{V}(t)$ in (2B.12)–(2B.22) has the form

$$\dot{V}(e_x, K_I, t) = W_1(e_x, K_I, t) + W_2(e_x, K_I, t) \tag{2B.35}$$

where

$$W_1(e_x, K_I, t) \triangleq f_1 + f_2 + f_3 + f_4 \tag{2B.36}$$

$$W_1(e_x, K_I, t) \leq \tilde{W}_1(e_x, K_I) \leq 0 \tag{2B.37}$$

$$W_2(e_x, K_I, t) = f_5 + f_6 + f_7 + f_8 + f_9 + f_{10} \tag{2B.38}$$

$$\lim_{t \to \infty} W_2(e_x, K_I, t) = 0. \tag{2B.39}$$

According to the modified invariance principle, since all solutions $e(t)$, $K(t)$ are bounded, all trajectories of (2.254) and (2.257) approach asymptotically the set

$$\{(e_x,\ K_I) \mid \tilde{W}_1(e_x, K_I) \equiv 0\} \tag{2B.40}$$

which, from (2B.12)–(2B.22) is equivalent with the set

$$\{(e_x,\ K_I) \mid (e_y(t) \equiv 0\} \tag{2B.41}$$

or in other words, the output error vanishes asymptotically.

Appendix 2C

Poles, Zeros and Relative Degree in Multivariable Systems

Let $G(s)$ be a $m \times m$ square transfer matrix

$$G(s) = \begin{bmatrix} g_{11}(s) & g_{12}(s) & g_{13}(s)\dots & g_{1m}(s) \\ g_{21}(s) & g_{22}(s) & g_{23}(s)\dots & g_{2m}(s) \\ g_{31}(s) & g_{32}(s) & g_{33}(s)\dots & g_{3m}(s) \\ \dots & \dots & \dots\ \dots & \dots \\ g_{m1}(s) & g_{m2}(s) & g_{m3}(s)\dots & g_{mm}(s) \end{bmatrix} \tag{2C.1}$$

with minimal realization [95]:

$$G(s) = C(sI - A)^{-1}B + D = D + \frac{CB}{s} + \frac{CAB}{s^2} + \frac{CA^2B}{s^3} + \dots \tag{2C.2}$$

If the matrix D is nonsingular, then $G(s)$ is proper but not strictly proper, and it is said to be of relative degree *zero*, because it has the same number of poles and zero.

When D is zero, the transfer function is strictly proper, and it has more poles than zeros. The ASPR property requires $G(s)$ to be of relative degree m. In other words, if the plant has n poles, it must have $n - m$ zeros. The easiest way to check if the relative degree is m in this case is to test if the matrix product CB is nonsingular (i.e., det $[CB] \neq 0$). However, since the ASPR lemmas require a test of the positive definiteness of CB the relative degree is a byproduct of this test.

To test if all zeros are in the open left half plane, recall that the poles of $G(s)$ are the roots of the polynomial det $(sI - A) = 0$, and that the zeros of $G(s)$ are those values of s that reduce the rank of the matrix [95]

$$P(s) = \begin{bmatrix} sI - A & B \\ -C & D \end{bmatrix}. \tag{2C.3}$$

These values of s are the roots of the polynomial

$$\det\{P(s)\} = 0. \tag{2C.4}$$

In general, if the transfer function $G(s)$ has the irreducible right matrix fraction description (MFD)

$$G(s) = N(s)D^{-1}(s) \tag{2C.5}$$

where $N(s)$ and $D(s)$ are polynomial matrices, then the poles of $G(s)$ are the roots of $\det[D(s)] = 0$, and the zeros of $G(s)$ are the roots of $det\,[N(s)] = 0$.

To locate the poles and zeros of multivariable systems, it may be convenient to write $G(s) = \frac{1}{d(s)} N(s)$, and perform elementary operations on the polynomial matrix $N(s)$ to bring it to the Smith form [95]:

$$N(s) = \begin{bmatrix} \lambda_1(s) & 0 & 0 \ldots & 0 \\ 0 & \lambda_2(s) & 0 \ldots & 0 \\ 0 & 0 & \lambda_3(s) \ldots & 0(s) \\ \ldots & \ldots & \ldots\;\ldots & \ldots \\ 0 & 0 & 0 \ldots & \lambda_m(s) \end{bmatrix} \tag{2C.6}$$

where $\lambda_1(s)$ divides $\lambda_{i+1}(s)$.

Then define the Smith–McMillan form

$$M(s) = \begin{bmatrix} \frac{\lambda_1(s)}{d(s)} & 0 & 0 \ldots & 0 \\ 0 & \frac{\lambda_2(s)}{d(s)} & 0 \ldots & 0 \\ 0 & 0 & \frac{\lambda_3(s)}{d(s)} \ldots & 0(s) \\ \ldots & \ldots & \ldots\;\ldots & \ldots \\ 0 & 0 & 0 \ldots & \frac{\lambda_m(s)}{d(s)} \end{bmatrix}. \tag{2C.7}$$

After cancellation of any common factor in $\lambda_i(s)$ and $d(s)$, this becomes:

$$M(s) = \begin{bmatrix} \frac{\epsilon_1(s)}{\psi_1(s)} & 0 & 0 \ldots & 0 \\ 0 & \frac{\epsilon_2(s)}{\psi_2(s)} & 0 \ldots & 0 \\ 0 & 0 & \frac{\epsilon_3(s)}{\psi_3(s)} \ldots & 0(s) \\ \ldots & \ldots & \ldots\;\ldots & \ldots \\ 0 & 0 & 0 \ldots & \frac{\epsilon_m(s)}{\psi_m(s)} \end{bmatrix} \tag{2C.8}$$

where now the pairs $\{\epsilon_i(s),\ \psi_i(s)\}$ are respectively coprime. The zeros of $G(s)$ are the roots of the polynomials $\epsilon_i(s)$, and the poles of $G(s)$ are the roots of the polynomials $\psi_i(s)$ [95].

Chapter 3

Extensions of the Basic Adaptive Algorithm: Alleviation of the SPR Constraints

Because the algorithms developed and discussed in Chapter 2, require the satisfaction of certain positivity constraints if asymptotic model following is to be guaranteed, various modifications have been developed so that the algorithms can be used in a much wider class of systems. Section 3.1 shows how a feedforward compensator can easily be used so that the resulting augmented system satisfies the positivity constraints.

However since the resulting controlled output is a combination of the original output and the feedforward augmentation, further modification might be needed if the actual output is to track the reference model output. To this effect, Section 3.3 introduces a supplementary feedforward compensator into the reference model, and Section 3.4 develops a unified approach to supplementary dynamics by introducing the concept of a metasystem.

Finally systems with a specific type of nonlinearity of known form are considered in Section 3.5.

3.1 Parallel Feedforward and Stability Considerations

The previous chapter developed adaptive controllers that guarantee stability of any system that satisfy the so–called almost strict positive realness (ASPR) conditions. It was also shown that any $m \times m$ minimum–phase transfer function with n poles and n or $n - m$ zeros is ASPR. This chapter will show how basic stabilizability properties of general plants can be used to augment either the controlled plants or the adaptive algorithms, in order to satisfy the desired ASPR conditions and thus, guarantee stability of more realistic systems.

To introduce the relationship between stabilizability and almost positivity, consider again Rohrs' example of Section 2.3.1 As a starting point, assume that the plant can be stabilized by some constant feedback, and

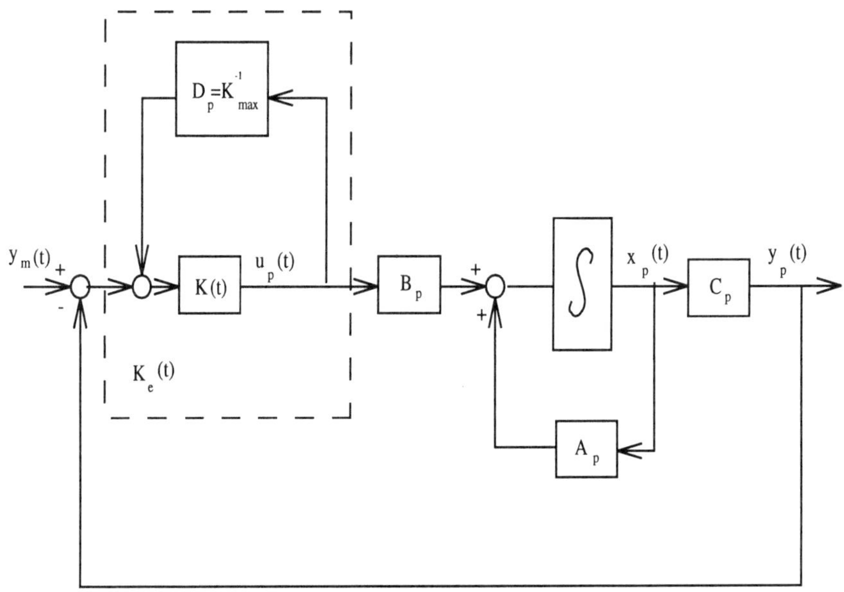

FIGURE 3.1. Bounded Effective Adaptive Gain.

that $K_{\max}$ is an estimate of the maximal admissible gain. Any estimate of a stabilizing gain will do, but it is called $K_{\max}$ because it will be used to guarantee that the effective adaptive gains that affect the plant do not become too large. For convenience of presentation, consider only the adaptive gain feedback $K_e(t)$. Instead of imposing a hard bound on the adaptive gain, as was suggested in Chapter 2, one can as in Figure 3.1 use the inverse $D = K_{\max}^{-1}$ in feedback around the adaptive control gain $K_e(t)$, such that the effective controller gain is $K_c(t) = [I + K_e(t)D]^{-1} K_e(t)$.

This way, if the adaptive gain $K(t)$ varies from 0 to ∞, the effective gain $K_c(t)$ varies *smoothly* from 0 to $K_{\max}$. As the adaptive gain increases, the effect of $D = K_{\max}^{-1}$ in feedback becomes relevant and strongly opposes the tendency of the effective gain to reach high and possible dangerous values. Therefore, the effective adaptive gain $K_c(t)$ has $K_{\max}$ as a soft bound.

Now without affecting the plant, the branch $D = K_{\max}^{-1}$ is represented as parallel feedforward as shown in Figure 3.2 in order to emphasize the control problem from the point of view of the adaptive controller $K_e(t)$. It can be seen from Figure 3.2 that now, instead of directly controlling the original plant $G_p(s)$, the adaptive gain $K_e(t)$ controls an 'augmented' plant $G_a(s) = G_p(s) + 1/K_{\max}$. The augmented plant $G_a(s)$ has an equal number of poles and zeros. Furthermore all zeros are minimum phase. As shown in Appendix 3A this augmented plant is ASPR, and therefore, adaptive control can be used with confidence.

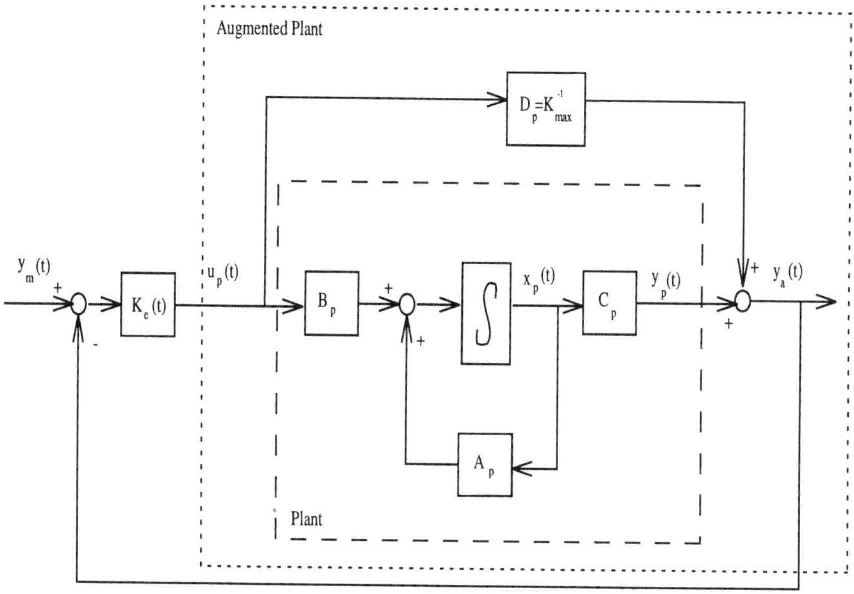

FIGURE 3.2. The Equivalent Control Scheme With Parallel Feedforward

It is also possible to perform similar useful augmentations with parallel feedforward in plants that are not stabilizable by constant feedback. Consider, for illustration, the simplified model of a non–linear generator–voltage control system (also discussed in Section 1.2) [27]. The nominal transfer function of the open–loop system is

$$G_p(s) = \frac{1.5}{(s+1.17)(s-2.9)}. \tag{3.1}$$

This plant cannot be stabilized by any constant gain. Yet, a PD controller of the form $H(s) = K(1 + s/s_0)$, with sufficiently high gain, can indeed stabilize this plant. A root locus plot for the system with controller $H(s) = K(s+2.5)$ is shown in Figure 3.3. In this case one can see that there exists some gain K, sufficiently large, such as K_2 or even K_3, that stabilizes the plant, and perhaps provides satisfactory behavior.

The dynamic stabilizing PD controller $H(s)$ will now be used to play a similar role to that played by the fixed stabilizing gain in the previous example. In other words, instead of implementing the (noncausal) stabilizing configuration $H(s)$ in series with the plant, it suggested to use its inverse,

$$D(s) = H^{-1}(s) = D/(1 + s/s_o) \tag{3.2}$$

as parallel feedforward around the plant, so as to obtain an augmented plant with certain desirable passivity properties, as shown in Figure 3.4.

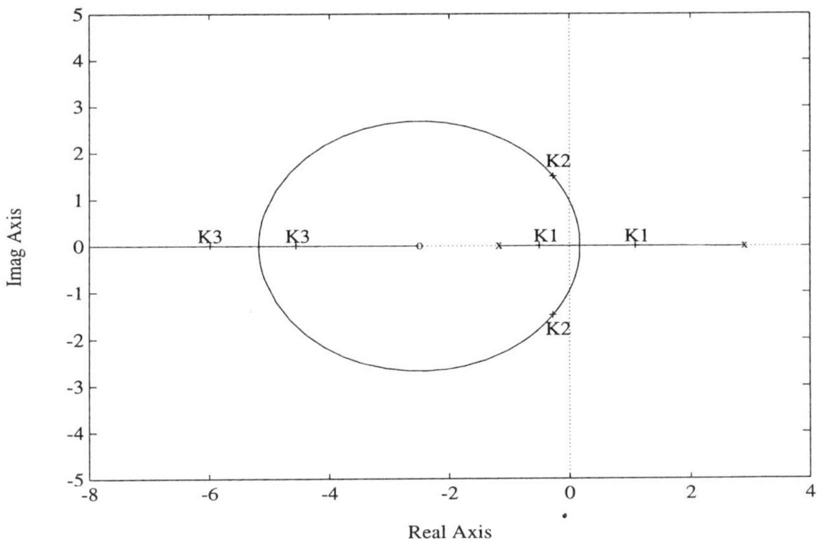

FIGURE 3.3. Root–Locus Diagram for the Example with PD Controller.

As shown in Section 2.3, this configuration is ASPR. Furthermore if a large gain K can be used for PD stabilization then $D = K^{-1}$ will be small, and the contribution of the resulting feedforward loop to the augmented output might be negligible relative to that of the actual plant output.

As an illustration for the particular gain value $K = 40$, the parallel feedforward is $D(s) = H^{-1}(s) = 0.025/(s + 2.5)$. The transfer function of the augmented plant is then

$$G_a(s) = G_p(s) + H^{-1} = \frac{1.5}{(s+1.17)(s-2.9)} + \frac{0.025}{s+2.5} \tag{3.3}$$

or as in (1.7),

$$G_a(s) = \frac{0.025(s^2 + 58.217s + 146.607)}{(s+1.17)(s-2.9)(s+2.5)}. \tag{3.4}$$

Note that $G_a(s)$, the augmented plant is minimum–phase and of relative degree 1, having two zeros and three poles. It thus has those ASPR properties that are so useful for the proofs of stability of the adaptive control systems (Appendix 3A) [70], [96], [19], [31], [51], [66]. Note also that the relative degree of the augmented plant is identical to the (selected) relative degree of the parallel feedforward. Adaptive control can now be applied with confidence.

It is easy to show the effect of parallel feedforward and the ASPR property in SISO systems. Consider the transfer function $G_p(s) = B(s)/A(s)$,

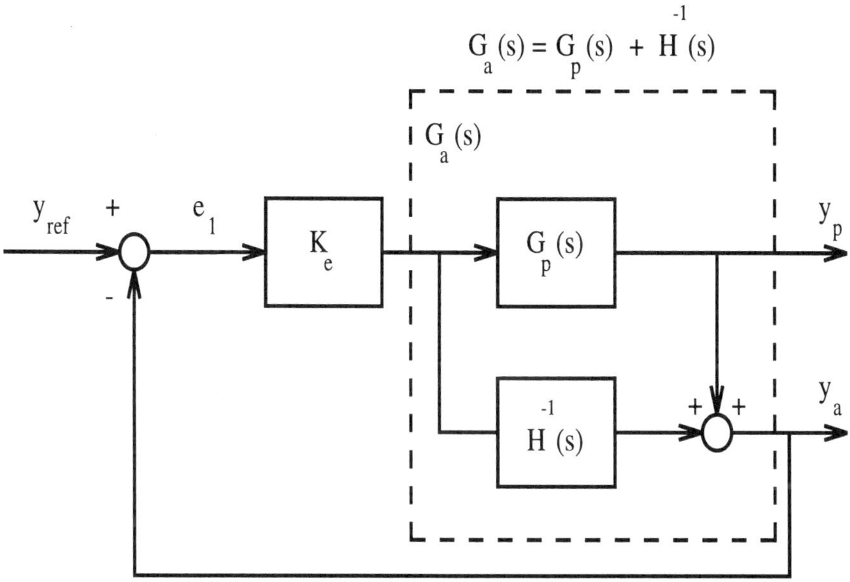

FIGURE 3.4. An Equivalent Representation of the Augmented Control System.

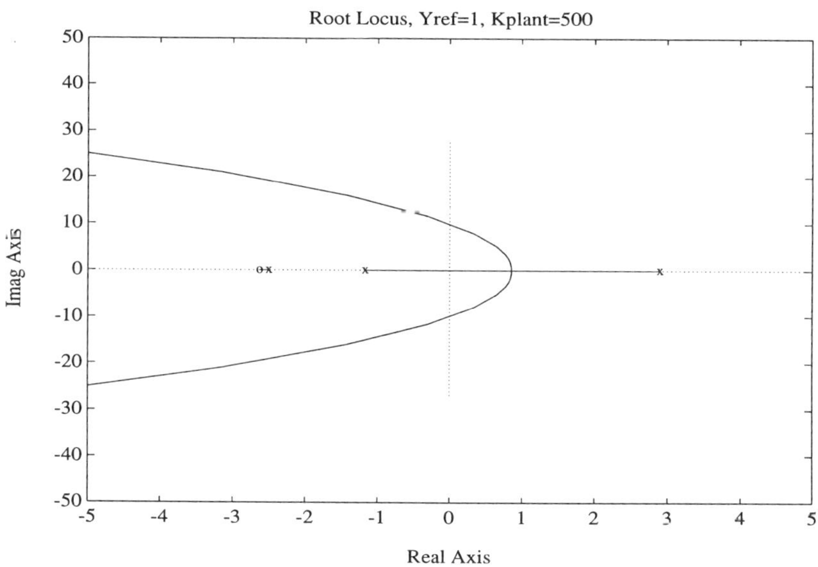

FIGURE 3.5. Root Locus of the Augmented ASPR Plant

and assume that it can be stabilized by some configuration $H(s) = KQ(s)/\ P(s)$, where $A(s)$, $B(s)$, $P(s)$, $Q(s)$ are polynomials, and $P(s)$ and $Q(s)$ are monic (i.e., the coefficient of highest degree term is unity). The closed loop system is $T(s) = G_p(s)/\ [1 + H(s)G_p(s)] = [B(s)P(s)]\,/\,[A(s)P(s) + KB(s)Q(s)]$. Because $H(s)$ is a stabilizing configuration, the polynomial $A(s)P(s) + KB(s)Q(s)$ is Hurwitz. If now, as suggested, one uses $H^{-1}(s)$ in parallel with the plant, one gets $G_a(s) = G_p(s) + H^{-1}(s) = B(s)/A(s) + P(s)/\,[KQ(s)]$, or $G_a(s) = [A(s)P(s) + KB(s)Q(s)]\,/\,[KA(s)\,Q(s)]$. It is clear that $G_a(s)$ is minimum–phase, and if the relative degree is unity, it is ASPR (Appendix 3A).

Generalizations of this parallel feedforward technique to multivariable systems are shown in detail in Appendix 3A. It is shown that, if a plant with the $m \times m$ transfer function $G_p(s)$ can be stabilized by some configuration $H(s)$, then the augmented plant $G_a(s) = G_p(s) + H^{-1}(s)$ is ASPR if $H(s)$ is selected such that the relative degree of $G_a(s)$ is m or zero. This is not a restrictive condition, because any factorization of $H(s) = H_1(s)H_2(s)$ can be used to implement the augmented system $G_a(s) = H_2(s)G_p(s) + H_1^{-1}(s)$ and satisfy the relative degree condition.

Use of parallel feedforward in adaptive control assumes that the designer can pick some stabilizing configuration before implementation of adaptive control. It is reasonable to assume, usually, that out of all existing methods for robust control with uncertainty, one can at least design some stabilizing configuration.

The PD controller $H(s) = K(1 + s/s_o)$ has been mentioned so often in this context because of its wide applicability and because of its ease of application. Although any gain will do, it is desired to estimate the highest gain $K = K_{\max}$ that maintains stability. The parallel configuration in this case is $H^{-1}(s) = K_{\max}^{-1}/(1+s/s_o)$. It is advised to evaluate the highest finite gain that still maintains stability for any general stabilizing configuration $H(s)$, so that the inverse $H^{-1}(s)$ has small gains, and its output signal $y_s(t)$ remains small compared to the plant output $y_p(t)$. This way, although it will be shown that the addition of $H^{-1}(s)$ may dramatically improve the stabilizability properties of the adaptive system, the measured augmented output maintains $y_a(t) \cong y_p(t)$ for all practical purposes. A root locus plot showing the high gain stability of the augmented open loop ASPR plant is shown in Figure 3.5.

3.2 Feedforward Around Plant

3.2.1 Adaptive Control with Basic Feedforward Augmentation

In general, to avoid algebraic loops, the feedforward compensation discussed in the previous section will be defined by the strictly proper transfer function matrix $R_p(s)$ with realization:

$$\dot{s}_p(t) = A_s s_p(t) + B_s u_p(t) \tag{3.5}$$

$$r_p(t) = D_s s_p(t). \tag{3.6}$$

The augmented output to be controlled is then

$$z_p(t) = y_p(t) + r_p(t) \tag{3.7}$$

In Appendix 3A, it is shown that the augmented system

$$G_a(s) = G_p(s) + R_p(s) \tag{3.8}$$

is ASPR provided that:

- $R_p(s)$ is such that the relative degree of $G_a(s)$ is m.
- $R_p^{-1}(s)$ stabilizes the closed loop output feedback system with transfer function $\left[I + G_p(s)R_p^{-1}(s)\right]^{-1} G_p(s)$.

Augmenting the plant with a feedforward compensator $R_p(s)$, which satisfies the above conditions, results in the configuration shown in Figure 3.6. Given minimal prior knowledge, a stabilizing configuration must be designed. For example, if a plant *is* stabilizable by a PD controller $R_p^{-1}(s) = K(1 + s/s_0)$, all that is needed is some estimation of the highest gain $K = K_{\max}$ that maintains stability. The PD controller is mentioned in particular because of its wide applicability and ease of application. The parallel feedforward configuration in this case is $R_p(s) = K_{\max}^{-1}/(1 + s/s_0)$. It is important to note that only stability and not closed loop performance is the goal of this intermediate procedure.

For any general stabilizing configuration it is important to find the highest gain that still maintains stability. This way, $R_p^{-1}(s)$ will have small gains, and its output $r_p(t)$ will remain small relative to the plant output $y_p(t)$. Although the addition of $R_p(s)$ may dramatically improve the stabilizability properties of the adaptive system, it is desired that the measured augmented output $z_p(t)$ remain approximately equal to the plant output $y_p(t)$. General procedures for designing such feedforward compensators are presented in Chapter 6.

3.2.2 Summary of MRAC Using Plant Feedforward

PLANT:

$$\dot{x}_p(t) = A_p x_p(t) + B_p u_p(t) \tag{3.9}$$

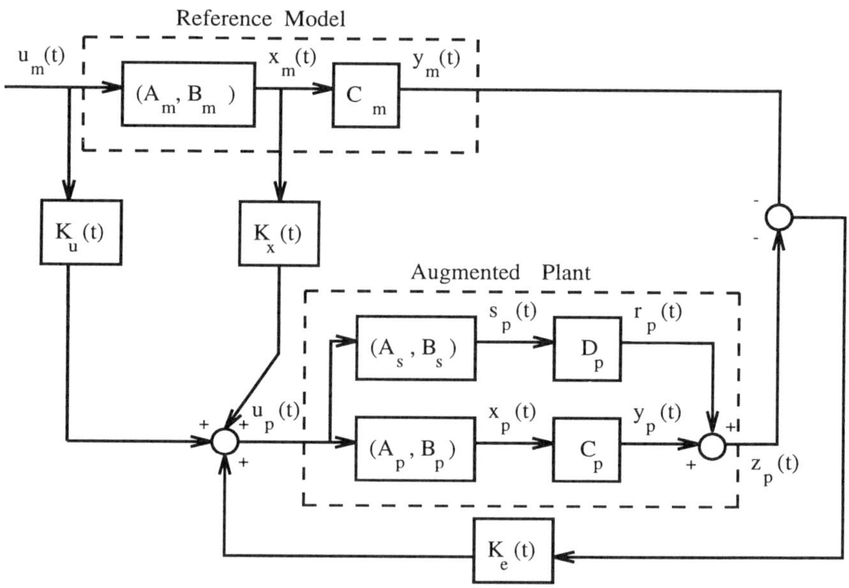

FIGURE 3.6. MRAC with Plant Feedforward

$$y_p(t) = C_p x_p(t) \tag{3.10}$$

SUPPLEMENTARY FEEDFORWARD:

$$\dot{s}_p(t) = A_s s_p(t) + B_s u_p(t) \tag{3.11}$$

$$r_p(t) = D_p s_p(t) \tag{3.12}$$

AUGMENTED MEASURED OUTPUT:

$$z_p(t) = y_p(t) + r_p(t) \tag{3.13}$$

MODEL:

$$\dot{x}_m(t) = A_m x_m(t) + B_m u_m(t) \tag{3.14}$$

$$y_m(t) = C_m x_m(t) \tag{3.15}$$

AUGMENTED TRACKING ERROR:

$$e_y(t) = y_m(t) - z_p(t) = y_m(t) - y_p(t) - D_p s_p(t) \tag{3.16}$$

ADAPTIVE ALGORITHM:

$$r^T(t) \triangleq \left[e_y^T(t) \;\; x_m^T(t) \;\; u_m^T(t)\right] \tag{3.17}$$

$$K(t) \triangleq \left[K_e(t) \;\; K_x(t) \;\; K_u(t)\right] \tag{3.18}$$

$$\dot{K}_I(t) = e_y(t) r^T(t) T \qquad\qquad T > 0 \tag{3.19}$$

$$K_p(t) = e_y(t) r^T(t) \bar{T} \qquad\qquad \bar{T} \geq 0 \tag{3.20}$$

$$K(t) = K_p(t) + K_I(t) \tag{3.21}$$

$$u_p(t) = K(t) r(t) \tag{3.22}$$

3.2.3 Illustrative Examples

To illustrate model reference adaptive control with feedforward augmentation, two examples will be considered. The first which was also considered in [40] is defined as follows:

The plant was given by:

$$\frac{y_p(s)}{u_p(s)} \doteq \frac{2}{(s+1)} \frac{229}{(s^2 + as + 229)} \tag{3.23}$$

where either $a = 30$ or $a = 30 + 5\sin(2t)$

The output of this plant was required to follow the output of the reference model:

$$\frac{y_m(s)}{u_m(s)} = \frac{1}{1 + s/3}. \tag{3.24}$$

The reference command u_m was a square wave with magnitude of 0.3 units and a period of 20 seconds, and all initial conditions were set to zero.

The feedforward augmentation was defined as

$$R_p(s)/U_p(s) = D/(\tau s + 1). \tag{3.25}$$

Thus with respect to 3.11 and 3.12,

$$A_s = -1/\tau \tag{3.26}$$

$$B_s = 1/\tau \tag{3.27}$$

$$D_p = D. \tag{3.28}$$

For simplicity, T and $\bar{T}$ were selected as

$$T = T_I \, I_3 \tag{3.29}$$

and

$$\bar{T} = T_p \, I_3 \tag{3.30}$$

where I_3 denotes a (3×3) identity matrix, and T_I and T_p are scalars.

Of interest are the effects of changing the design parameters

$$T_I, \; T_p, \; D, \; \tau.$$

and the effects of the sinusoidal variation on the coefficient a of s in the second order factor. In all cases, to satisfy the positive real constraints, D must be non–negative. Results, shown in Figures 3.7–3.10 for a fixed at 30 and for variations in the design parameters, show that:

- As D was increased, the steady state tracking error increased (Figure 3.7).
- As T_p was increased, the overshoot decreased. (Figure 3.8).
- As T_I was increased, the rate of adaptation increased, and the steady state error decreased somewhat. (Figure 3.9). Note that although the steady state error is about the same for T_I and $T_p = 10.$, more cycles are needed for error reduction for $T_I = 1.$
- As τ was changed within small multiples of the reference model time constant (.333), the response was not noticeably affected (Figure 3.10).

Finally, as shown in Figure 3.11, the sinusoidal variation in the coefficient a did not present any tracking problems.

Representative plots for the gains are shown in Figure 3.12.

The second example considered was an unstable modification of another example from [40]; namely the single input single output system defined by:

$$\frac{y_p(s)}{u_p(s)} = \frac{200}{(s-1)(s^2 + as + 100)} \tag{3.31}$$

where either $a = 8.0$ or $a = 8 + 6 \sin\,(2t)$

The reference model, feedforward compensator, and command were the same as for the first example.

However, in this example, in order to satisfy the positive real condition, for $\tau = 0.2$ and the sinusoidal variation; namely,

$$2. \leq a \leq 14.$$

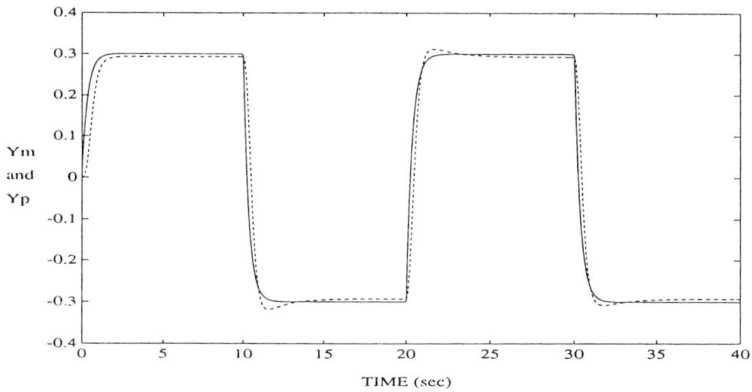

(a) $D_p = 0.05$

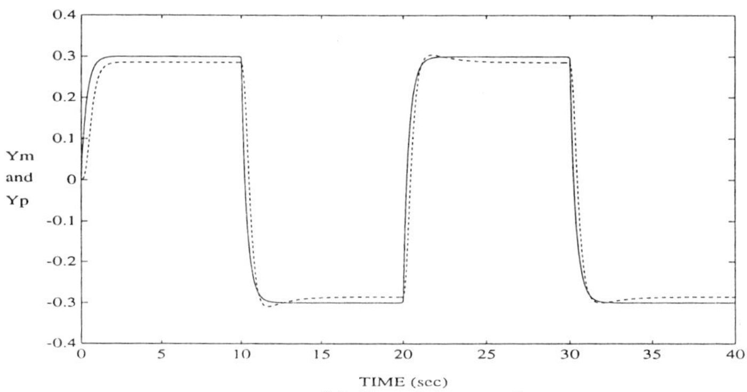

(b) $D_p = 0.1$

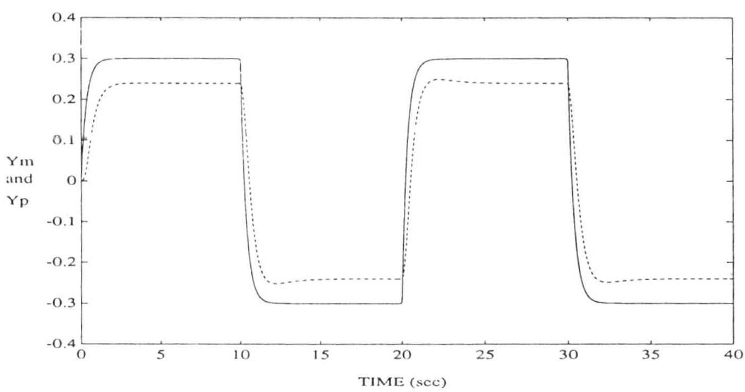

(c) $D_p = 0.5$

FIGURE 3.7. Plant and Model Responses for First Example, Feedforward in Plant, $T_I = T_p = 10.$, $\tau = 0.2$, $a = 30$

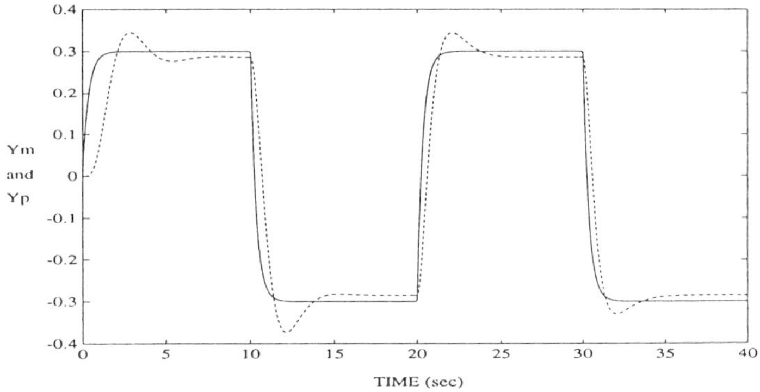

(a) $\bar{T} = 0$

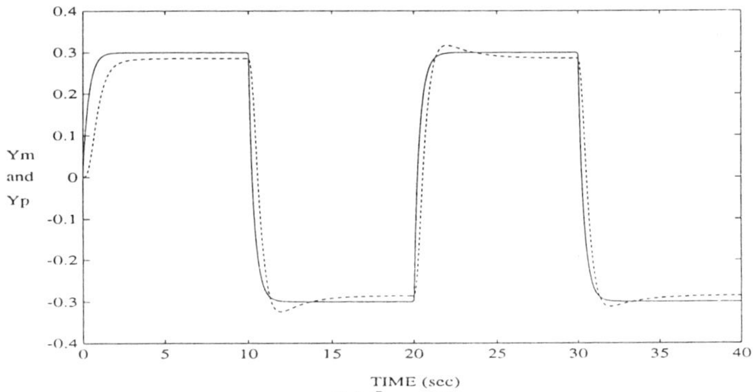

(b) $\bar{T} = 5I$

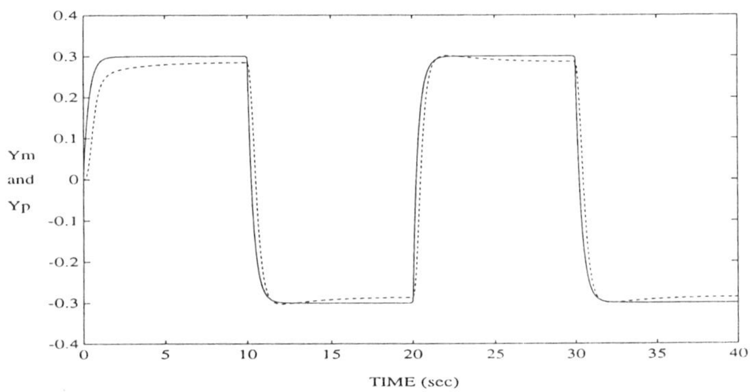

(c) $\bar{T} = 10I$

FIGURE 3.8. Plant and Model Responses for First Example, Feedforward in Plant, $T_I = 50$, $\tau = 0.2$, $D = 0.1$, $a = 30$

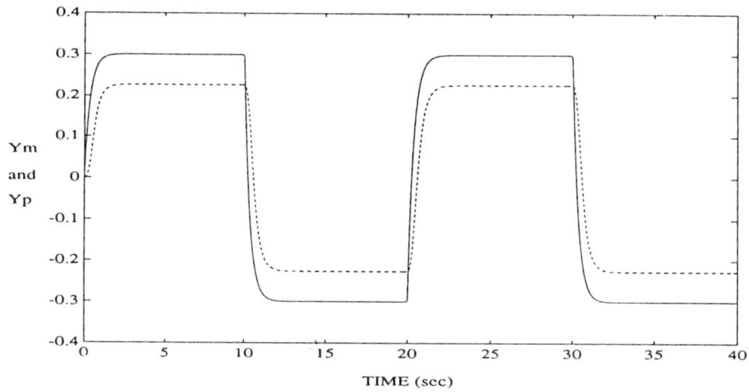

(a) $T = 0$

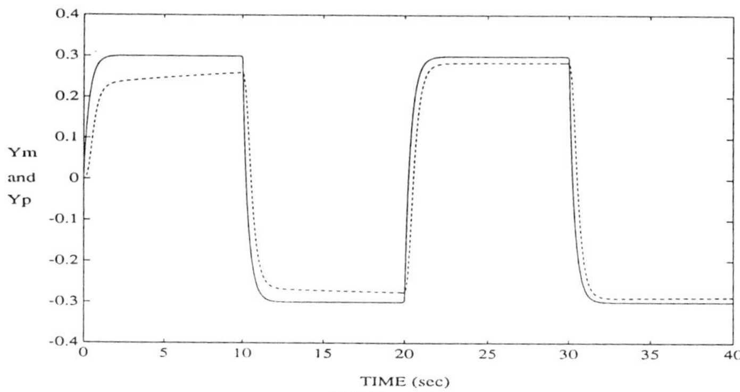

(b) $T = 1I$

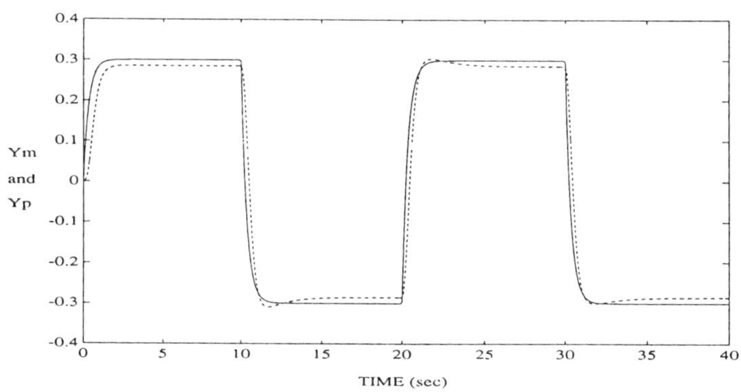

(c) $T = 10I$

FIGURE 3.9. Plant and Model Responses for First Example, Feedforward in Plant, $T_p = 10.$, $\tau = 0.2$, $D = 0.1, a = 30$

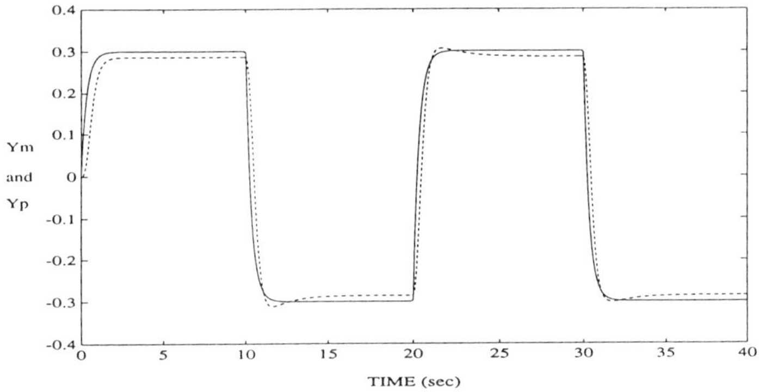

(a) $\tau = 0.1$

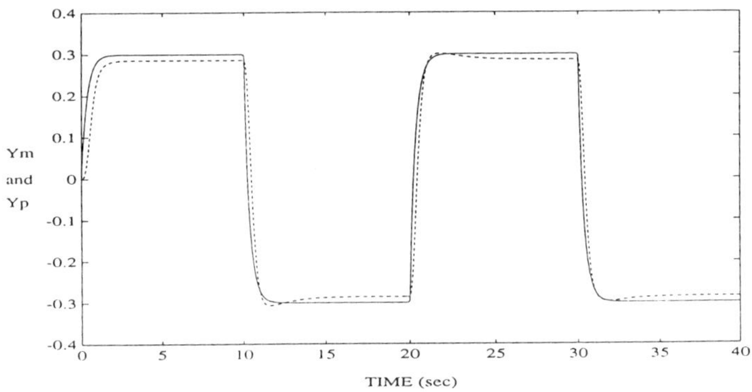

(b) $\tau = 0.5$

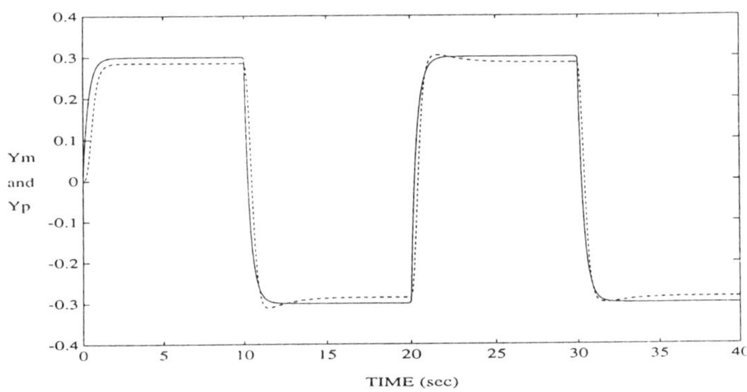

(c) $\tau = 1.0$

FIGURE 3.10. Plant and Model Responses for First Example, Feedforward in Plant, $T_I = T_p = 10.$, $D = 0.1, a = 30$

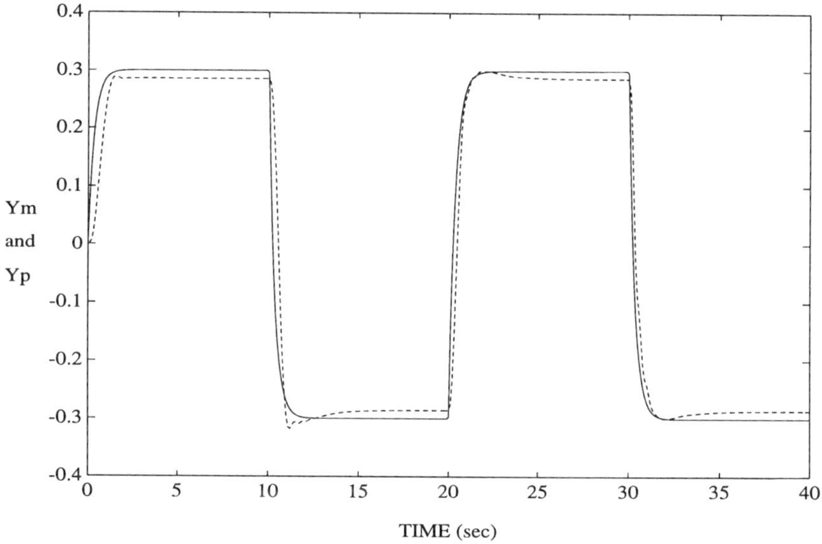

FIGURE 3.11. Plant and Model Responses for First Example, Feedforward in Plant, Sinusoidal Variation in a, $T_I = T_p = 10.$, $\tau = 0.2$, $D = 0.1$

it is sufficient to select D such that

$$0.86 < D < 2.$$

For $a = 8.0$, and $\tau = 0.2$,

$$0 < D \leq 2.$$

Representative results are shown for various D values in Figure 3.13 for $T_I = 7500.$,, $T_p = 3750.$, $\tau = .2$, $a = 8$. Note that although lower values of D result in smaller steady state errors, the corresponding transients can contain significant high frequency oscillations. As shown in Figures 3.14 and 3.15, these oscillations can be reduced by increasing T_p or by increasing the feedforward time constraint τ. Tracking in the presence of the sinusoidal variation in a is shown in Figure 3.16.

3.3 Feedforward in Both Plant and Model

3.3.1 Modifications to Insure Asymptotic Model Following

Although it is fairly easy to select a value for $R_p(s)$ in (3.8) so as to satisfy the ASPR condition, the resulting adaptive controller will in general result

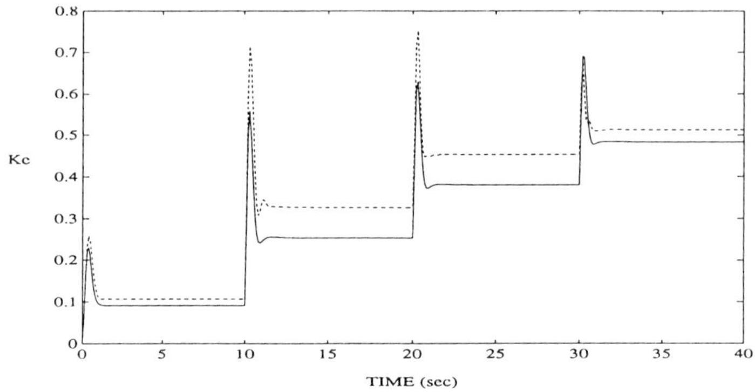

(a) $K_e(t)$

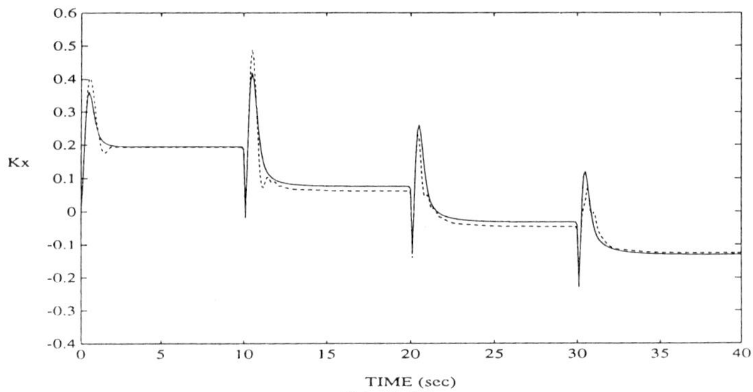

(b) $K_x(t)$

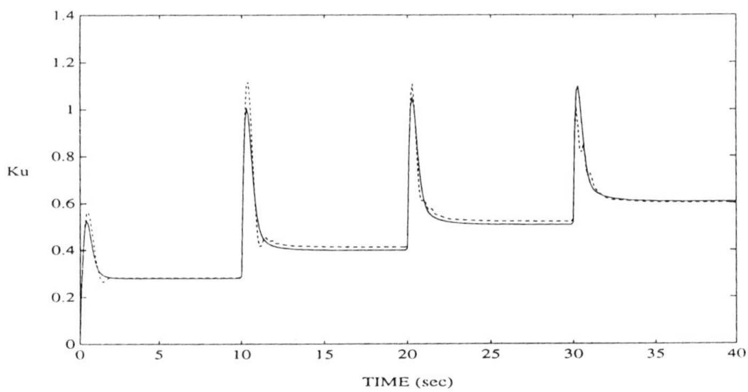

(c) $K_u(t)$

FIGURE 3.12. Gain Responses for First Example, Feedforward in Plant, $T_I = T_1 = 10.$, $\tau = 0.2$, $D = 0.1$

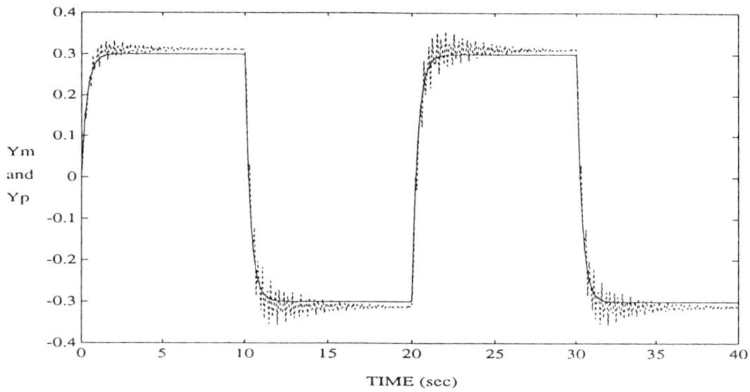

(a) $D_p = 0.075$

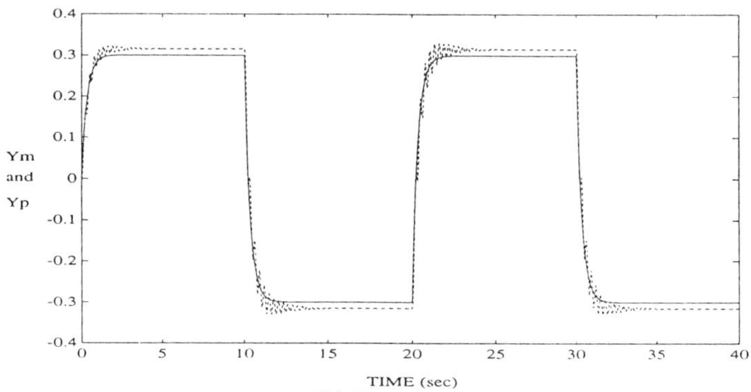

(b) $D_p = 0.095$

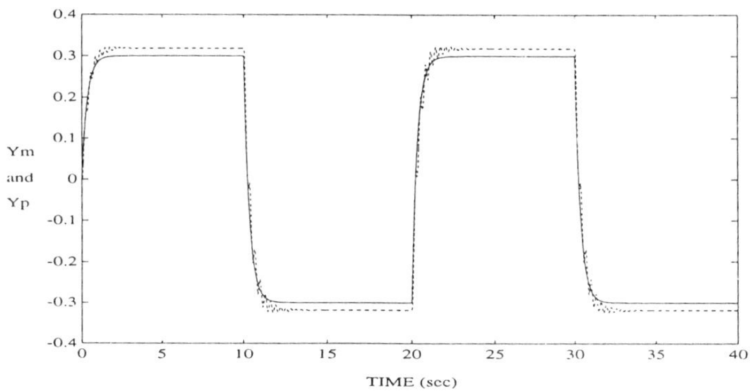

(c) $D_p = 0.115$

FIGURE 3.13. Plant and Model Responses for Second Example, Feedforward in Plant, $T_I = 7500$, $T_p = 3750$, $\tau = 0.2, a = 8.0$

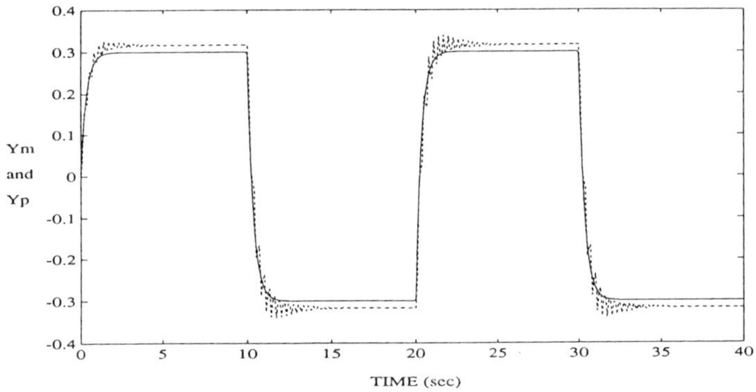

(a) $\bar{T} = 2500I$

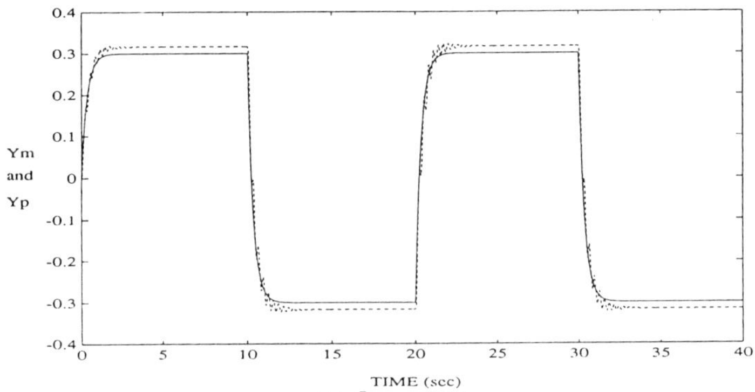

(b) $\bar{T} = 5000I$

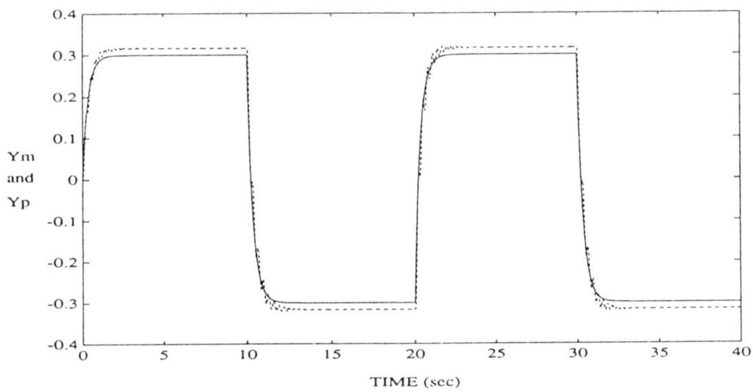

(c) $\bar{T} = 7500I$

FIGURE 3.14. Plant and Model Responses for Second Example, Feedforward in Plant, $T_I = 7500$, $\tau = 0.2$, $D = 0.105, a = 8.0$

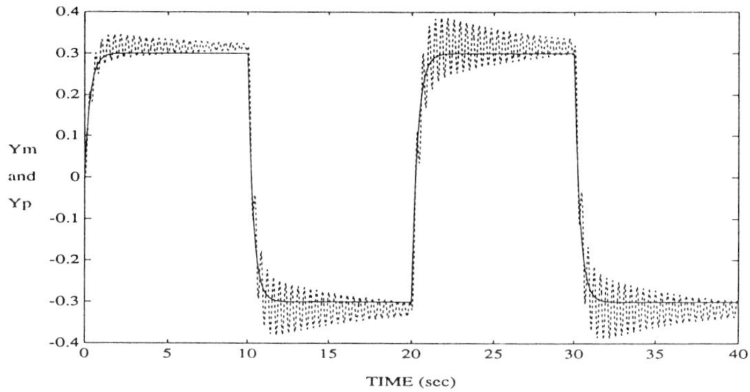

(a) $\tau = 0.1$

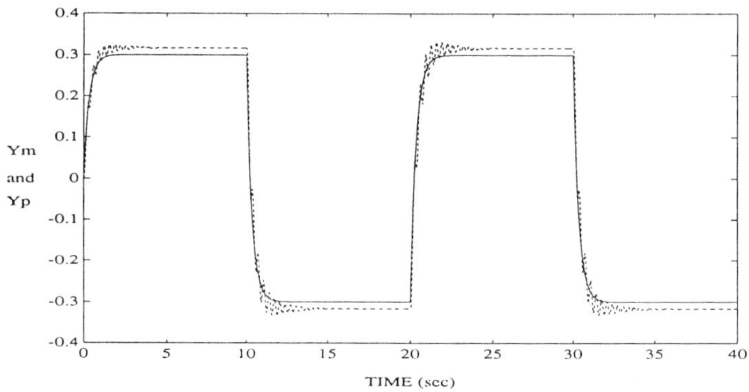

(b) $\tau = 0.15$

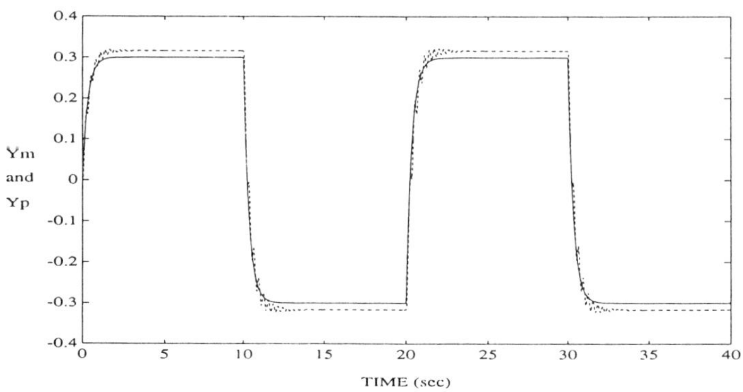

(c) $\tau = 0.2$

FIGURE 3.15. Plant and Model Responses for Second Example, Feedforward in Plant, $T_I = 7500$, $T_p = 3750$, $D = 0.105, a = 8.0$

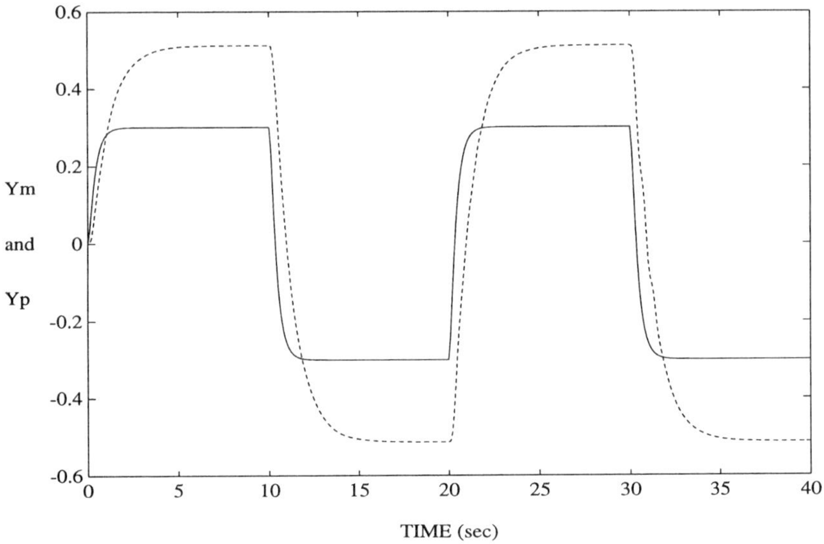

FIGURE 3.16. Plant and Model Responses for Second Example, Feedforward in Plant, Sinusoidal Variation in a, $T_I = 7500$, $T_p = 3750$, $\tau = 0.2$, $D = 0.105$

in a model following error that is bounded but not zero in steady state. To alleviate this situation, a modification is made that incorporates the supplementary feedforward of (3.8) into the reference model output as well as the plant output. To generalize this approach, the feedforward of (3.8) will be used so that the modified plant output $z_p(t)$ is again:

$$z_p(t) = y_p(t) + r_p(t) \tag{3.32}$$

where $r_p(t)$ is the result of the operator defined by $R_p(s)$ acting on u_p.
Thus,

$$z_p(t) = y_p(t) + R_p u_p(t) \tag{3.33}$$

where R_p is a linear operator on u_p, chosen such that the transfer matrix $G_a(s)$ defined by

$$z_p(s) = G_a(s) u_p(s) \tag{3.34}$$

is almost strictly positive real (ASPR) [20].

In a similar manner, define an augmented or modified reference model output z_m such that

$$z_m(t) = y_m(t) + r_m(t) \tag{3.35}$$

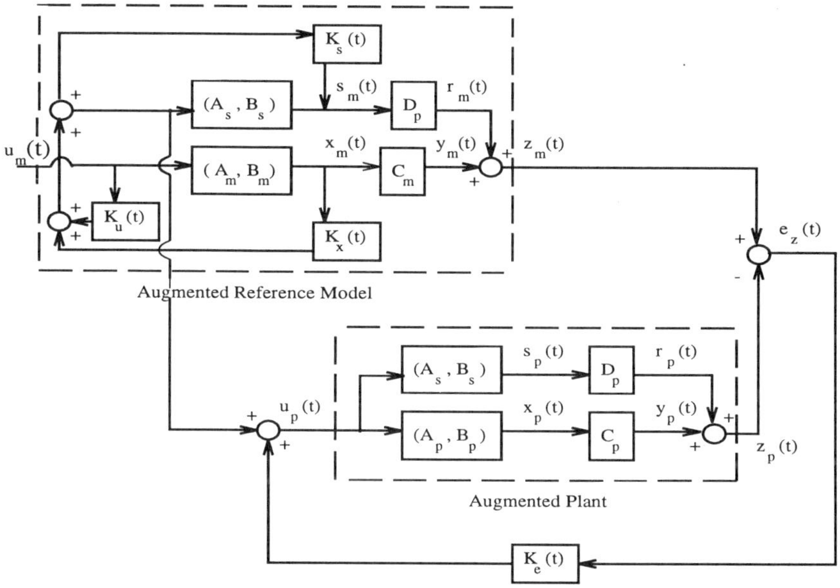

FIGURE 3.17. MRAC with Plant and Reference Model Feedforward

where the new feedforward component r_m is defined by:

$$\dot{s}_m(t) = A_s s_m(t) + B_s[u_p(t) - K_e(t)(z_m(t) - z_p(t))] \tag{3.36}$$

$$r_m(t) = D_p s_m(t) \tag{3.37}$$

and where $K_e(t)$ is an adaptive gain which will be subsequently defined. This configuration is shown in Figure 3.17. It is useful to note that since in (3.36) the error feedback from

$$e_z(t) = z_m(t) - z_p(t) \tag{3.38}$$

is subtracted out from u_p, the model augmentation r_m corresponds to the output of a system forced by the feedforward component of the adaptive controller $u_p(t)$. Thus:

$$\begin{aligned} \dot{z}_m(t) - \dot{z}_p(t) &= \dot{y}_m(t) - \dot{y}_p(t) + \dot{r}_m(t) - \dot{r}_p(t) \\ &= \dot{y}_m(t) - \dot{y}_p(t) + D_p(A_s(s_m(t) - \\ & s_p(t)) - B_s K_e(t) e_z). \end{aligned} \tag{3.39}$$

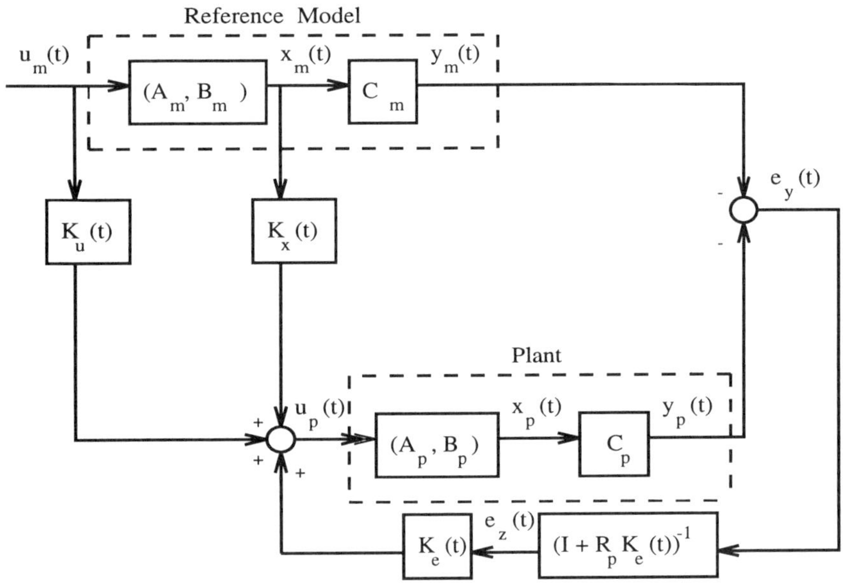

FIGURE 3.18. Equivalent Representation of MRAC with Feedforward

or

$$\dot{e}_z(t) = \dot{e}_y(t) + D_pA_se_s(t) - D_pB_sK_e(t)e_z(t) \tag{3.40}$$

where $e_s(t) = s_m(t) - s_p(t)$. But,

$$(s_m(t) - s_p(t)) = D_p^{-1}(r_m(t) - r_p(t)) = D_p^{-1}(e_z(t) - e_y(t)\,). \tag{3.41}$$

Thus, substitution of (3.41) into (3.40) gives:

$$\dot{e}_z(t) = \dot{e}_y(t) + D_pA_sD_p^{-1}[e_z(t) - e_y(t)] - D_pB_sK_e(t)e_z(t). \tag{3.42}$$

It is useful to note that (3.42) implies that incorporation of reference model feedforward is equivalent to defining $e_z(t)$ as the output of a time varying filter operation on $e_y(t)$ as shown in Figure 3.18.

Therefore, if the MRAC is designed such that $e_z(t)$ and $\dot{e}_z(t)$ both approach zero, (3.42) becomes:

$$0 = \dot{e}_y(t) - D_pA_sD_p^{-1}e_y(t). \tag{3.43}$$

Thus, if the eigenvalues of A_s are in the left half plane, $e_y(t)$ and $\dot{e}_y(t)$ will as desired both approach zero.

The proposed adaptive controller would then be

$$u_p(t) = K_e(t)e_z(t) + K_x(t)x_m(t) + K_s(t)s_m(t) + K_u(t)u_m(t) \qquad (3.44)$$

or

$$u_p(t) = K_r(t)\ r(t) \qquad (3.45)$$

where

$$K_r(t) = [K_e(t), K_x(t), K_s(t), K_u(t)] \qquad (3.46)$$

and

$$r^T(t) = [e_z^T(t), x_m^T(t), s_m^T(t), u_m^T(t)]. \qquad (3.47)$$

In a manner similar to (3.18)–(3.22), the corresponding adaptive controller becomes:

$$K_r(t) = K_p(t) + K_I(t) \qquad (3.48)$$

$$K_p(t) = e_z(t)r^T(t)\bar{T}\ ,\ \bar{T} \geq 0 \qquad (3.49)$$

$$\dot{K}_I(t) = e_z(t)r^T(t)T\ ,\ T > 0. \qquad (3.50)$$

It is important to note that in implementation of (3.49) and (3.50), it is preferable to compute $e_z(t)$ from (3.42) rather than from direct subtraction of $z_p(t)$ from $z_m(t)$.

The proof that these equations define an asymptotically stable adaptive controller is given in the next section.

3.3.2 Stability Proof

In order to prove stability, it is necessary to augment the plant and model equations. Combining (3.9) and (3.10) with (3.11), (3.12), and (3.13) gives as the augmented plant:

$$\dot{q}_p(t) = \begin{pmatrix} \dot{x}_p(t) \\ \dot{s}_p(t) \end{pmatrix} = \begin{pmatrix} A_p & 0 \\ 0 & A_s \end{pmatrix} \begin{pmatrix} x_p(t) \\ s_p(t) \end{pmatrix} + \begin{pmatrix} B_p \\ B_s \end{pmatrix} u_p(t). \qquad (3.51)$$

$$z_p(t) = [C_p \quad D_p] \begin{pmatrix} x_p(t) \\ s_p(t) \end{pmatrix}. \tag{3.52}$$

Thus, the plant can be considered as the time invariant process:

$$\dot{q}_p(t) = F_p q_p(t) + G_p u_p(t) \tag{3.53}$$

$$z_p(t) = H_p q_p(t). \tag{3.54}$$

However, the augmented model will be time varying because of the coupling with the adaptive gains. To see this, substitute (3.44) into (3.36) giving:

$$\dot{s}_m(t) = A_s s_m(t) + B_s[K_x(t)x_m + K_s(t)s_m + K_u(t)u_m]. \tag{3.55}$$

Combining this with (3.14) and (3.15) gives as the augmented model:

$$\dot{q}_m(t) = \begin{pmatrix} \dot{x}_m(t) \\ \dot{s}_m(t) \end{pmatrix} =$$

$$\begin{pmatrix} A_m & 0 \\ B_s K_x(t) & A_s + B_s K_s(t) \end{pmatrix} \begin{pmatrix} x_m(t) \\ s_m(t) \end{pmatrix} + \begin{pmatrix} B_m \\ B_s K_u(t) \end{pmatrix} u_m(t) \tag{3.56}$$

or

$$\dot{q}_m(t) = F_m(t) q_m(t) + G_m(t) u_m(t) \tag{3.57}$$

$$z_m(t) = H_m q_m(t). \tag{3.58}$$

Figure 3.19 shows the algorithm in this representation. Note the similarity between this structure and the original algorithm shown in Figure 2.1.

Because the augmented model is time varying, the model reference algorithm controller of Abida and Kaufman [97] will be used. This controller is based upon the existence of a command generator tracker (CGT) controller defined by the equations:

$$\begin{bmatrix} q_p^*(t) \\ u_p^*(t) \end{bmatrix} = \begin{bmatrix} S_{11}(t) & S_{12}(t) \\ S_{21}(t) & S_{22}(t) \end{bmatrix} \begin{bmatrix} q_m(t) \\ u_m(t) \end{bmatrix} \tag{3.59}$$

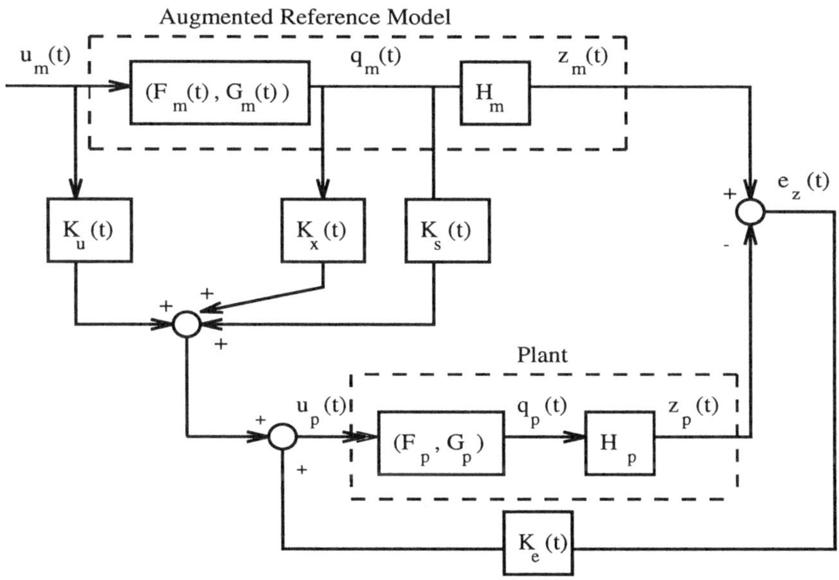

FIGURE 3.19. MRAC with Feedforward, Augmented Representation

where $q_p^*(t)$ is an ideal state vector satisfying

$$\dot{q}_p^*(t) = F_p q_p^*(t) + G_p u_p^*(t) \tag{3.60}$$

and $u_p^*(t)$ is an ideal controller such that if at some time:

$$q_p(t_1) = q_p^*(t_1) \tag{3.61}$$

and

$$u_p(t) = u_p^*(t) \qquad t \geq t_1. \tag{3.62}$$

Then:

$$q_p(t) \equiv q_p^*(t) \qquad t \geq t_1 \tag{3.63}$$

and

$$z_p(t) \equiv z_m(t) \qquad t \geq t_1. \tag{3.64}$$

As shown in [97] and under mild conditions, the $S_{ij}(t)$ matrices exist and satisfy:

$$\dot{S}_{11}(t) = F_p\ S_{11}(t) - S_{11}(t)\ F_m(t) + G_p\ S_{21}(t) \tag{3.65}$$

$$\dot{S}_{12}(t) = F_p\ S_{12}(t) - S_{11}(t)\ G_m(t) + G_p\ S_{22}(t) \tag{3.66}$$

$$H_p S_{11}(t) = H_m \tag{3.67}$$

$$H_p S_{12}(t) = 0. \tag{3.68}$$

In order to study the properties of the solution, it is useful to partition the above S_{ij} matrices according to their association with either the original states (x_p, x_m) or the augmentation (s_p, s_m). To this effect let:

$$S_{11}(t) = \begin{pmatrix} S_{11_{xx}}(t) & S_{11_{xs}}(t) \\ S_{11_{sx}}(t) & S_{11_{ss}}(t) \end{pmatrix} \tag{3.69}$$

$$S_{12}(t) = \begin{pmatrix} S_{12_x}(t) \\ S_{12_s}(t) \end{pmatrix} \tag{3.70}$$

$$S_{21}(t) = \left(S_{21_x}(t)\ S_{21_s}(t) \right). \tag{3.71}$$

With these partitions, (3.65) becomes:

$$\dot{S}_{11_{xx}}(t) = A_p S_{11_{xx}}(t) - S_{11_{xx}}(t) A_m - S_{11_{xs}}(t) B_s K_x(t) + B_p S_{21_x}(t) \tag{3.72}$$

$$\dot{S}_{11_{xs}}(t) = A_p S_{11_{xs}}(t) - S_{11_{xs}}(t)(A_s + B_s K_s(t)) + B_p S_{21_s}(t) \tag{3.73}$$

$$\dot{S}_{11_{sx}}(t) = A_s S_{11_{sx}}(t) - S_{11_{sx}}(t) A_m - S_{11_{ss}}(t) B_s K_x(t) + B_s S_{21_x}(t) \tag{3.74}$$

$$\dot{S}_{11_{ss}}(t) = A_s S_{11_{ss}}(t) - S_{11_{ss}}(t)(A_s + B_s K_s(t)) + B_s S_{21_s}(t). \tag{3.75}$$

Similarly, (3.66) becomes:

$$\dot{S}_{12_x}(t) = A_p S_{12_x}(t) - S_{11_{xx}}(t) B_m - S_{11_{xs}}(t) B_s K_u(t) + B_p S_{22}(t) \tag{3.76}$$

$$\dot{S}_{12_s}(t) = A_s S_{12_s}(t) - S_{11_{sx}}(t) B_m - S_{11_{ss}}(t) B_s K_u(t) + B_s S_{22}(t). \tag{3.77}$$

(3.67) becomes:

$$C_p S_{11_{xx}}(t) + D_p S_{11_{sz}}(t) = C_m \tag{3.78}$$

$$C_p S_{11_{xs}}(t) + D_p S_{11_{ss}}(t) = D_p. \tag{3.79}$$

Finally, (3.68) becomes:

$$C_p S_{12_x}(t) + D_p S_{12_s}(t) = 0. \tag{3.80}$$

Observe that (3.73), (3.74), (3.75), (3.77), and (3.79) can be satisfied by setting:

$$S_{21_s}(t) = K_s(t) = 0 \tag{3.81}$$

$$S_{11_{ss}}(t) = I \text{ (i.e } \dot{S}_{11_{ss}} = 0) \tag{3.82}$$

$$S_{11_{xs}}(t) = 0 \tag{3.83}$$

$$S_{11_{sx}}(t) = 0 \tag{3.84}$$

$$S_{21_x}(t) = K_x(t) \tag{3.85}$$

$$S_{22}(t) = K_u(t) \tag{3.86}$$

$$S_{21_s}(t) = 0. \tag{3.87}$$

Thus, (3.72), (3.76), (3.78), (3.80) respectively become:

$$S_{11_{xx}}(t) = A_p S_{11_{xx}}(t) - S_{11_{xx}}(t) A_m + B_p S_{21_x}(t) \tag{3.88}$$

$$\dot{S}_{12x}(t) = A_p S_{12_x}(t) - S_{11_{xx}}(t) B_m + B_p S_{22}(t) \tag{3.89}$$

$$C_p S_{11_{xx}}(t) = C_m \tag{3.90}$$

$$C_p S_{12_x}(t) = 0. \tag{3.91}$$

It is important to note that these equations were shown by Abida [97] to have a solution under rather mild conditions. With the above values, given in (3.81)–(3.87):

$$s_p^*(t) = s_m(t) \tag{3.92}$$

and

$$C_p x_p^*(t) = C_m x_m(t). \tag{3.93}$$

Furthermore, since $K_s(t) = S_{21_s}(t) = 0$, it is unnecessary to include $s_m(t)$ in (3.44) and $K_s(t)$ in (3.46). This in turn means that in (3.55), the $K_s(t)$ term can be removed from adaptation so that

$$\dot{s}_m(t) = A_s s_m(t) + B_s[K_x(t)x_m(t) + K_u(t)u_m(t)]. \tag{3.94}$$

Thus, since $K_s(t) = 0$, the adaptive control law as in [97] is chosen to be:

$$u_p(t) = K_e(t)(z_m(t) - z_p(t)) + K_x(t)x_m(t) + K_u(t)u_m(t) \tag{3.95}$$

or

$$u(t) = K(t)r(t) \text{ where } K(t) = [K_e(t), K_x(t), K_u(t)] \tag{3.96}$$

and

$$r^T(t) = \left[z_m^T(t) - z_p^T(t), x_m^T(t), u_m^T(t)\right]. \tag{3.97}$$

$K(t)$ is then generated according to the following adaptive rule:

$$K(t) = K_I(t) + K_p(t) \tag{3.98}$$

$$\dot{K}_I(t) = (z_m(t) - z_p(t))r^T(t)T, \quad K_I(0) = K_{I0}, \ T > 0 \tag{3.99}$$

$$K_p(t) = (z_m(t) - z_p(t))r^T(t)\bar{T}, \quad \bar{T} \geq 0. \tag{3.100}$$

To show stability, define the error $e(t)$ as the difference between the ideal and actual augmented state; namely, $e(t) = q_p^*(t) - q_p(t)$. Then the error dynamics become:

$$\begin{aligned} \dot{e}(t) = F_p e(t) \quad &+ \quad G_p[S_{21}(t)q_m(t) + S_{22}(t)u_m(t) \\ &- \quad K_p(t)r(t) - K_I(t)r(t)] \end{aligned} \tag{3.101}$$

where $S_{21}(t) = [S_{21_x}(t)\ 0]$. This error equation is used in Appendix 3B to derive the following main theorem:

Theorem 3.1: Consider the adaptive controller defined by:

$$u_p(t) = K_e(t)(z_m(t) - z_p(t)) + K_x(t)x_m(t) + K_u(t)u_m(t) \tag{3.102}$$

or

$$u_p(t) = K(t)r(t) \tag{3.103}$$

where

$$K(t) = [K_e(t), K_x(t), K_u(t)] \tag{3.104}$$

and

$$r^T(t) = \left[z_m^T(t) - z_p^T(t), x_m^T(t), u_m^T(t)\right]. \tag{3.105}$$

$K(t)$ is then generated according to the following adaptive rule:

$$K(t) = K_I(t) + K_p(t) \tag{3.106}$$

$$\dot{K}_I(t) = (z_m(t) - z_p(t))r^T(t)T,\ K_I(0) = K_{10}, T > 0 \tag{3.107}$$

$$K_p(t) = (z_m(t) - z_p(t))r^T(t)\bar{T},\ \bar{T} \geq 0 \tag{3.108}$$

where z_m, z_p are as defined in (3.32)–(3.35). Then provided that:

- G_a as defined in (3.8) is ASPR,
- $u_m(t)$ is a step or constant command,
- a solution exists to (3.65)–(3.68),
- and all eigenvalues of A_s are in left half plane,

the output error $(y_m - y_p)$ will asymptotically vanish, and all states and gains will be bounded.

3.3.3 Summary of Constraints and Design Rules

PLANT:

$$\dot{x}_p(t) = A_p x_p(t) + B_p u_p(t) \tag{3.109}$$

$$y_p(t) = C_p x_p(t) \tag{3.110}$$

MODEL:

$$\dot{x}_m(t) = A_m x_m(t) + B_m u_m(t) \tag{3.111}$$

$$y_m(t) = C_m x_m(t) \tag{3.112}$$

SUPPLEMENTARY FEEDFORWARD:

$$\dot{s}_p(t) = A_s s_p + B_s u_p(t) \tag{3.113}$$

$$r_p(t) = D_p s_p(t) \tag{3.114}$$

$$z_p(t) = y_p(t) + r_p(t) \tag{3.115}$$

$$\dot{s}_m(t) = A_s s_m(t) + B_s \left[u_p(t) - K_e(t)(z_m(t) - z_p(t))\right] \tag{3.116}$$

$$r_m(t) = D_p s_m(t) \tag{3.117}$$

$$z_m(t) = y_m(t) + z_m(t) \tag{3.118}$$

ADAPTIVE CONTROL LAWS:

$$u_p(t) = K(t) r(t) \tag{3.119}$$

where:

$$K(t) = K_p(t) + K_I(t) \tag{3.120}$$

$$r^T(t) = \left[z_m^T(t) - z_p^T(t), x_m^T(t), u_m^T(t)\right] \tag{3.121}$$

$$\dot{K}_I(t) = (z_m(t) - z_p(t)) r^T(t) T \tag{3.122}$$

$$K_p(t) = (z_m(t) - z_p(t)) r^T(t) \bar{T} \tag{3.123}$$

CONSTRAINTS:

- $T > 0, \bar{T} \geq 0$
- G_a as defined in (3.8) is ASPR,
- $u_m(t)$ is a step or constant command,
- a solution exists to (3.65)–(3.68),
- All eigenvalues of A_s are in the left half plane

3.3.4 Illustrative Examples

In order to demonstrate that the use of feedforward in both plant and reference model yields asymptotic tracking, the same two examples of Section 3.2.3 were considered. The first is again defined by the plant.

$$\frac{Y_p(s)}{U_p(s)} = \frac{2}{(s+1)} \frac{229}{(s^2 + as + 229)} \tag{3.124}$$

where either $a = 30$ or $a = 30 + 25\sin(2t)$

The output of this plant is required to follow the output of the reference model

$$\frac{Y_m(s)}{U_m(s)} = \frac{1}{1 + s/3}. \tag{3.125}$$

As before the reference command u_m was a square wave with magnitude of 0.3 units and a period of 20 seconds, and all initial conditions were set to zero.

The feedforward augmentation for both plant and model is again defined as:

$$R_p(s)/U_p(s) = D/(\tau s + 1). \tag{3.126}$$

For simplicity, T and $\bar{T}$ were selected as

$$T = T_I \, I_3 \tag{3.127}$$

and

$$\bar{T} = T_p \, I_3 \tag{3.128}$$

where I_3 denotes a (3×3) identity matrix and T_I and T_p are scalar.

Of interest again are the effects of changing the design parameters

$$T_I, \; T_p, \; D, \; \tau.$$

and the effects of the sinusoidal variation on the coefficient a of s in the second order term. In all cases, to satisfy the position of real constraints, D must be non–negative. Results shown in Figures 3.20–3.23, for variations in plant parameters and a fixed at 30, show that in all cases the steady state error is zero and furthermore that:

- As D was increased, the overshoot increased (Figure 3.20).
- As T_p was increased, the overshoot decreased. (Figure 3.21).
- As T_I was increased, the rate of adaptation increased. (Figure 3.22).

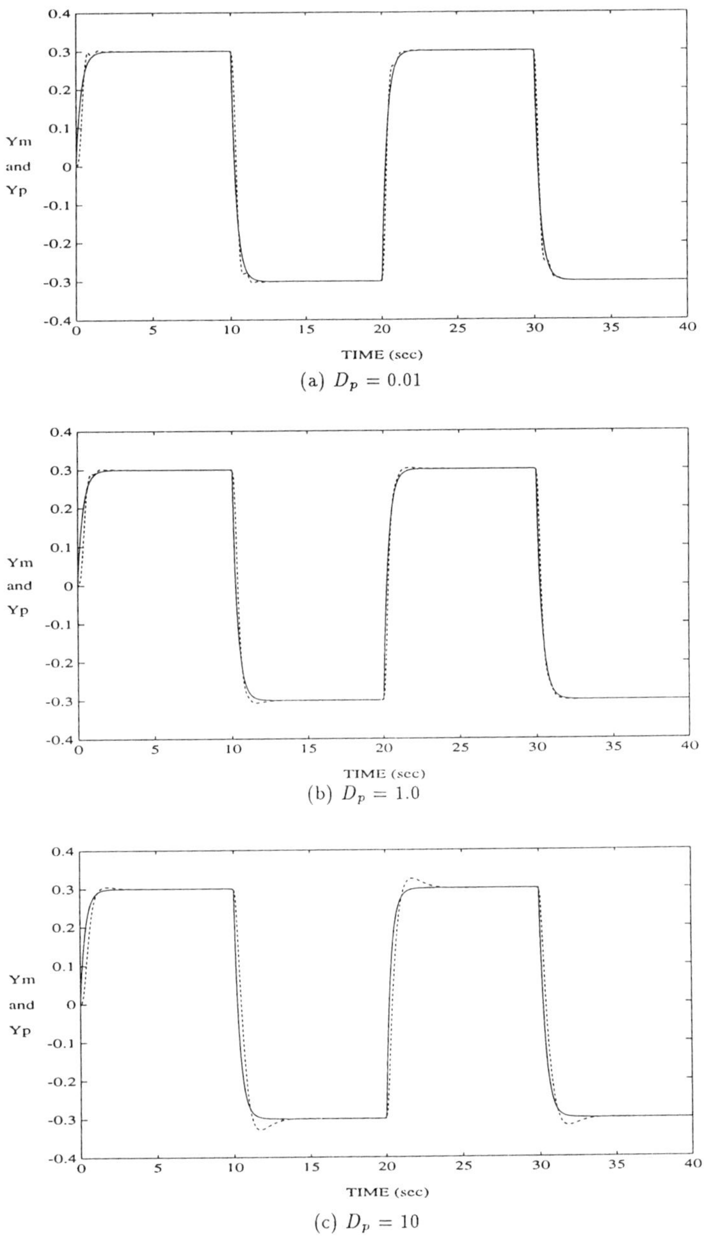

FIGURE 3.20. Plant and Model Responses for First Example, Feedforward in Plant and Model, $T_I = 50.0$, $T_p = 25.0$, $\tau = 0.3, a = 30$.

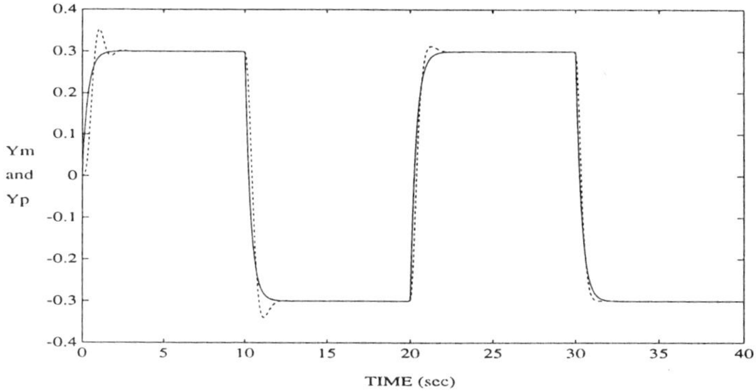

(a) $\bar{T} = 10I$

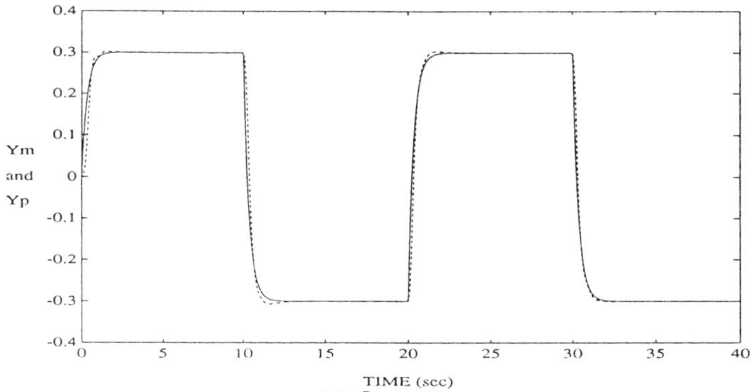

(b) $\bar{T} = 25I$

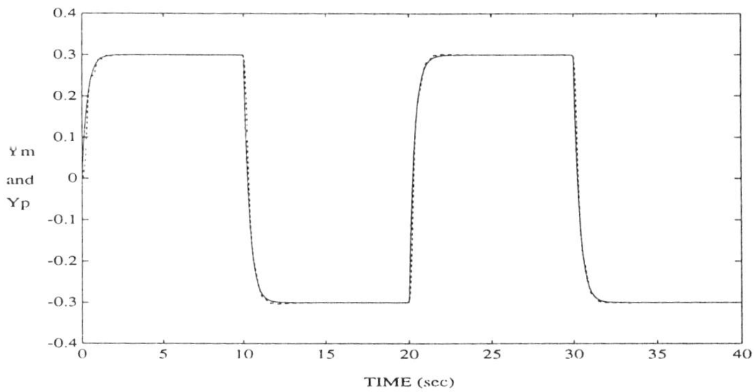

(c) $\bar{T} = 50I$

FIGURE 3.21. Plant and Model Responses for First Example, Feedforward in Plant and Model, $T_I = 50.0$, $\tau = 0.3$, $D = 1.0, a = 30$

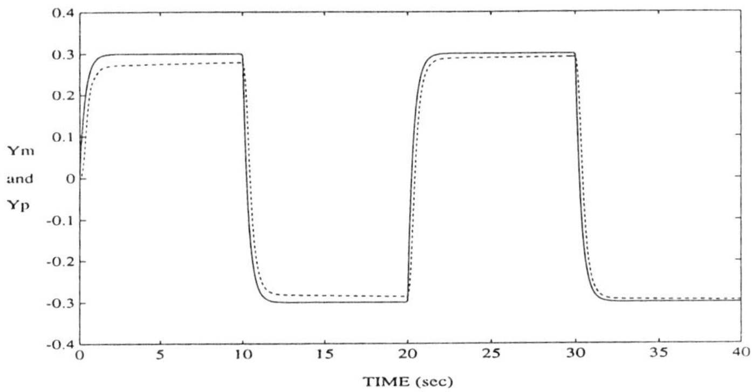

(a) $T = 1I$

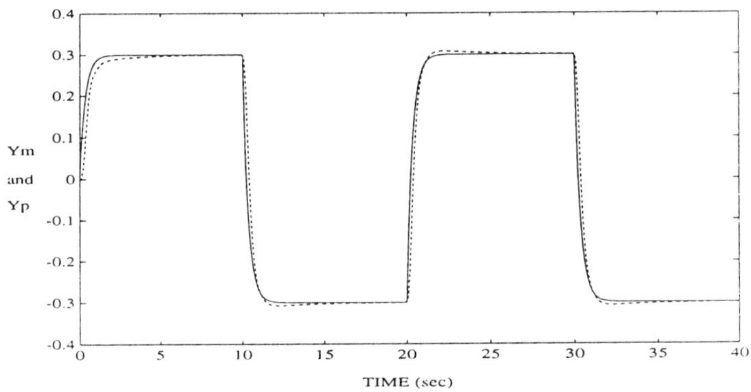

(b) $T = 10I$

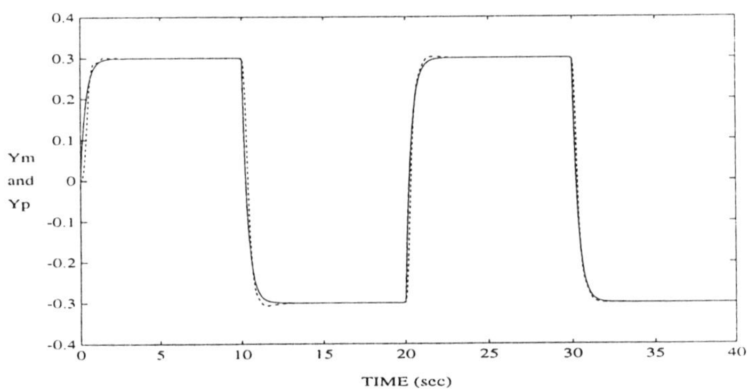

(c) $T = 50I$

FIGURE 3.22. Plant and Model Responses for First Example, Feedforward in Plant and Model, $T_p = 25.0$, $\tau = 0.3, a = 30$.

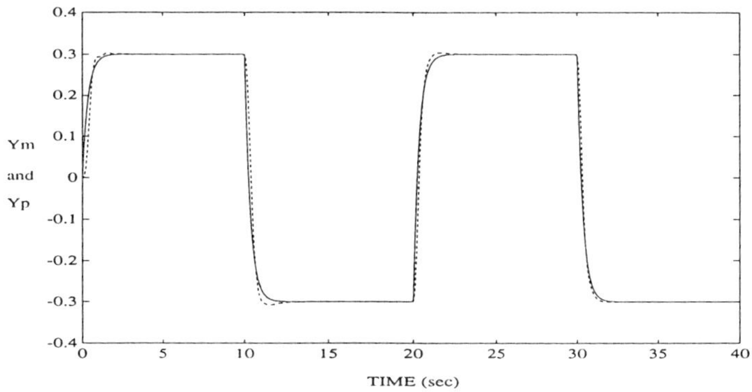

(a) $\tau = 0.2$

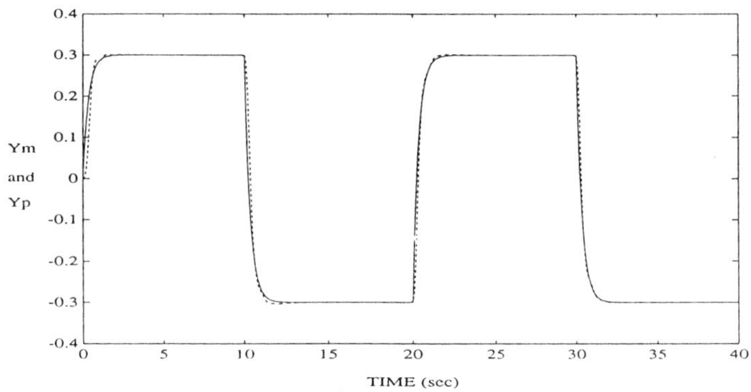

(b) $\tau = 0.5$

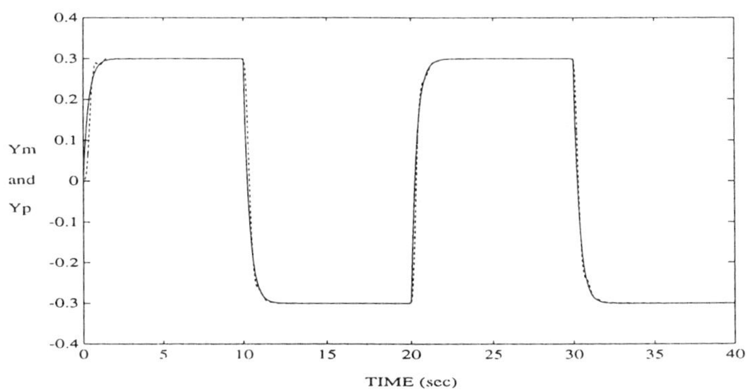

(c) $\tau = 1.0$

FIGURE 3.23. Plant and Model Responses for First Example, Feedforward in Plant and Model, $T_I = 50.0$, $T_p = 25.0$, $D = 1.0, a = 30$

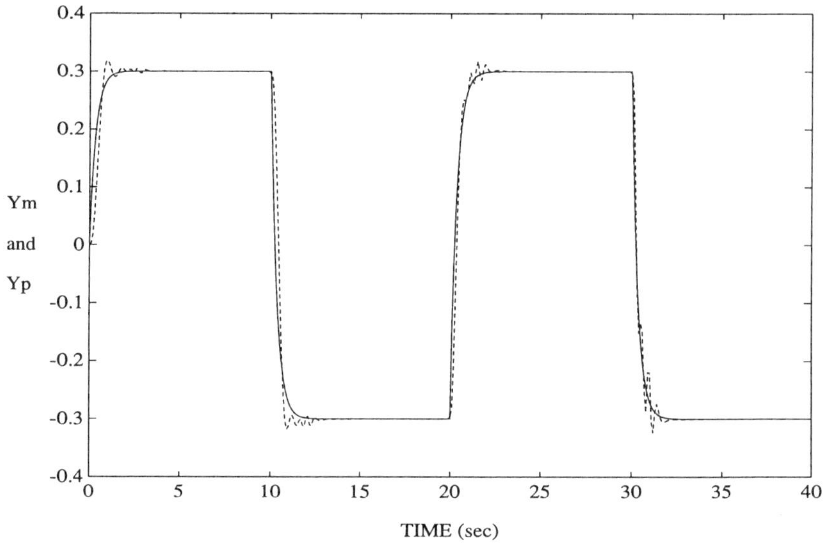

FIGURE 3.24. Plant and Model Responses for First Example, Feedforward in Plant and Model, Sinusoidal Variation in a, $T_I = 50$, $T_p = 25$, $\tau = 0.3$, $D = 1.0$

- As τ was changed within small multiples of the reference model time constraint (.333), the overshoot was slightly decreased (Figure 3.23).

Finally, as shown in Figure 3.24, the sinusoidal variation in the coefficient a did not present any tracking problems.

Representative plots for the gains are shown in Figure 3.25.

The second example considered was again the single input single output unstable system defined by:

$$\frac{y_p(s)}{u_p(s)} = \frac{200}{(s-1)(s^2+as+100)} \tag{3.129}$$

where either $a = 8.0$ or $a = 8 + 6\sin(2t)$.

The reference model, feedforward compensator, and command were the same as for the first example.

However, in this example, in order to satisfy the positive real condition for $\tau = 0.2$ and the sinusoidal variation; namely,

$$2. \leq a \leq 14,$$

it is sufficient to select D such that

$$0.86 < D < 2.$$

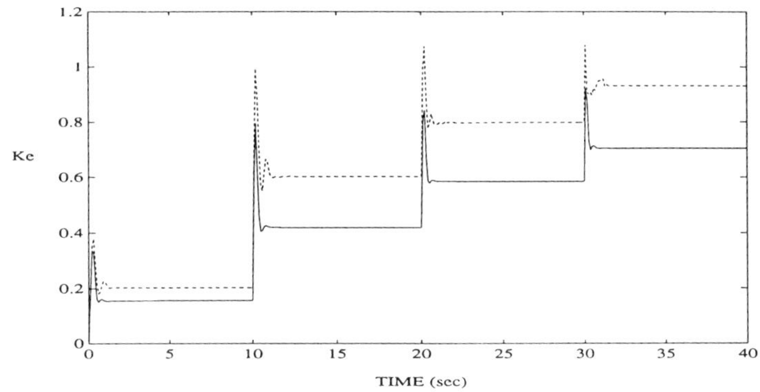

(a) $K_e(t)$

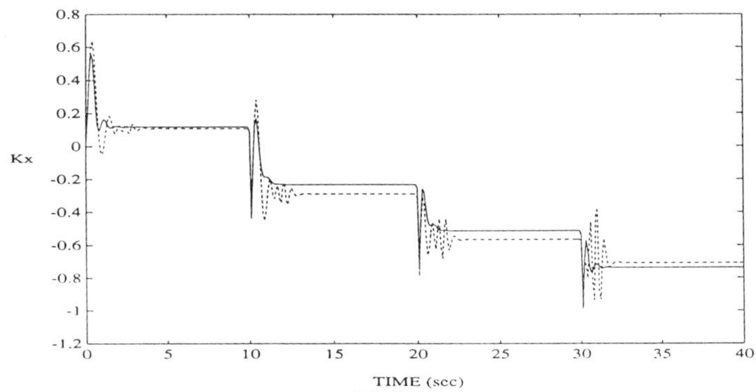

(b) $K_x(t)$

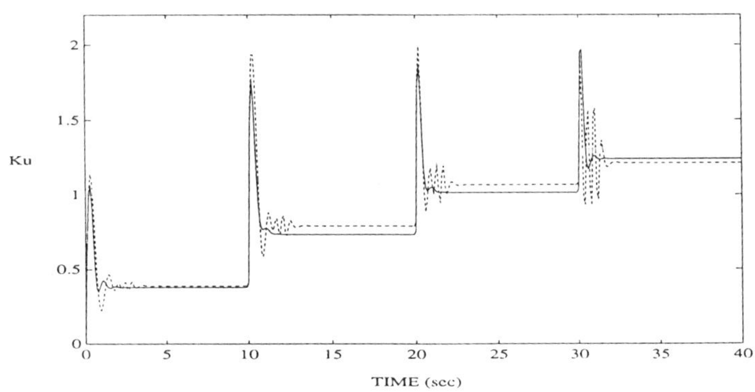

(c) $K_u(t)$

FIGURE 3.25. Gain responses for First Example, Feedforward in Plant and Model, $T_I = 50.0$, $T_p = 25.0$, $\tau = 0.3$, $D = 1.0$

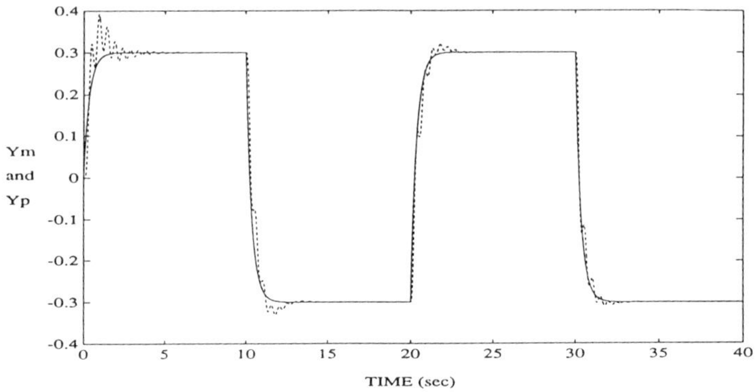

(a) $D_p = 0.5$

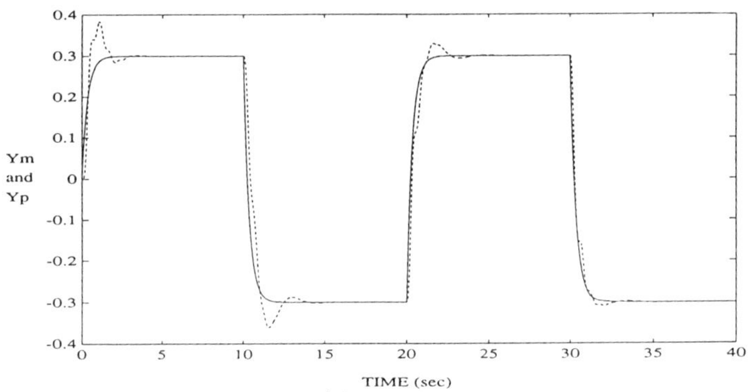

(b) $D_p = 1.0$

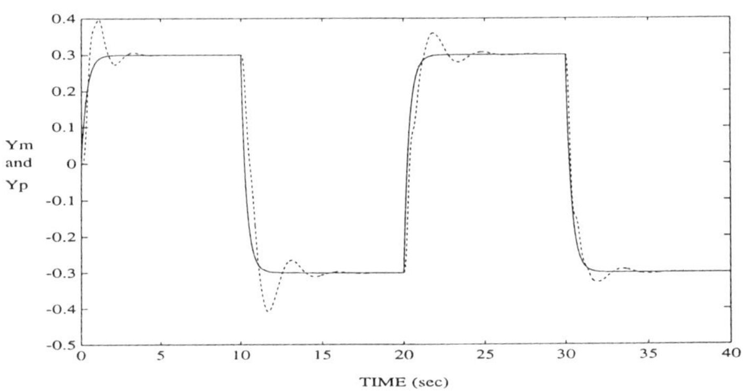

(c) $D_p = 1.5$

FIGURE 3.26. Plant and Model Responses for Second Example, Feedforward in Plant and Model, $a = 8.0$, $T_{I_e} = 50,000.$, $T_{I_x} = T_{I_u} = 50.0$, $T_p = 0$.

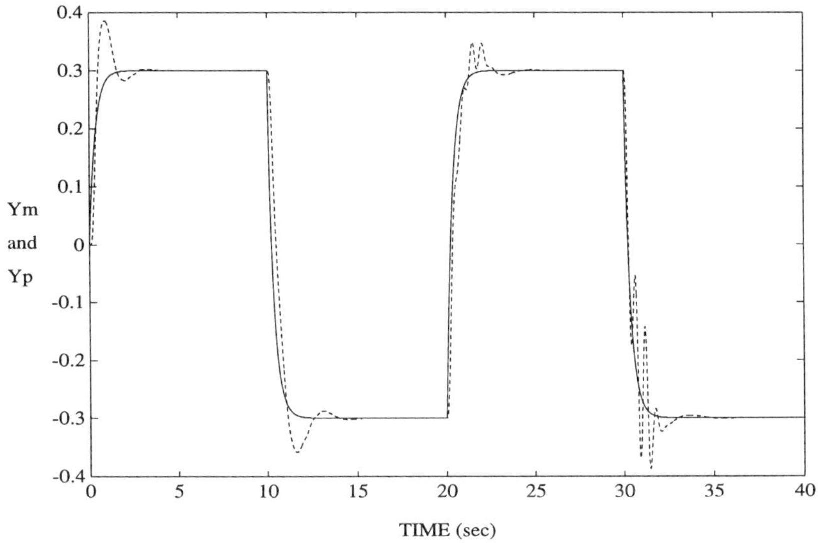

FIGURE 3.27. Plant and Model Responses for Second Example, Feedforward in Plant and Model, Sinusoidal Variation in a, $\tau = 0.3$, $D = 1.$, $T_{I_e} = 50,000.$, $T_{I_x} = T_{I_u} = 50.0$, $T_p = 0$.

For $a = 8.0$, and $\tau = 0.2$,

$$0 < D \leq 2.$$

Representative results are shown for various D values in Figure 3.26 for $T_{I_e} = 50,000.$, $T_{I_x} = T_{I_u} = 500.$, $T_p = 0.$. Note that low values of D can result in significant high frequency transient oscillations. Tracking in the presence of the sinusoidal variation in a is shown in Figure 3.27.

3.3.5 Conclusions and Recommendations

A proof for a new generalized modification to a direct multiple input-multiple output model reference adaptive controller has been developed. This modification, which involves augmenting the reference model as well as the plant with a supplementary feedforward filter, eliminates the bias in the steady state error from feedforward only in the plant. Such an advancement allows for the increased utility of the algorithm over a much wider class of processes.

3.4 A Unified Approach to Supplementary Dynamics

In this section, three algorithms for non–ASPR plants are described which are characterized by the insertion of supplementary dynamics in different locations in the adaptive loop. These locations include a feedback path, a parallel path, and a cascade path with the plant. A metasystem is obtained by concatenating the plant with the supplementary dynamics, and all three algorithms are shown to be special cases of the metasystem representation of the adaptive system. These algorithms differ from those in Sections 3.1 and 3.2 because of the introduction of adaptive gains for the supplementary dynamics in addition to adaptive gains for the controller.

A theorem is presented which describes sufficient conditions for all three algorithms to yield a bounded error. Furthermore, if the reference model input is constant for $t \geq t_1$ with t_1 arbitrary, then another theorem describes sufficient conditions for an asymptotically vanishing error when the supplementary dynamics are in a parallel or cascade path with the plant.

3.4.1 Theory

Problem Formulation

The plant and reference model were described in Section 2.2. The state representation for the supplementary dynamics is given by

$$\dot{x}_f(t) \;=\; A_f x_f(t) \;+\; B_f u_f(t) \tag{3.130}$$

$$y_f(t) \;=\; C_f x_f(t) \tag{3.131}$$

where $x_f(t)$ is a $(n_f \times 1)$ vector, $y_f(t)$ is a $(l_f \times 1)$ vector, and $u_f(t)$ is a $(m \times 1)$ vector. The form for $u_f(t)$ will depend upon whether the supplementary dynamics are inserted in a parallel path, feedback path, or a cascade path with the plant. The constant matrices A_f, B_f and C_f are chosen by design rules which will be discussed later.

Metasystem Equations and Adaptive Control Signal

The block diagrams for the adaptive algorithms with the supplementary dynamics in a feedback path (Algorithm 1), a parallel path (Algorithm 2), and a cascade path (Algorithm 3) are shown in Figures 3.28, 3.29 and 3.30, respectively. The different expressions for the plant input $u_p(t)$ and the supplementary input $u_f(t)$ are substituted into the equations for the plant and supplementary dynamics. Then, the plant and supplementary dynamics are concatenated to form an $(n + n_f)$ order metasystem.

In Algorithm 1, supplementary dynamics are inserted into the output feedback path as part of the adaptive mechanism. As shown in Figure 3.28, the supplementary dynamics process the output error to produce a signal

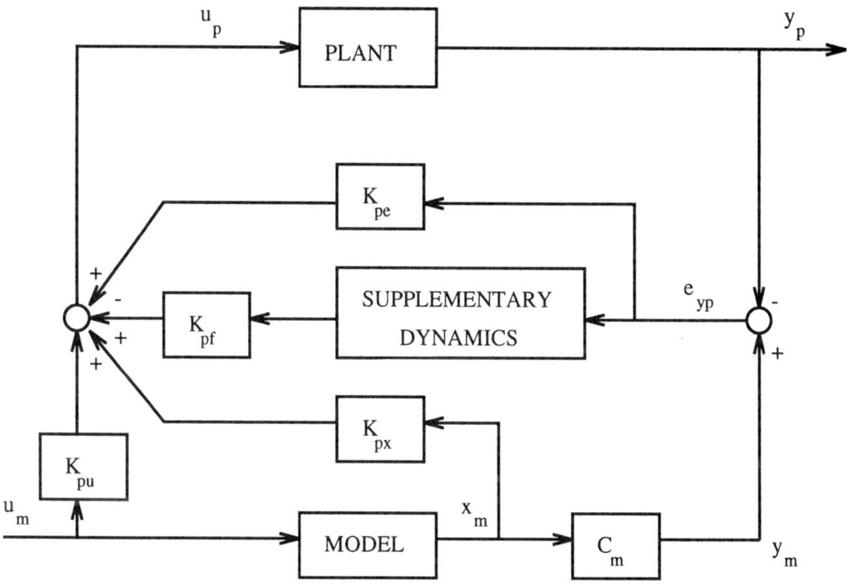

FIGURE 3.28. Supplemental Dynamics Algorithm 1

which is multiplied by a new adaptive gain $K_f(t)$, such that the control law is of the form,

$$u_p(t) = K_e(t)e_{yp}(t) - K_f(t)y_f(t) + K_x(t)x_m(t) + K_u(t)u_m(t) \quad (3.132)$$

$$u_f(t) \;=\; e_{yp} = y_m(t) \;-\; y_p(t) \quad (3.133)$$

where $K_e(t), K_f(t), K_x(t)$, and $K_u(t)$ are adaptive gain matrices. The plant and supplementary dynamics are concatenated to form an $(n + n_f)$ order metasystem as follows:

$$\begin{bmatrix} \dot{x}_p(t) \\ \dot{x}_f(t) \end{bmatrix} = \begin{bmatrix} A_p & 0 \\ -B_f C_p & A_p \end{bmatrix} \begin{bmatrix} x_p(t) \\ x_f(t) \end{bmatrix} +$$

$$\begin{bmatrix} B_p \\ 0 \end{bmatrix} [K_e(t), K_f(t), K_x(t), K_u(t)] \begin{bmatrix} e_{yp}(t) \\ -y_f(t) \\ x_m(t) \\ u_m(t) \end{bmatrix} + \begin{bmatrix} 0 \\ B_f y_m(t) \end{bmatrix} \quad (3.134)$$

$$\begin{bmatrix} y_p(t) \\ y_f(t) \end{bmatrix} = \begin{bmatrix} C_p & 0 \\ 0 & C_f \end{bmatrix} \begin{bmatrix} x_p(t) \\ x_f(t) \end{bmatrix}. \quad (3.135)$$

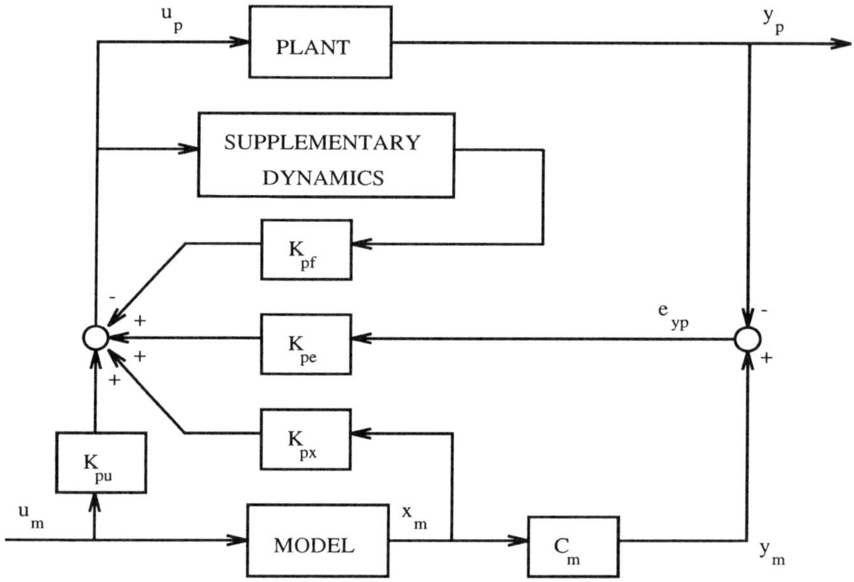

FIGURE 3.29. Supplemental Dynamics Algorithm 2

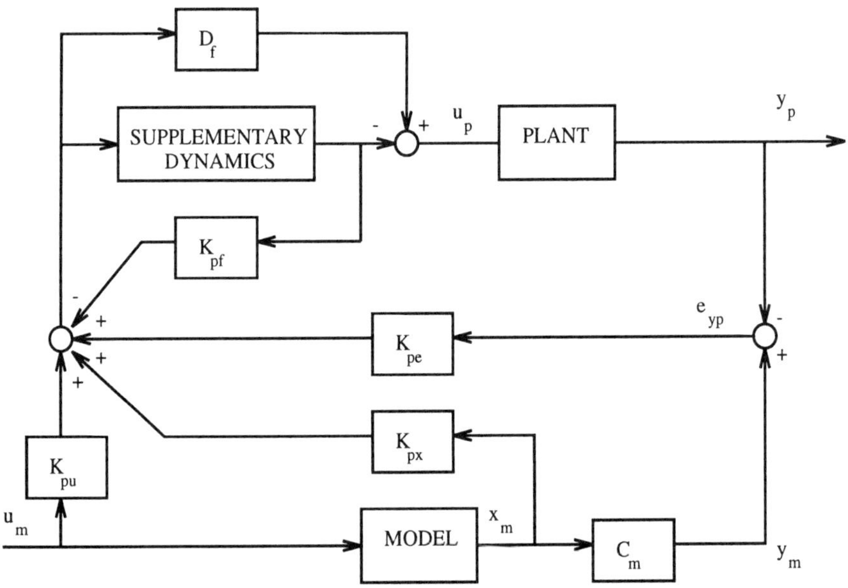

FIGURE 3.30. Supplemental Dynamics Algorithm 3

Algorithm 2 is characterized by supplementary dynamics which are parallel with the plant. The control law is of the form:

$$u_p(t) = u_f(t) = K_e(t)e_{yp}(t) - K_f(t)y_f(t) + K_x(t)x_m(t) + K_u(t)u_m(t) \tag{3.136}$$

where $K_e(t), K_f(t), K_x(t)$, and $K_u(t)$ are adaptive gain matrices. The plant and supplementary dynamics are concatenated to form a metasystem as follows

$$\begin{bmatrix} \dot{x}_p(t) \\ \dot{x}_f(t) \end{bmatrix} = \begin{bmatrix} A_p & 0 \\ 0 & A_f \end{bmatrix} \begin{bmatrix} x_p(t) \\ x_f(t) \end{bmatrix} \tag{3.137}$$

$$+ \begin{bmatrix} B_p \\ B_f \end{bmatrix} [K_e(t), K_f(t), K_x(t), K_u(t)] \begin{bmatrix} e_{yp}(t) \\ -y_f(t) \\ x_m(t) \\ u_m(t) \end{bmatrix}$$

$$\begin{bmatrix} y_p(t) \\ y_f(t) \end{bmatrix} = \begin{bmatrix} C_p & 0 \\ 0 & C_f \end{bmatrix} \begin{bmatrix} x_p(t) \\ x_f(t) \end{bmatrix}. \tag{3.138}$$

Algorithm 3 is characterized by supplementary dynamics which are in cascade with the plant. The control law is of the form:

$$u_p(t) = D_f u_f(t) - y_f(t) = D_f u_f(t) - C_f x_f(t) \tag{3.139}$$

where:

$$u_f(t) = K_e(t)e_{yp}(t) - K_f(t)y_f(t) + K_x(t)x_m(t) + K_u(t)u_m(t). \tag{3.140}$$

The metasystem for Algorithm 3 is of the form:

$$\begin{bmatrix} \dot{x}_p(t) \\ \dot{x}_f(t) \end{bmatrix} = \begin{bmatrix} A_p & -B_pC_f \\ 0 & A_f \end{bmatrix} \begin{bmatrix} x_p(t) \\ x_f(t) \end{bmatrix} + \begin{bmatrix} B_pD_f \\ B_f \end{bmatrix} [K_e(t), K_f(t), K_x(t), K_u(t)] \begin{bmatrix} e_{yp}(t) \\ -y_f(t) \\ x_m(t) \\ u_m(t) \end{bmatrix} \tag{3.141}$$

$$\begin{bmatrix} y_p(t) \\ y_f(t) \end{bmatrix} = \begin{bmatrix} C_p & 0 \\ 0 & C_f \end{bmatrix} \begin{bmatrix} x_p(t) \\ x_f(t) \end{bmatrix}. \tag{3.142}$$

Metasystem Summary

The metasystem is formed by concatenating the plant dynamics with the supplementary dynamics. These metasystem equations may be written more compactly as:

Plant Equations

$$\dot{x}(t) = Ax(t) + BK(t)r(t) + d(t) = Ax(t) + Bu(t) + d(t) \quad (3.143)$$

$$y(t) = Cx(t) \quad (3.144)$$

where

$$x(t) = \begin{bmatrix} x_p(t) \\ x_f(t) \end{bmatrix}, \quad y(t) = \begin{bmatrix} y_p(t) \\ y_f(t) \end{bmatrix}, \quad C = \begin{bmatrix} C_p & 0 \\ 0 & C_f \end{bmatrix} \quad (3.145)$$

and where for Algorithm 1:

$$A = \begin{bmatrix} A_p & 0 \\ -B_f C_p & A_f \end{bmatrix} \quad B = \begin{bmatrix} B_p \\ 0 \end{bmatrix} \quad d(t) = \begin{bmatrix} 0 \\ B_f y_m(t) \end{bmatrix} \quad (3.146)$$

for Algorithm 2:

$$A = \begin{bmatrix} A_p & 0 \\ 0 & A_f \end{bmatrix}, \quad B = \begin{bmatrix} B_p \\ B_f \end{bmatrix} \quad (3.147)$$

and for Algorithm 3:

$$A = \begin{bmatrix} A_p & -B_p C_f \\ 0 & A_f \end{bmatrix}, \quad B = \begin{bmatrix} B_p D_f \\ B_f \end{bmatrix}. \quad (3.148)$$

Adaptive gain matrix

$$K(t) = [K_a(t), K_x(t), K_u(t)] \quad (3.149)$$

where

$$K_a(t) = [K_e(t), K_f(t)] \quad (3.150)$$

$$r(t) = \begin{bmatrix} e_{yv}(t) \\ x_m(t) \\ u_m(t) \end{bmatrix} \quad (3.151)$$

and where

$$e_{yv}(t) = \begin{bmatrix} e_{yp}(t) \\ -y_f(t) \end{bmatrix}. \tag{3.152}$$

The adaptive gains which form $K(t)$ are chosen as before to be a combination of proportional and integral terms. The gain matrices are shown below:

$$K(t) = K_p(t) + K_I(t) \tag{3.153}$$

where

$$K_p(t) = v(t)r^T(t)\bar{T}, \quad \bar{T} \text{ symmetric and positive semi–definite} \tag{3.154}$$

$$\dot{K}_I(t) = v(t)r^T(t)T, \quad T \text{ symmetric and positive definite} \tag{3.155}$$

and where the initial integral gains are given by

$$K_I(0) = [K_{eI}(0), \; K_{fI}(0), \; K_{xI}(0), K_{uI}(0)]. \tag{3.156}$$

The signal $v(t)$ is chosen based upon a subsequent Lyapunov stability analysis and is given by

$$v(t) = Qe_{yv}(t) + GK(t)r(t) \tag{3.157}$$

where

$$Q = [Q_p \quad Q_f]. \tag{3.158}$$

Error Equation Summary
The plant state error is defined as

$$e_{xp}(t) = x_p^*(t) - x_p(t) \tag{3.159}$$

where x_p^* is the target state defined in (2.111), and the plant output error is:

$$e_{yp}(t) = y_m(t) - y_p(t) = y_p^*(t) - y_p(t) = C_p e_{xp}(t). \tag{3.160}$$

The supplementary dynamics state error is defined as

$$e_{xf}(t) = x_f^*(t) - x_f(t) \tag{3.161}$$

where $x_f^*(t)$ is the supplementary state trajectory when the plant output $y_p(t)$ is equal to the model output $y_m(t)$. Then, the metastate error is of the form

$$e_x(t) = \begin{bmatrix} e_{xp}(t) \\ e_{xf}(t) \end{bmatrix} = x^*(t) - x(t) \tag{3.162}$$

where

$$x^*(t) = \begin{bmatrix} x_p^*(t) \\ x_f^*(t) \end{bmatrix} \tag{3.163}$$

and the error derivative will be

$$\dot{e}_x(t) = \dot{x}^*(t) - \dot{x}(t) = \dot{x}^*(t) - Ax(t) - BK(t)r(t) - d(t). \tag{3.164}$$

Furthermore, suppose that there exists a constant matrix $\tilde{K}$ given by

$$\tilde{K} = \left[\tilde{K}_a,\ \tilde{K}_x,\ \tilde{K}_u\right] \tag{3.165}$$

such that

$$A_c = A - B\tilde{K}_a C \tag{3.166}$$

is a stability matrix.

Next, use (3.162) and (3.164) to obtain

$$\begin{aligned} \dot{e}_x(t) = \dot{x}^*(t) - A\left[x^*(t) - e_x(t)\right] - B\left[(K(t) - \tilde{K})r(t)\right] \\ -B\tilde{K}r(t) - d(t). \end{aligned} \tag{3.167}$$

Define $z(t) = \left[K(t) - \tilde{K}\right] r(t)$. Then, (3.167) becomes

$$\dot{e}_x(t) = \dot{x}^*(t) - Ax^*(t) + Ae_x(t) - Bz(t) - B\tilde{K}r(t) - d(t). \tag{3.168}$$

Substitute $\tilde{K}r(t) = \tilde{K}_a e_{yv}(t) + \tilde{K}_x x_m(t) + \tilde{K}_u u_m(t)$ into (3.168) to obtain

$$\begin{aligned} &\dot{e}_x(t) = \dot{x}^*(t) - Ax^*(t) + Ae_x(t) - Bz(t) - \\ &B\left[\tilde{K}_a e_{yv}(t) + \tilde{K}_x x_m(t) + \tilde{K}_u u_m(t)\right] - d(t). \end{aligned} \tag{3.169}$$

Define

$$e_v(t) = \begin{bmatrix} e_{xp}(t) \\ -x_f(t) \end{bmatrix} \tag{3.170}$$

and

$$x_0^*(t) = \begin{bmatrix} 0 \\ x_f^*(t) \end{bmatrix} \tag{3.171}$$

such that

$$e_{yv}(t) = Ce_v(t) \tag{3.172}$$

and

$$e_v(t) = e_x(t) - x_0^*(t). \tag{3.173}$$

Substitute (3.166), (3.172) and (3.173) into (3.169) to obtain

$$\dot{e}_x(t) = A_c e_x(t) - Bz(t) - F_1(t) \tag{3.174}$$

where

$$\begin{aligned} F_1(t) &= -\dot{x}^*(t) + Ax^*(t) - B\tilde{K}_a C x_0^*(t) + \\ &B\left[\tilde{K}_x x_m(t) + \tilde{K}_u u_m(t)\right] + d(t). \end{aligned} \tag{3.175}$$

Note that $F_1(t)$ is bounded because $\dot{x}^*(t), x^*(t), x_m(t), x_0^*(t)$, and $u_m(t)$, are bounded and $\tilde{K}_a, \tilde{K}_x$, and $\tilde{K}_u$ are constant.

Stability Analysis

Theorem 3.2: Consider the metasystem representation given by (3.143)–(3.157) with the controllable and observable LTI plant described by (3.9) and (3.10). Further, suppose that there exists a real symmetric positive definite matrix P and real matrices $J, L, W, \tilde{K}_a$, and R, with $(R + R^T) > 0$, such that

$$P(A - B\tilde{K}_a C) + (A - B\tilde{K}_a C)^T P = -LL^T - R < 0 \tag{3.176}$$

$$PB = C^T(Q^T + \tilde{K}_a^T G^T) - LW \tag{3.177}$$

$$W^T W = J + J^T \tag{3.178}$$

$$J + J^T + G + G^T < 0 \tag{3.179}$$

where the matrices T and $\bar{T}$ are positive definite symmetric and positive semi–definite symmetric, respectively. Then, all states, gains, and errors in the adaptive system are bounded.
Proof: See Appendix 3C.

The three MRAC algorithms are special cases of the metasystem which is described by (3.143)–(3.157). Thus, the stability of these algorithms is ensured if the sufficient conditions of (3.176)–(3.179) are satisfied.

It is important to observe that the first three sufficient conditions, given by (3.176)–(3.178), are equivalent to requiring that the transfer matrix given by

$$H(s) = J + (Q + G\tilde{K}_a)C(sI - A + B\tilde{K}_aC)^{-1}B \tag{3.180}$$

be strictly positive real (SPR).

Corollary 3.1.1
The sufficient conditions, (3.176)–(3.179), can be satisfied by Algorithm 1 for any controllable and observable plant.
Proof: If the plant is controllable and observable then there exists a compensator, denoted by the quadruple (A_c, B_c, C_c, D_c), which stabilizes the plant. Thus, the composite plant/compensator system is represented by

$$A_{\text{comp}} = \begin{bmatrix} A_p + B_pD_cC_p & B_pC_c \\ B_cC_p & A_c \end{bmatrix} \tag{3.181}$$

where A_{comp} is a stability matrix. Next, from (3.132) and (3.133) it can be seen that the composite system for Algorithm 1 is given by

$$A_1 = A - B\tilde{K}_aC = \begin{bmatrix} A_p - B_p\tilde{K}_eC_p & -B_p\tilde{K}_fC_f \\ -B_fC_p & A_f \end{bmatrix} \tag{3.182}$$

which is required to be a stability matrix for some gain $\tilde{K}_a = \left[\tilde{K}_e, \ \tilde{K}_f\right]$. Thus, by comparing (3.181) and (3.182) it is observed that the choice

$$\tilde{K}_e = -D_c \tag{3.183}$$

$$\tilde{K}_fC_f = -C_c \tag{3.184}$$

$$B_f = -B_c \tag{3.185}$$

$$A_f = A_c \tag{3.186}$$

will result in A_1 being a stability matrix. Now, it is necessary to show that (3.179) and (3.180) can be satisfied. Let $C_f = I$ and let $\tilde{K}_a = \left[\tilde{K}_e,\ \tilde{K}_f\right] = \left[-D_c,\ -C_c\right]$. Choose $G = -\gamma_1 I$ and $J = \gamma_2 I$, where $\gamma_1 > \gamma_2$ are positive scalars, so that (3.79) is satisfied, i.e.,

$$J + J^T + G + G^T = 2\gamma_2 I - 2\gamma_1 I < 0. \tag{3.187}$$

Then, let $Q = \gamma_1 \tilde{K}_a$ such that (3.180) reduces to $J = \gamma_2 I > 0$. Hence, there exists $Q, G, \tilde{K}_a$, and J which satisfy (3.179) and (3.180).

Corollary 3.1.2

The sufficient conditions, (3.179) and (3.180), can be satisfied by Algorithm 2 for the class of plants which are output stabilizable with a proper, but not strictly proper compensator, denoted by (A_c, B_c, C_c, D_c), which satisfies the mild restriction that

$$\text{Rank}\,[B_c] = \text{Rank}\begin{bmatrix} B_c \\ D_C \end{bmatrix}. \tag{3.188}$$

Proof: The composite system consisting of the plant with the compensator (A_c, B_c, C_c, D_c) is given by the stability matrix A_{comp} of (3.181). It should be observed from (3.135) and (3.136) that the composite system for Algorithm 2 is given by

$$A_2 = A - B\tilde{K}_a C = \begin{bmatrix} A_p - B_p\tilde{K}_e C_p & -B_p\tilde{K}_f C_f \\ -B_f\tilde{K}_e C_p & A_f - B_f\tilde{K}_f C_f \end{bmatrix} \tag{3.189}$$

which is required to be a stability matrix for some gain $\tilde{K}_a = \left[\tilde{K}_e, \tilde{K}_f\right]$. Thus, by comparing (3.181) and (3.189), it is seen that the choice

$$\tilde{K}_e = -D_c \neq 0 \tag{3.190}$$

$$C_f = I,\ \ \tilde{K}_f = -C_c \tag{3.191}$$

$$-B_f\tilde{K}_e = B_c \ \ (\text{ or } B_f D_c = B_c\ ) \tag{3.192}$$

and

$$A_f = A_c - B_f C_c \tag{3.193}$$

will result in A_2 being a stability matrix. (3.192) has a solution for B_f if (3.188) is satisfied. Then, Q, G, and J can be chosen to satisfy (3.179) and

(3.180) by using the constructive method shown in the proof of Corollary 3.1.1.

Corollary 3.1.3
The sufficient conditions, (3.179) and (3.180), can be satisfied by Algorithm 3 for the class of plants which are output stabilizable with a proper but not strictly proper compensator, denoted by (A_c, B_c, C_c, D_c), which satisfies the mild restriction that

$$\text{Rank } [\, B_c \,] = \text{Rank} \begin{bmatrix} B_c \\ D_c \end{bmatrix}. \tag{3.194}$$

Proof: The composite system consisting of the plant with the compensator (A_c, B_c, C_c, D_c) is given by the stability matrix A_{comp} of (3.181). It should be observed from (3.139) and (3.140) that the composite system for Algorithm 3 is given by

$$A_3 = A - B\tilde{K}_a C = \begin{bmatrix} A_p - B_p D_f \tilde{K}_e C_p & -B_p(I + D_f \tilde{K}_f) C_f \\ -B_f \tilde{K}_e C_p & A_f - B_f \tilde{K}_f C_f \end{bmatrix} \tag{3.195}$$

which is required to be a stability matrix for some gain $\tilde{K}_a = \left[\tilde{K}_e, \tilde{K}_f\right]$. Thus, by comparing (3.181) and (3.195) it is seen that the choice

$$D_f = I, \;\; \tilde{K}_e = -D_c \tag{3.196}$$

$$C_f = I, \;\; \tilde{K}_f = -C_c - I \tag{3.197}$$

$$-B_f \tilde{K}_e = B_c \;\; (\text{or } B_f D_c = B_c) \tag{3.198}$$

$$A_f = A_c - B_f (C_c + I) \tag{3.199}$$

will result in A_3 being a stability matrix where D_c must not be a zero matrix and where (3.198) has a solution for B_f if (3.194) is satisfied. Then Q, G, and J can be chosen to satisfy (3.179) and (3.180) by using the constructive method shown in the proof of Corollary 3.1.1.

It should be noted that the class of plants which can be controlled by Algorithms 2 and 3 is more restrictive than the class of plants which can be controlled by Algorithm 1. However, Algorithms 2 and 3 may be the preferred controllers because it will be subsequently shown that Algorithms 2 and 3 yield an asymptotically vanishing error provided that some additional conditions are satisfied.

Asymptotic Output Tracking

The stability analysis of Theorem 3.1 only ensures that all signals in the adaptive system are bounded. In this section, conditions are derived for Algorithms 2 and 3 under which the output error is asymptotically vanishing provided that the model input is constant for $t \geq t_1$. This new result is based upon extending O'Brien and Broussard's [76] command generator tracker (CGT) for model following control of known plants to the metasystem described by (3.143)–(3.157).

Command Generator Tracker for the Metasystem

When perfect tracking occurs (i.e., $e_{yp}(t) = \dot{e}_{yp}(t) = 0$), the corresponding metasystem state and control trajectories are defined to be the ideal metasystem state and ideal metasystem control trajectories, respectively. These ideal trajectories will be denoted by $x^*(t)$ and $u_p^*(t)$, respectively, where

$$x^*(t) = \begin{bmatrix} x_p^*(t) \\ x_f^*(t) \end{bmatrix} \tag{3.200}$$

and where $x_p^*(t)$ is the ideal plant state and $x_f^*(t)$ is the ideal supplementary dynamics state. These correspond with the CGT vector defined in Section 2.2.1

By definition, the ideal metasystem is such that it satisfies the same dynamics as the real metasystem. In addition, the output of the ideal plant is defined to be identically equal to the output of the model. Mathematically;

$$\dot{x}^*(t) = Ax^*(t) + Bu_p^*(t) \quad \text{for all } t \geq t_0 \tag{3.201}$$

and

$$y_p^*(t) = y_m(t) \text{ or } [C_p,\ 0]\, x^* = C_m x_m(t). \tag{3.202}$$

Hence, when perfect tracking occurs, the real metasystem trajectories become the ideal metasystem trajectories, and the real plant output becomes the ideal output, which is defined to be the model output.

Assume that the model input u_m is constant and that the ideal trajectories are linear functions of the model state $x_m(t)$ and the model input u_m. Mathematically:

$$\begin{bmatrix} x^*(t) \\ u_p^*(t) \end{bmatrix} = \begin{bmatrix} S_1 & S_2 \\ S_{31} & S_{32} \end{bmatrix} \begin{bmatrix} x_m(t) \\ u_m \end{bmatrix}. \tag{3.203}$$

where

$$S_1 = \begin{bmatrix} S_{11} \\ S_{21} \end{bmatrix} \qquad S_2 = \begin{bmatrix} S_{12} \\ S_{22} \end{bmatrix} \tag{3.204}$$

Thus,

$$x_p^*(t) = S_{11}x_m(t) + S_{12}u_m \tag{3.205}$$

$$x_f^*(t) = S_{21}x_m(t) + S_{22}u_m \tag{3.206}$$

$$u_p^*(t) = S_{31}x_m(t) + S_{32}u_m. \tag{3.207}$$

Combining the ideal metasystem, (3.201), with the ideal plant output, (3.202), yields:

$$\begin{bmatrix} \dot{x}^*(t) \\ y_p^*(t) \end{bmatrix} = \begin{bmatrix} A & B \\ [C_p, 0] & 0 \end{bmatrix} \begin{bmatrix} x^*(t) \\ u_p^*(t) \end{bmatrix}. \tag{3.208}$$

Substituting (3.203) into (3.208) gives:

$$\begin{bmatrix} \dot{x}^*(t) \\ y_p^*(t) \end{bmatrix} = \begin{bmatrix} A & B \\ [C_p, 0] & 0 \end{bmatrix} \begin{bmatrix} S_1 & S_2 \\ S_{31} & S_{32} \end{bmatrix} \begin{bmatrix} x_m(t) \\ u_m \end{bmatrix}. \tag{3.209}$$

Now, differentiate $x^*(t)$ in (3.203) to yield:

$$\dot{x}^*(t) = S_1\dot{x}_m(t) \tag{3.210}$$

where again it is assumed that u_m is constant.

Next, substitute the equation of the model dynamics into (3.210) to obtain

$$\dot{x}^*(t) = S_1A_mx_m(t) + S_1B_mu_m. \tag{3.211}$$

Concatenate (3.211) with the reference model output to obtain

$$\begin{bmatrix} \dot{x}^* \\ y_m(t) \end{bmatrix} = \begin{bmatrix} S_1A_m & S_2B_m \\ C_m & 0 \end{bmatrix} \begin{bmatrix} x_m(t) \\ u_m(t) \end{bmatrix}. \tag{3.212}$$

Comparing (3.209) and (3.212) gives;

$$\begin{bmatrix} S_1A_m & S_1B_m \\ C_m & 0 \end{bmatrix} = \begin{bmatrix} A & B \\ [C_p, 0] & 0 \end{bmatrix} \begin{bmatrix} S_1 & S_2 \\ S_{31} & S_{32} \end{bmatrix}. \tag{3.213}$$

Defining

$$\begin{bmatrix} \Omega_{11} & \Omega_{12} \\ \Omega_{21} & \Omega_{22} \end{bmatrix} = \begin{bmatrix} A & B \\ [C_p, 0] & 0 \end{bmatrix}^{-1} \tag{3.214}$$

results in

$$\begin{bmatrix} S_1 & S_2 \\ S_{31} & S_{32} \end{bmatrix} = \begin{bmatrix} \Omega_{11} & \Omega_{12} \\ \Omega_{21} & \Omega_{22} \end{bmatrix} \begin{bmatrix} S_1 A_m & S_1 B_m \\ C_m & 0 \end{bmatrix}. \tag{3.215}$$

O'Brien and Broussard [76] have shown that an equation of the type given by (3.215) has a solution for S_1, S_2, S_{31}, and S_{32} if (i) u_m is a constant, (ii) $\dim [y_p(t)] = \dim [u_p(t)]$, and (iii) no eigenvalue of Ω_{11} is equal to the inverse of an eigenvalue of A_m.

Theorem 3.3: Let the adaptive controller be that defined by Algorithm 2 or 3. The adaptive control algorithm described by Theorem 3.2 yields an asymptotically vanishing output error if the conditions of Theorem 3.2 are satisfied and if (i) u_m is constant for $t \geq t_1$, (ii) a solution exists for the matrices S_1, S_2, S_{31}, and S_{32} in (3.215), and (iii) there exists a real matrix E of dimension $m \times n_f$ such that

$$G = -EB_f. \tag{3.216}$$

and

$$Q_f C_f = EA_f \tag{3.217}$$

Furthermore, if $\det [A_f] \neq 0$, then (3.216) and (3.217) reduce to

$$G = -Q_f C_f A_f^{-1} B_f \tag{3.218}$$

Proof: See Appendix 3D.

Corollary 3.2.1

Let the adaptive controller be Algorithm 2 or 3 and let the plant belong to the class described by Theorem 3.3 with the additional restriction that $A_c = 0$. Suppose that u_m is constant for $t \geq t_1$ and a solution exists for the matrices S_1, S_2, S_{31}, and S_{32} in (3.213). If there exists a real matrix E of dimension $m \times n_f$ such that

$$EB_f > 0, \tag{3.219}$$

then the sufficient conditions for a bounded and asymptotically vanishing error can be satisfied.

Proof: Let the supplementary dynamics and the gain $\tilde{K}_a$ be chosen as in Corollary 3.1.2 (Algorithm 2), or Corollary 3.1.3 (Algorithm 3). Then, choose Q_f and G to satisfy Theorem 3.2. That is,

$$Q_f = EA_f \tag{3.220}$$

$$G = -EB_f. \tag{3.221}$$

Choose

$$Q_p = -EB_f D_c \tag{3.222}$$

and use (3.220) and (3.221) to obtain

$$\begin{aligned} Q + G\tilde{K}_a &= [Q_p, \ \ Q_f] - \left[EB_f\tilde{K}_e, \ EB_f\tilde{K}_f\right] \\ &= \left[-EB_f D_c - EB_f\tilde{K}_e, \ \ EA_f - EB_f\tilde{K}_f\right] \\ &= \left[-EB_f(D_c + \tilde{K}_e), \ \ E(A_f - B_f\tilde{K}_f)\right]. \end{aligned} \tag{3.223}$$

Using either Corollary 3.1.2 or 3.1.3, gives: $\tilde{K}_e = -D_c$ and $A_c = A_f - B_f\tilde{K}_f = 0$. Thus,

$$Q + G\tilde{K}_a = [-EB_f(D_c - D_c), \ \ EA_c] = 0 \tag{3.224}$$

and the sufficient condition given by (3.179) and (3.178) reduces to

$$H(s) = J > 0 \tag{3.225}$$

$$J + J^T + G + G^T < 0. \tag{3.226}$$

Choose

$$J = \gamma_3 EB_f \tag{3.227}$$

with the scalar γ_3 such that $0 < \gamma_3 < 1$.

Then,

$$J = \gamma_3 EB_f > 0 \tag{3.228}$$

and

$$\begin{aligned} J + J^T + G + G^T &= \gamma_3(EB_f + B_f^T E^T) - (EB_f + B_f^T E^T) \\ &= (\gamma_3 - 1)(EB_f + B_f^T E^T) < 0. \end{aligned} \tag{3.229}$$

Hence, the sufficient conditions for a bounded error given by (3.179) and (3.180) and for an asymptotically vanishing error given by (3.216) and (3.217) are satisfied.

A trivial solution for Corollary 3.2.1 is to choose $E = B_f^T$ such that $EB_f = B_f^T B_f > 0$. Furthermore, if $\det[A_f] \neq 0$, Q_f can be chosen such that $G = -Q_f A_f^{-1} B_f < 0$, and $Q_p = GD_c$. This is a special case of Corollary 3.2.1 with $E = -Q_f A_f^{-1}$

Corollary 3.2.1 shows that it is possible to satisfy all of the sufficient conditions for a rather large class of plants. One example of a compensator which satisfies the assumptions of Corollary 3.2.1 is a proportional plus integral (PI) controller.

3.4.2 Summary of Constraints and Design Rules

Objective: To find a control signal by means of adaptive computation such that the plant, which is not ASPR, will follow the reference model with either a bounded or asymptotically vanishing output error. The plant, model, and control law used in Algorithms 1, 2, and 3 are described below.

PLANT:

$$\dot{x}_p(t) = A_p x_p(t) + B_p u_p(t) \tag{3.230}$$

$$y_p(t) = C_p x_p(t) \tag{3.231}$$

MODEL:

$$\dot{x}_m(t) = A_m x_m(t) + B_m u_m(t) \tag{3.232}$$

$$y_m(t) = C_m x_m(t) \tag{3.233}$$

SUPPLEMENTARY DYNAMICS:

$$\dot{x}_f(t) = A_f x_f(t) + B_f u_f(t) \tag{3.234}$$

$$y_f(t) = C_f x_f(t) \tag{3.235}$$

ADAPTIVE CONTROL LAW:

$$u_p(t) = K(t)r(t) \text{ for Algorithms 1 and 2} \tag{3.236}$$

$$u_p(t) = D_f K(t)r(t) - y_f(t) \text{ for Algorithm 3} \tag{3.237}$$

where

$$K(t) = K_p(t) + K_I(t) \tag{3.238}$$

$$K_p(t) = v(t)r^T(t)\bar{T} \tag{3.239}$$

$$\dot{K}_I(t) = v(t)r^T(t)T \tag{3.240}$$

$$v(t) = Qe_{yv}(t) + GK(t)r(t) \tag{3.241}$$

$$r^T(t) = \left[e_{yv}^T(t) \;\; x_m^T(t) \;\; u_m^T(t)\right] \tag{3.242}$$

$$e_{yv}^T(t) = \left[(y_m(t) - y_p(t))^T \quad -y_f^T(t)\right] \tag{3.243}$$

$$T > 0 \text{ and } \bar{T} \geq 0. \tag{3.244}$$

Algorithm 1: (Bounded Tracking Error)

Constraints

(i) There exists a compensator (A_c, B_c, C_c, D_c) which stabilizes the plant.

(ii)

$$Q + G\tilde{K}_a = 0 \tag{3.245}$$

(iii)

$$J + J^T + G + G^T < 0 \tag{3.246}$$

where J is a positive definite matrix.

Design Method

(i) Choose $C_f = I$.

(ii) Choose $\tilde{K}_a = \left[\tilde{K}_e, \tilde{K}_f\right] = [-D_c, -C_c]$.

(iii) Choose $B_f = -B_c$.

(iv) Choose $A_f = A_c$.

(v) Choose $G + G^T < 0$.

(vi) Choose $Q = [Q_p, Q_f] = G[D_c, C_c]$.

Algorithm 2: (Bounded Tracking Error)

Constraints

(i) There exists a proper, but not strictly proper, compensator (A_c, B_c, C_c, D_c) which stabilizes the plant and satisfies

$$\text{Rank } [B_c] = \text{Rank} \begin{bmatrix} B_c \\ D_c \end{bmatrix} \tag{3.247}$$

(ii)

$$Q + G\tilde{K}_a = 0 \tag{3.248}$$

(iii)

$$J + J^T + G + G^T < 0. \tag{3.249}$$

Design Method

(i)

$$\text{Choose } C_f = I. \tag{3.250}$$

(ii)

$$\text{Choose } \tilde{K}_a = \left[\tilde{K}_e,\ \tilde{K}_f\right] = [-D_c,\ -C_c]\,. \tag{3.251}$$

(iii)

$$\text{Solve } B_f D_c = B_c \text{ for } B_f. \tag{3.252}$$

(iv)

$$\text{Choose } A_f = A_c - B_f C_c. \tag{3.253}$$

(v)

$$\text{Choose } G + G^T < 0. \tag{3.254}$$

(vi)

$$\text{Choose } Q = [Q_p,\ Q_f] = G\,[D_c,\ C_c]\,. \tag{3.255}$$

Algorithm 2: (asymptotically vanishing tracking error)

Constraints

(i) There exists a proper, but not strictly proper, compensator (A_c, B_c, C_c, D_c) which stabilizes the plant and satisfies

$$\text{Rank } [B_c] = \text{Rank} \begin{bmatrix} B_c \\ D_c \end{bmatrix} \tag{3.256}$$

(ii)

$$Q + G\tilde{K}_a = 0 \tag{3.257}$$

(iii)

$$J + J^T + G + G^T < 0. \tag{3.258}$$

(iv)

$$\begin{aligned} &G = -EB_f \text{ and } Q_f C_f = EA_f; \\ &\text{or } G = -Q_f C_f A_f^{-1} B_f \text{ if } \det[A_f] \neq 0 \end{aligned} \tag{3.259}$$

where J is a positive definite matrix.

Design Method

(i)

$$\text{Choose } C_f = I. \tag{3.260}$$

(ii)

$$\text{Choose } \tilde{K}_a = \left[\tilde{K}_e, \tilde{K}_f\right] = [-D_c, -C_c]. \tag{3.261}$$

(iii)

$$\text{Solve } B_f D_c = B_c \text{ for } B_f \tag{3.262}$$

(iv)

$$\text{Choose } A_f = -B_f C_c \tag{3.263}$$

(v)

$$\text{Choose } G = -EB_f < 0 \tag{3.264}$$

(vi)

$$\text{Choose } Q = [Q_p, Q_f] = G[D_c, C_c] \tag{3.265}$$

If $\det[A_f] \neq 0$, (3.264) and (3.265) become $G = -Q_f A_f^{-1} B_f < 0$ and $Q_p = GD_c$.

Algorithm 3: (Bounded Tracking Error)

Constraints

(i) There exists a proper, but not strictly proper, compensator (A_c, B_c, C_c, D_c) which stabilizes the plant and satisfies

$$\text{Rank } [B_c] = \text{Rank} \begin{bmatrix} B_c \\ D_c \end{bmatrix}. \tag{3.266}$$

(ii)

$$Q + G\tilde{K}_a = 0. \tag{3.267}$$

(iii)

$$J + J^T + G + G^T < 0. \tag{3.268}$$

where J is a positive definite matrix.

Design Method

(i)

$$\text{Choose } C_f = I \text{ and } D_f = I. \tag{3.269}$$

(ii)

$$\text{Choose } \tilde{K}_a = \left[\tilde{K}_e,\ \tilde{K}_f\right] = [-D_c,\ -C_c - I]. \tag{3.270}$$

(iii)

$$\text{Solve } B_f D_c = B_c \text{ for } B_f. \tag{3.271}$$

(iv)

$$\text{Choose } A_f = A_c - B_f(C_c + I). \tag{3.272}$$

(v)

$$\text{Choose } (G + G^T) < 0. \tag{3.273}$$

(vi)

$$\text{Choose } Q = [Q_p,\ Q_f] = G\,[D_c,\ C_c]. \tag{3.274}$$

Algorithm 3: (Asymptotically Vanishing Tracking Error)

Constraints

(i) There exists a proper, but not strictly proper, compensator (A_c, B_c, C_c, D_c) which stabilizes the plant and satisfies

$$\text{Rank } [B_c] = \text{Rank} \begin{bmatrix} B_c \\ D_c \end{bmatrix}. \tag{3.275}$$

(ii)

$$Q + G\tilde{K}_a = 0. \tag{3.276}$$

(iii)

$$J + J^T + G + G^T < 0. \tag{3.277}$$

(iv)

$$G = -EB_f \text{ and } Q_f C_f = EA_f;$$
$$\text{or } G = -Q_f C_f A_f^{-1} B_f \text{ if } \det[A_f] \neq 0. \tag{3.278}$$

where J is a positive definite matrix.

Design Method

(i)

$$\text{Choose } C_f = I \text{ and } D_f = I. \tag{3.279}$$

(ii)

$$\text{Choose } \tilde{K}_a = \left[\tilde{K}_e,\ \tilde{K}_f\right] = [-D_c,\ -C_c - I]. \tag{3.280}$$

(iii)

$$\text{Solve } B_f D_c = B_c \text{ for } B_f. \tag{3.281}$$

(iv)

$$\text{Choose } A_f = -B_f(C_c + I). \tag{3.282}$$

(v)

$$\text{Choose } G = -EB_f < 0. \tag{3.283}$$

(vi)

$$\text{Choose } Q = [Q_p,\ Q_f] = G\,[D_c,\ C_c + I]. \tag{3.284}$$

If $\det[A_f] \neq 0$, (3.283) and (3.284) become $G = -Q_f A_f^{-1} B_f < 0$ and $Q_p = GD_c$.

3.4.3 Illustrative Examples

In order to demonstrate the preceding developments, the same examples, of Section 3.1.3 are again considered.

Example 1: (Rohrs' Example)

The plant is given by

$$\frac{y_p(s)}{u_p(s)} = \frac{2}{(s + 1)} \frac{229}{(s^2 + 30s + 229)}. \tag{3.285}$$

The output of this plant is required to follow the output of the reference model which is shown below:

$$\frac{y_m(s)}{u_m(s)} = \frac{1}{1 + s / 3}. \tag{3.286}$$

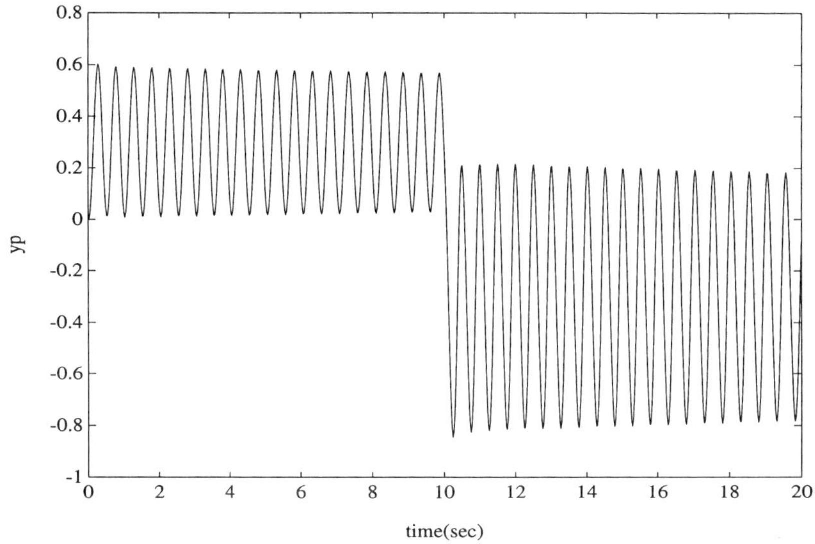

FIGURE 3.31. Rohrs' Example: Output with Non–Adaptive Compensator

Consider a stabilizing PI compensator described by:

$$G_c(s) = \frac{-(10s+35)}{s}. \tag{3.287}$$

Although the compensator described by (3.287) is stabilizing, it yields a closed loop system with a settling time greater than 500 seconds and a damping ratio for the dominant poles of less than 0.001. The time response of the output to a square wave input is shown in Figure 3.31 and is completely unacceptable.

Note that the choicc of some stabilizing compensator which is needed for computation of the matrices A_f, B_f, Q, and G is a much easier task than the computation of a compensator which satisfies stringent performance specifications for an unknown or poorly known plant.

The compensator described by (3.287) has a state space realization given by $A_c = 0$, $B_c = 10$, $C_c = -3.5$, and $D_c = -10$. Adaptive controllers are designed using each of the three algorithms for the purpose of comparing the behavior of the different algorithms. The reference command is a square wave with magnitude of 0.3 units and a period of 20 seconds. The supplementary dynamics measurement matrix C_f is chosen to be unity and the weighting matrices T and $\bar{T}$ are chosen to be identity matrices.

<u>Algorithm 1</u>

Use (3.183)–(3.186) to obtain

$$\tilde{K}_e = -D_c = 10 \tag{3.288}$$

$$\tilde{K}_f = -C_c = 3.5 \tag{3.289}$$

$$B_f = -B_c = -10 \tag{3.290}$$

$$A_f = A_c = 0. \tag{3.291}$$

Therefore, the supplementary dynamics, which are inserted into the output feedback path, are described by

$$\dot{x}_f(t) = -10e_{yf}(t) \tag{3.292}$$

$$y_f(t) = x_f(t). \tag{3.293}$$

Next, choose

$$Q = [57.14,\ 20] \tag{3.294}$$

$$G = -5.714 \tag{3.295}$$

such that $Q + G\tilde{K}_a = 0$ is satisfied.

The time responses of the plant and model outputs were obtained from a digital computer simulation and are shown in Figure 3.32. Observe that the plant output exhibits excessive oscillations. Note the output error can only be guaranteed to be asymptotically vanishing when using either Algorithm 2 or 3. Therefore, Algorithm 1 is not the algorithm of choice.

Algorithm 2

Use (3.190)–(3.193) to obtain

$$\tilde{K}_e = -D_c = 10 \tag{3.296}$$

$$\tilde{K}_f = -C_c = 3.5 \tag{3.297}$$

$$B_f D_c = B_c \Rightarrow B_f = B_c/D_c = -1 \tag{3.298}$$

$$A_f = -B_f C_c = -3.5. \tag{3.299}$$

Therefore, the parallel supplementary dynamics are described by

$$\dot{x}_f(t) = -3.5x_f(t) - u_f(t) \tag{3.300}$$

$$y_f(t) = x_f(t). \tag{3.301}$$

Since A_f is non–singular, Q_f can be arbitrarily selected as

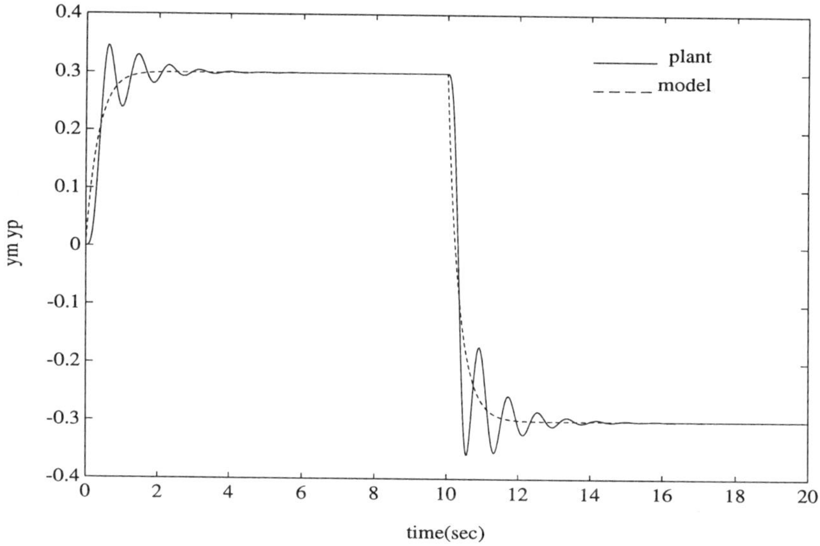

FIGURE 3.32. Rohrs' Example: Plant and Model Outputs, (Algorithm 1)

$$Q_f = 20 \tag{3.302}$$

and then G and Q_p are chosen to satisfy:

$$G = -Q_f A_f^{-1} B_f = -5.71428, \tag{3.303}$$

$$Q_p = GD_c = 57.1428. \tag{3.304}$$

The plant and model output time responses are shown in Figure 3.33. Of interest is the excellent transient behavior with zero error in approximately 2 seconds. The condition $G = -Q_f A_f^{-1} B_f$ is required for achieving an asymptotically vanishing output error. To demonstrate this, consider the time responses for Algorithm 2 with $G = -10$. The plant and model output time responses for these conditions are shown in Figure 3.34 from which it is observed that an asymptotically vanishing output error is no longer achieved.

Algorithm 3

Use (3.196)–(3.199) to obtain

$$D_f = 1 \tag{3.305}$$

$$\tilde{K}_e = -D_c = 10 \tag{3.306}$$

$$\tilde{K}_f = -C_c - 1 = 2.5 \tag{3.307}$$

$$B_f D_c = B_c \Rightarrow B_f = B_c / D_c = -1 \tag{3.308}$$

$$A_f = -B_f(C_c + I) = -2.5. \tag{3.309}$$

Therefore, the cascade supplementary dynamics are described by

$$\dot{x}_f(t) = -2.5 x_f(t) - u_f(t) \tag{3.310}$$

$$y_f(t) = x_f(t) + u_f(t). \tag{3.311}$$

Since A_f is non–singular, Q_f can be arbitrarily selected as

$$Q_f = 20 \tag{3.312}$$

and the G and Q_p selected to satisfy:

$$G = -Q_f A_f^{-1} B_f = -8.0 \tag{3.313}$$

$$Q_p = G D_c = 80.0. \tag{3.314}$$

The plant and model output time responses are shown in Figure 3.35. Observe again the excellent transient behavior with zero error in approximately 2 seconds. The condition $G = -Q_f A_f^{-1} B_f$ is again required to achieve an asymptotically vanishing output error. To demonstrate this, consider the time responses for Algorithm 3 with $G = -20$. The plant and model output time responses for these values are shown in Figure 3.36 from which it can be observed that an asymptotically vanishing output error is no longer achieved.

Upon comparing the responses for the three algorithms, it is observed that Algorithm 1 yields an unacceptable decaying oscillatory response whereas Algorithms 2 and 3 yield excellent responses with approximately zero error after 2 seconds. Furthermore, note the importance of choosing $G = -Q_f A_f^{-1} B_f$ when an asymptotically vanishing error is desired. Nevertheless, the output error is guaranteed to be bounded even when the choice of G does not satisfy $G = -Q_f A_f^{-1} B_f$.

Example 2: (Unstable Plant Using Algorithm 2 Parallel Dynamics)
Consider the unstable single–input single–output plant with transfer function given by

$$\frac{y_p(s)}{u_p(s)} = \frac{200}{(s - 1)(s^2 + 8s + 100)}. \tag{3.315}$$

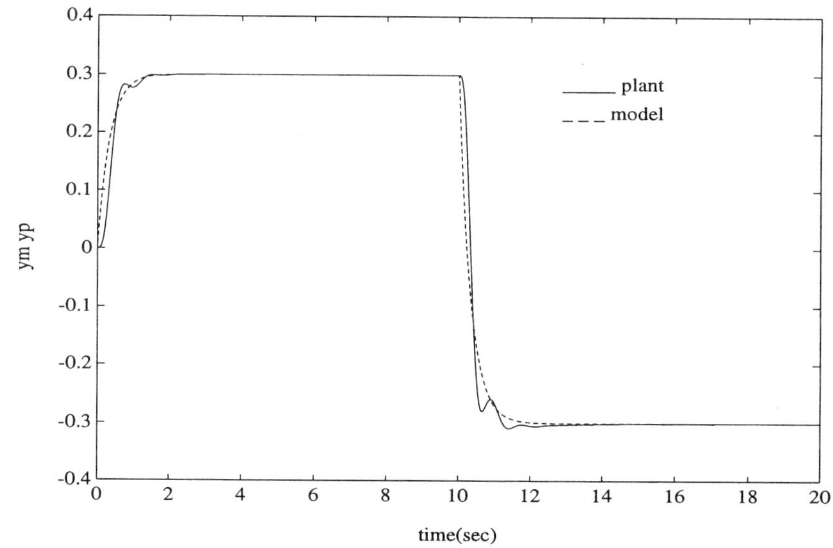

FIGURE 3.33. Rohrs' Example: Plant and Model Outputs, Ideal G, (Algorithm 2)

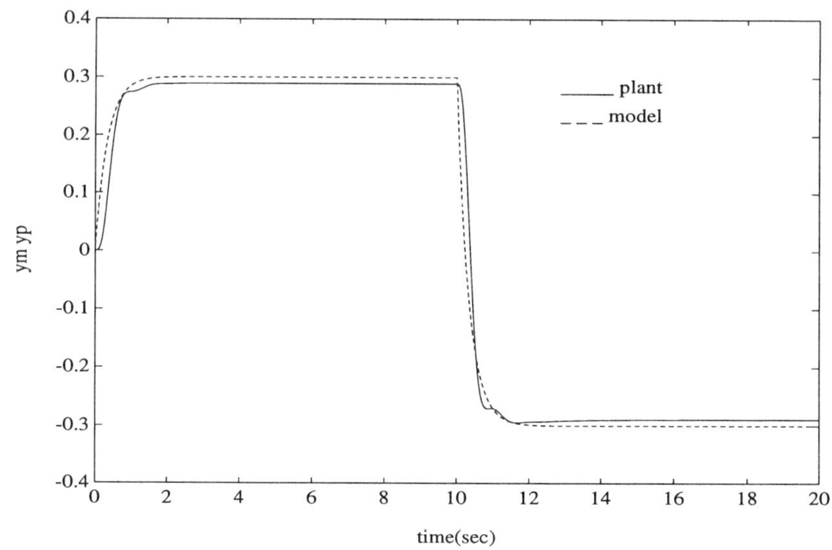

FIGURE 3.34. K_e Rohrs' Example, Plant and Model Outputs, Mismatched G (Algorithm 2)

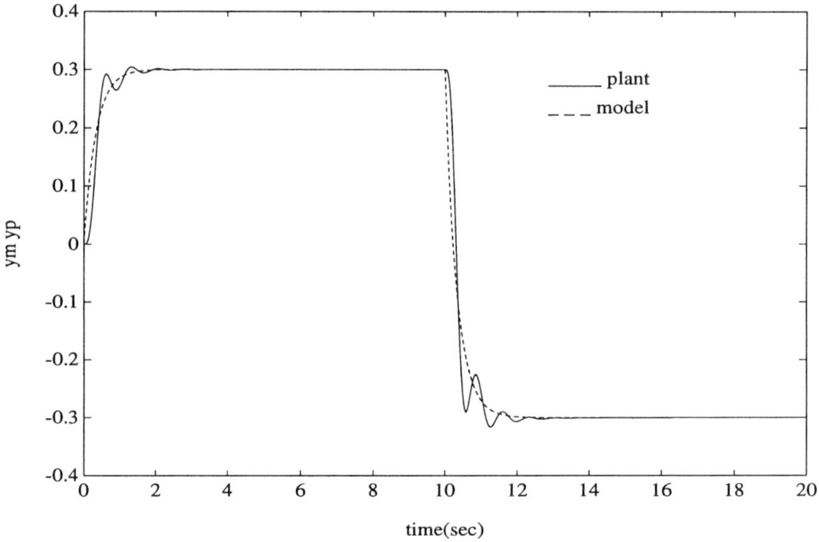

FIGURE 3.35. Rohrs' Example: Plant and Model Outputs, Ideal G (Algorithm 3)

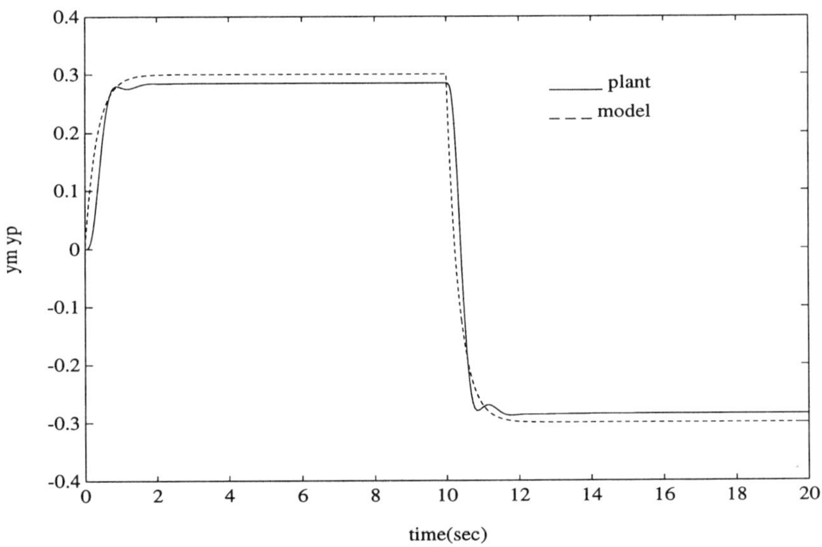

FIGURE 3.36. Rohrs' Example, Plant and Model Outputs, Mismatched G (Algorithm 3)

The output of the plant is required to follow the output of the reference model whose transfer function is given by (3.286).

Consider the stabilizing PI compensator

$$G_c(s) = \frac{-(3s+6)}{s}. \tag{3.316}$$

The compensator described by (3.316) has a state space realization given by $A_c = 0$, $B_c = 3$, $C_c = -2$, and $D_c = -3$. Use (3.190)–(3.193) to obtain

$$\tilde{K}_e = -D_c = 3 \tag{3.317}$$

$$C_f = I; \; \tilde{K}_f = -C_c = 2 \tag{3.318}$$

$$B_f D_c = B_c \Rightarrow B_f = B_c / D_c = -1 \tag{3.319}$$

$$A_f = -B_f C_c = -2. \tag{3.320}$$

Therefore, the parallel supplementary dynamics are described by

$$\dot{x}_f(t) = -2x_f(t) - u_p(t) \tag{3.321}$$

$$y_f(t) = x_f(t). \tag{3.322}$$

Since A_f is non–singular, Q_f can be arbitrarily selected as:

$$Q_f = 20 \tag{3.323}$$

and then G and Q_p selected to satisfy:

$$G = -Q_f A_f^{-1} B_f = -10.0 \tag{3.324}$$

$$Q_p = G D_c = 30.0. \tag{3.325}$$

The matrices T and $\bar{T}$ are chosen to be identity matrices. The plant and model outputs are shown in Figure 3.37 for a square wave reference command of magnitude 0.3 units and period of 40 seconds. Observe that the error is driven to zero in approximately 2 seconds albeit with a maximum overshoot of near 30 percent. The adaptive gains K_e and K_f are shown in Figures 3.38 and 3.39, respectively. Observe that the gains become momentarily large at $t = 0$ and $t = 20$ seconds in order to force the error to zero when the reference command is initially applied and when the reference command changes sign.

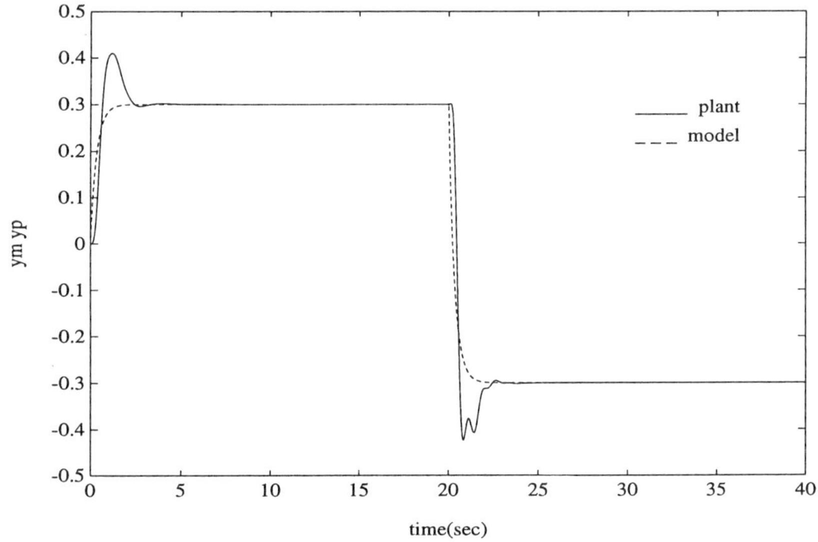

FIGURE 3.37. Unstable Plant: Plant and Model Outputs (Algorithm 2)

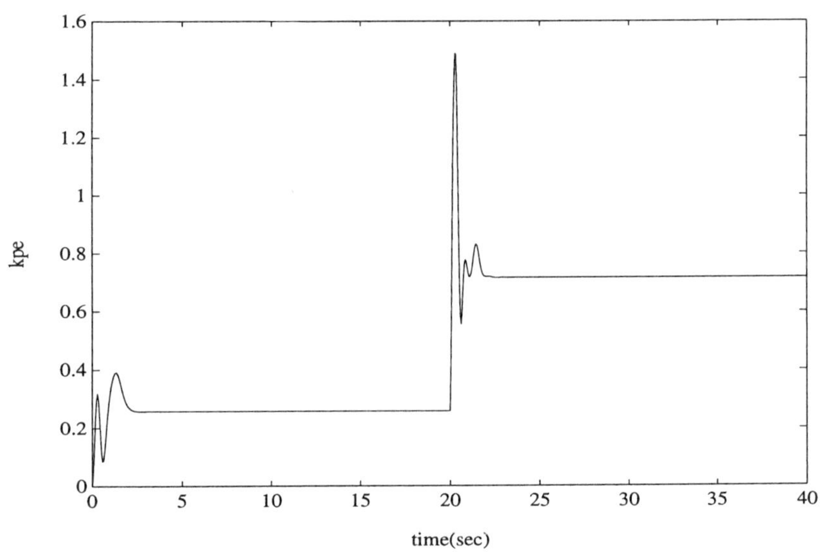

FIGURE 3.38. Unstable Plant: Adaptive Gain K_e (Algorithm 2)

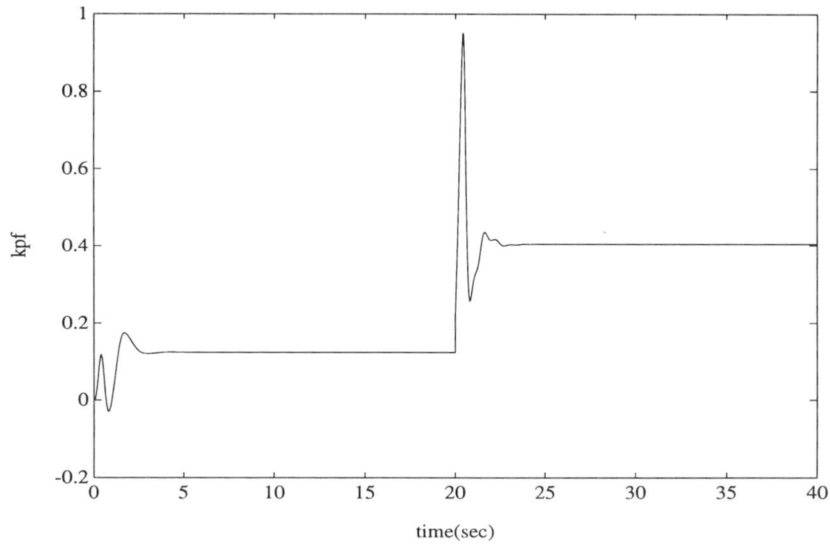

FIGURE 3.39. Unstable Plant: Adaptive Gain K_f (Algorithm 2)

3.5 Adaptive Control in the Presence of Nonlinearities

3.5.1 Theory

3.5.2 Adaptation for Nonlinearity of Known Form

This section presents an algorithm for plants with nonlinearities which are of known form but with unknown parameters. This approach does not require an assumption that the nonlinearities are bounded. First, the problem formulation and the adaptive control algorithm are described. Then, the command generator tracker is extended to include plants with nonlinearities of known form, but with unknown parameters. Next, the error equations for the adaptive algorithm are derived and a theorem is presented which describes the sufficient conditions for an asymptotically vanishing output error. In particular, the stability result shows that the linear part of the plant must be ASPR. A ship autopilot example is presented to illustrate the adaptive algorithm.

Problem Formulation

Let the nonlinear plant be described by

$$\dot{x}_p(t) = A_p x_p(t) + A_\gamma \gamma(C_p x_p, u_p, t) + B_p u_p(t) \tag{3.326}$$

$$y_p(t) = C_p x_p(t) \tag{3.327}$$

where $x_p(t)$ is the $(n \times 1)$ non–accessible plant state vector, $u_p(t)$ is the $(m \times 1)$ plant input, $y_p(t)$ is the $(m \times 1)$ plant output, $\gamma(C_p x_p, u_p, t)$ is a $(r \times 1)$ nonlinear function of known form. The matrices A_p, B_p, A_γ, and C_p are unknown and with appropriate dimensions.

The objective is to determine a control signal such that the plant tracks a desired trajectory which is generated by the linear time invariant reference model described by (3.14) and (3.15)

The choice for the adaptive control law, which includes a nonlinear term, is shown below.

$$u_p(t) = K_e(t)e_{yp}(t) + K_\gamma(t)\gamma(C_p x_p, u_p, t) + K_x(t)x_m(t) + K_u(t)u_m(t) \tag{3.328}$$

which may be written more compactly as $u(t) = K(t)r(t)$ where

$$e_{yp}(t) = y_m(t) - y_p(t) \tag{3.329}$$

$$K(t) = [K_e(t),\ K_\gamma(t), K_x(t), K_u(t)] \tag{3.330}$$

and

$$r(t) = \begin{bmatrix} e_{yp}(t) \\ \gamma(C_p x_p, u_p, t) \\ x_m(t) \\ u_m(t) \end{bmatrix}. \tag{3.331}$$

The adaptive gain $K(t)$ is computed as:

$$K(t) = K_p(t) + K_I(t) \tag{3.332}$$

$$K_p(t) = v(t)r^T(t)\bar{T} \tag{3.333}$$

$$\dot{K}_I(t) = v(t)r^T(t)T \tag{3.334}$$

where T and $\bar{T}$ are positive definite and positive semi-definite, respectively, and where the signal $v(t)$ is chosen as

$$v(t) = e_{yp}(t). \tag{3.335}$$

The initial integral gain is given by

$$K_I(0) = [K_{Ie}(0), K_{I\gamma}(0), K_{Ix}(0), K_{Iu}(0)]. \tag{3.336}$$

Asymptotic Output Tracking for Known Plants

O'Brien and Broussard's [76] command generator tracker is extended to the nonlinear plant by using the ideal plant state $x_p^*(t)$, ideal plant input $u_p^*(t)$, and ideal plant output $y_p^*(t)$. In the ideal situation, $y_p(t) = y_p^*(t) = y_m(t)$, and the ideal plant satisfies the following dynamics.

$$\dot{x}_p^*(t) = A_p x_p^*(t) + A_\gamma \gamma(C_p x_p^*, u_p^*, t) + B_p u_p^*(t) \tag{3.337}$$

$$y_p^*(t) = C_p x_p^*(t). \tag{3.338}$$

Next, the reference model command is constrained to be a constant and it is assumed that the form of the ideal state and ideal input are as follows:

$$x_p^*(t) = S_{11} x_m(t) + S_{12} u_m \tag{3.339}$$

$$u_p^*(t) = S_\gamma \gamma(C_p x_p^*, u_p^*, t) + S_{21} x_m(t) + S_{22} u_m. \tag{3.340}$$

Substitute (3.339) and (3.340) into (3.337) and (3.338) to obtain

$$\begin{aligned} \dot{x}_p^*(t) = {} & A_p \left[S_{11} x_m(t) + S_{12} u_m\right] + B_p \left[S_{21} x_m(t) + S_{22} u_m\right] \\ & + (A_\gamma + B_p S_\gamma)\gamma(C_p x_p^*, u_p^*, t). \end{aligned} \tag{3.341}$$

$$y_p^*(t) = C_p \left[S_{11} x_m(t) + S_{12} u_m\right] \tag{3.342}$$

Combine (3.341) and (3.342), and use the assumption that $A_\gamma + B_p S_\gamma = 0$ (which is satisfied if $\text{rank}\,[B_p] = \text{rank}\,[B_p | A_\gamma]$) to obtain

$$\begin{bmatrix} \dot{x}_p^*(t) \\ y_p^*(t) \end{bmatrix} = \begin{bmatrix} A_p & B_p \\ C_p & 0 \end{bmatrix} \begin{bmatrix} S_{11} & S_{12} \\ S_{21} & S_{22} \end{bmatrix} \begin{bmatrix} x_m(t) \\ u_m \end{bmatrix}. \tag{3.343}$$

Equation (3.343) is the same equation as in the linear time invariant case which was discussed in Section 2.2.1. Thus,

$$\begin{bmatrix} S_{11} A_m & S_{12} B_m \\ C_m & 0 \end{bmatrix} = \begin{bmatrix} A_p & B_p \\ C_p & 0 \end{bmatrix} \begin{bmatrix} S_{11} & S_{12} \\ S_{21} & S_{22} \end{bmatrix}. \tag{3.344}$$

For a known plant, a control law is chosen which is similar to the adaptive law of (3.328), but with constant gains. Thus,

$$u_p(t) = \tilde{K}_e e_{yp}(t) + \tilde{K}_\gamma \gamma(C_p x_p, u_p, t) + \tilde{K}_x x_m(t) + \tilde{K}_u u_m \tag{3.345}$$

where $\tilde{K}_e$, $\tilde{K}_\gamma$, $\tilde{K}_x$, and $\tilde{K}_u$ are constant gain matrices.

Next, it is shown that the control law of (3.345) is sufficient to yield perfect output tracking when (3.344) is satisfied. The state error derivative equation is described by

$$\dot{x}_p^*(t) - \dot{x}_p(t) = S_{11}\dot{x}_m(t) - A_p x_p(t) - A_\gamma \gamma(C_p x_p, u_p, t) - B_p u_p(t)$$

$$= S_{11}\left[A_m x_m(t) + B_m u_m\right] + A_p\left[x_p^*(t) - x_p(t)\right]$$

$$- A_p x_p^*(t) - A_\gamma \gamma(C_p x_p, u_p, t)$$

$$- B_p\left[\tilde{K}_e e_{yp}(t) + \tilde{K}_\gamma \gamma(C_p x_p, u_p, t) + \tilde{K}_x x_m(t) + \tilde{K}_u u_m\right]$$

$$= (A_p - B_p\tilde{K}_e C_p)\left[x_p^*(t) - x_p(t)\right] + \left[S_{11}A_m - A_p S_{11} - B_p\tilde{K}_x\right] x_m(t)$$

$$+ \left[S_{11}B_m - A_p S_{12} - B_p\tilde{K}_u\right] u_m - \left[A_\gamma + B_p\tilde{K}_\gamma\right]\gamma(C_p x_p, u_p, t). \quad (3.346)$$

Let $\tilde{K}_\gamma = S_\gamma$, $\tilde{K}_x = S_{21}$, $\tilde{K}_u = S_{22}$. Then, use (3.344) and (3.346) together with $A_\gamma + B_p S_\gamma = 0$ (which is satisfied if rank $[B_p]$ = rank $[B_p|A_\gamma]$) to obtain

$$\dot{x}_p^*(t) - \dot{x}_p(t) = (A_p - B_p\tilde{K}_e C_p)\left[x_p^*(t) - x_p(t)\right] \quad (3.347)$$

where $\tilde{K}_e$ is chosen such that $A_p - B_p\tilde{K}_e C_p$ is a stability matrix. Therefore, the output tracking error will asymptotically vanish. That is

$$C_p\left[x_p^*(t) - x_p(t)\right] = y_p^*(t) - y_p(t) = y_m(t) - y_p(t) \quad (3.348)$$

goes to zero as time approaches infinity.

Error Equations for the Adaptive System

Define the state and output errors as shown below.

$$e_{xp}(t) = x_p^*(t) - x_p(t) \quad (3.349)$$

$$e_{yp}(t) = y_m(t) - y_p(t) = y_p^*(t) - y_p(t) = C_p e_{xp}(t). \quad (3.350)$$

Then, the state error derivative is given by

$$\dot{e}_{xp}(t) = \dot{x}_p^*(t) - \dot{x}_p(t)$$

$$= \dot{x}_p^*(t) - A_p x_p(t) - A_\gamma \gamma(C_p x_p, u_p, t) - B_p u_p(t)$$

$$= \dot{x}_p^*(t) + A_p \left[x_p^*(t) - x_p(t)\right] - A_p x_p^*(t) - A_\gamma \gamma(C_p x_p, u_p, t)$$

$$-B_p \left[K(t) - \tilde{K}(t)\right] r(t)$$

$$-B_p \left[\tilde{K}_e e_{yp}(t) + \tilde{K}_\gamma \gamma(C_p x_p, u_p, t) + \tilde{K}_x x_m(t) + \tilde{K}_u u_m(t)\right]$$

$$= (A_p - B_p \tilde{K}_e C_p) e_{xp}(t) - B_p z(t) + \dot{x}_p^*(t) + \left[-A_p S_{11} - B_p \tilde{K}_x\right] x_m(t)$$

$$+ \left[-A_p S_{12} - B_p \tilde{K}_u\right] u_m(t) - \left[A_\gamma + B_p \tilde{K}_\gamma\right] \gamma(C_p x_p, u_p, t)$$

$$= A_c e_{xp}(t) - B_p z(t) - F_1(t) \tag{3.351}$$

where $A_c = A_p - B_p \tilde{K}_e C_p$ is a stability matrix

$$z(t) = \left[K(t) - \tilde{K}\right] r(t) \tag{3.352}$$

$$\tilde{K} = \left[\tilde{K}_e, \tilde{K}_\gamma, \tilde{K}_x, \tilde{K}_u\right] \tag{3.353}$$

and

$$F_1(t) = -\dot{x}_p^*(t) + \left[A_p S_{11} + B_p \tilde{K}_x\right] x_m(t) + \left[A_p S_{12} + B_p \tilde{K}_u\right] u_m(t) + \left[A_\gamma + B_p \tilde{K}_\gamma\right] \gamma(C_p x_p, u_p, t). \tag{3.354}$$

Stability Analysis of the Adaptive System

In the adaptive control problem, the plant is generally unknown or poorly known. Therefore, the adaptive control signal described by (3.328) is used. An asymptotically vanishing output error will be obtained if the conditions in the following theorem are satisfied.

Theorem 3.4: Consider the nonlinear plant represented by (3.326) and (3.327) and let the adaptive control algorithm be described by (3.28)–(3.34). If (i) there exists a real symmetric positive definite matrix P and real matrices L and $\tilde{K}_e$ such that

$$P(A_p - B_p\tilde{K}_e C_p) \;+\; (A_p - B_p\tilde{K}_e C_p)^T P \;=\; -LL^T \;<\; 0 \qquad (3.355)$$

$$PB_p \;=\; C_p^T \qquad (3.356)$$

which is equivalent to requiring that the linear part of the plant be ASPR; (ii) $T > 0, \bar{T} \geq 0$; (iii) $\text{rank}\,[B_p] = \text{rank}\,[B_p|A_\gamma]$; (iv) u_m is constant for $t \geq t_1$; (5) a solution exists for the matrices S_{11}, S_{12}, S_{21}, and S_{22} in (3.224); then all states and gains in the adaptive loop are bounded and the output error is asymptotically vanishing.

Proof: See Appendix 3E.

Example (Ship Autopilot)

The dynamics of the ship discussed by van Amerongen and Ten Cate [60] are described by the following third order equation.

$$\frac{d^3\theta(t)}{dt^3} + C\frac{d^2\theta(t)}{dt^2} + K\left(a\left(\frac{d\theta(t)}{dt}\right)^3 + b\frac{d\theta(t)}{dt}\right) = Ku_p(t) \qquad (3.357)$$

where $u_p(t)$ is the rudder angle, $\theta(t)$ is the ship heading or course, and $\frac{d\theta(t)}{dt}$, the component to be controlled, is the course angular velocity. The unknown parameters C, K, a, and b are related to the hydrodynamic coefficients and the mass of the ship.

Letting

$$x_{p1}(t) \;=\; \frac{d^2\theta(t)}{dt^2} \qquad (3.358)$$

and

$$x_{p2}(t) = \frac{d\theta(t)}{dt}, \qquad (3.359)$$

gives

$$\dot{x}_p(t) = \begin{bmatrix} -C & -Kb \\ 1 & 0 \end{bmatrix} x_p(t) + \begin{bmatrix} K \\ 0 \end{bmatrix} u_p(t) + \begin{bmatrix} -Ka \\ 0 \end{bmatrix} x_{p2}^3(t) \qquad (3.360)$$

$$y_p(t) \;=\; [0 \;\; 1]\, x_p(t) \qquad (3.361)$$

For illustrative purposes, let $a = 1.06$, $b = 4$, $C = 3.5$, and $K = 0.05$. Then, (3.326) becomes

$$\dot{x}_p(t) = \begin{bmatrix} -3.5 & -0.2 \\ 1 & 0 \end{bmatrix} x_p(t) + \begin{bmatrix} 0.05 \\ 0 \end{bmatrix} u_p(t) + \begin{bmatrix} -0.053 \\ 0 \end{bmatrix} x_{p2}^3(t). \tag{3.362}$$

This example will show that the output of the plant will asymptotically track the output of the linear reference model described by

$$\dot{x}_m(t) = \begin{bmatrix} -3 & -0.3 \\ 1 & 0 \end{bmatrix} x_m(t) + \begin{bmatrix} 0.1 \\ 0 \end{bmatrix} u_m(t) \tag{3.363}$$

$$y_m(t) = [0 \;\; 1]\, x_m(t). \tag{3.364}$$

The positive real procedure, described in Section 2.7, is used to determine if the plant output matrix defined in (3.361) satisfies the sufficient conditions described by (3.355) and (3.356). Let $C_p = [c_{p1},\ 1]$. Then, the characteristic equation of the linear part of the plant is

$$\lambda^2 + (3.5 + c_{p1}\tilde{k}_e)\lambda + (0.2 + 0.05\tilde{k}_e) = 0. \tag{3.365}$$

If the coefficients of this equation are always greater than zero, then the real part of the eigenvalues of $A_p - B_p\tilde{K}_eC_p$ will be negative as required in step one of the positive real procedure.

Thus, c_{p1} and $\tilde{k}_e$ must be chosen such that

$$3.5 + 0.05c_{p1}\tilde{k}_e > 0 \tag{3.366}$$

$$0.2 + 0.05\tilde{k}_e > 0. \tag{3.367}$$

Next, proceed by using the definition that

$$Z(s) = C_p(sI - A_p + B_p\tilde{K}_eC_p)^{-1}B_p \tag{3.368}$$

to obtain

$$Z(s) = \frac{0.05(1 + c_{p1}s)}{s^2 + (3.5 + c_{p1}\tilde{k}_e)s + (0.2 + 0.05\tilde{k}_e)} \tag{3.369}$$

and

$$Z(j\omega) = \frac{0.05(1 + j\omega c_{p1})}{\left[-\omega^2 + 0.2 + 0.05\tilde{k}_e\right] + j\left[(3.5 + 0.05c_{p1}\tilde{k}_e)\omega\right]}. \tag{3.370}$$

Then, use the definition that

$$F(\omega) = Z(j\omega) + Z^T(-j\omega) \tag{3.371}$$

to obtain

$$F(\omega) = \frac{0.1\{\left[(3.5 + 0.05c_{p1}\tilde{k}_e)c_{p1} - 1\right]\omega^2 + \left[0.2 + 0.05\tilde{k}_e\right]\}}{\left[-\omega^2 + 0.2 + 0.05\tilde{k}_e\right]^2 + \left[(3.5 + 0.05c_{p1}\tilde{k}_e)\omega\right]^2}. \tag{3.372}$$

If the coefficients of ω^2 and ω^0 in the numerator of (3.372) are positive, then $F(\omega)$ is positive for all ω. Therefore, it is required that

$$(3.5 + 0.05c_{p1}\tilde{k}_e)c_{p1} > 1 \tag{3.373}$$

$$0.2 + 0.05\tilde{k}_e > 0. \tag{3.374}$$

Furthermore, with these conditions

$$\lim_{\omega \to \infty} \omega^2 F(\omega) > 0.$$

The selection of $C_p = [c_{p1},\ 1]$ must be such that (3.365), (3.367), (3.373) and (3.374) are satisfied. Suppose $\tilde{k}_e$ is chosen to be zero. Then, it is easily shown that c_{p1} must satisfy $c_{p1} \geq 0.286$. Furthermore, it is seen that $\text{Rank}\,[B_p] = \text{Rank}\,[B_p|A_\gamma] = 1$ so that the third sufficient condition of the theorem is satisfied.

Digital simulations have been performed with a square wave reference model input which has an amplitude of 45° and a period of 20 seconds. The initial plant states are zero, and the gain weighting matrices T and $\bar{T}$ are $2000I$. First, the plant output matrix is chosen to be $C_p = [0.286, 1]$. The plant and model angular velocities are shown in Figure 3.36 where it is seen that the velocity of the ship perfectly follows the output of the linear time invariant reference model. In Figure 3.37 it is observed that the acceleration tracking error vanishes asymptotically even though the measurement is $0.286x_{p1}(t) + x_{p2}(t)$.

Next, the plant output matrix is chosen to be $C_p = [0,\ 1]$. This is the matrix which was given in the description of the plant for the ship autopilot problem. Note that this choice for the plant output matrix does not satisfy the sufficient conditions of the theorem. The plant and model angular velocities are shown in Figure 3.42 where it is seen that the velocity of the ship perfectly follows the output of the linear time invariant reference model. In Figure 3.43, it is observed that the acceleration tracking error vanishes asymptotically even though only the velocity $x_{p2}(t)$ is measured.

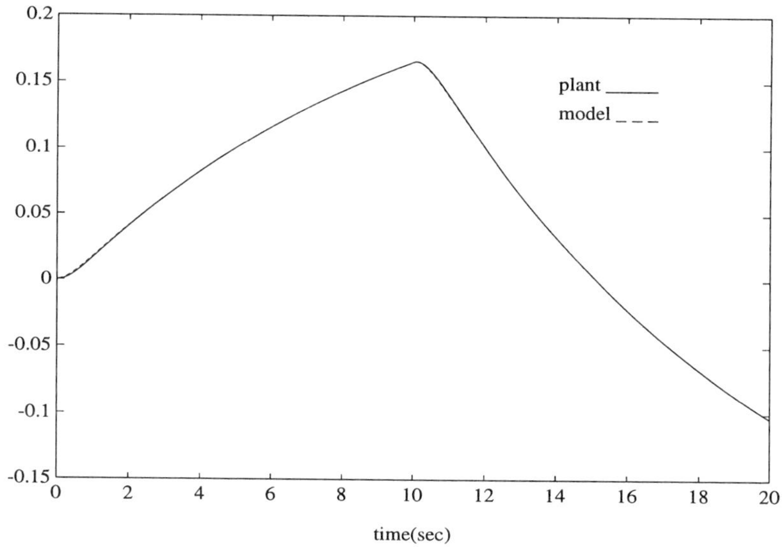

FIGURE 3.40. Plant and Model Angular Velocities $C_p = [0.286,\ 1]$

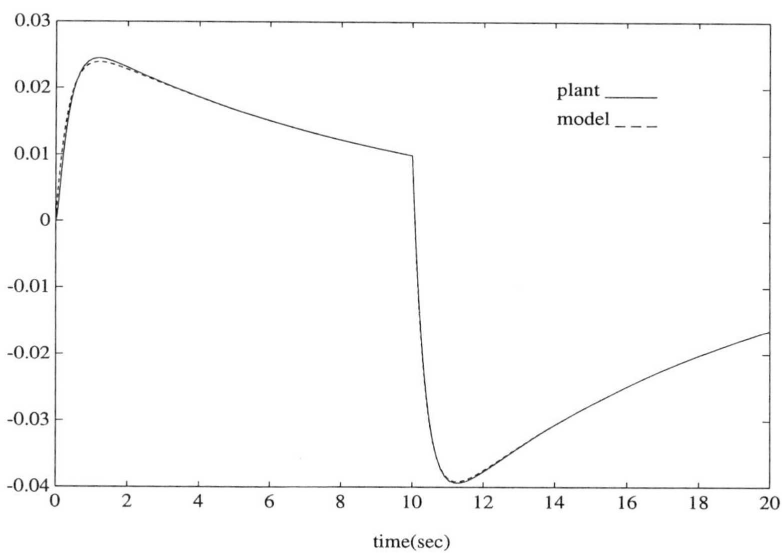

FIGURE 3.41. Plant and Model Angular Accelerations $C_p = [0.286,\ 1]$

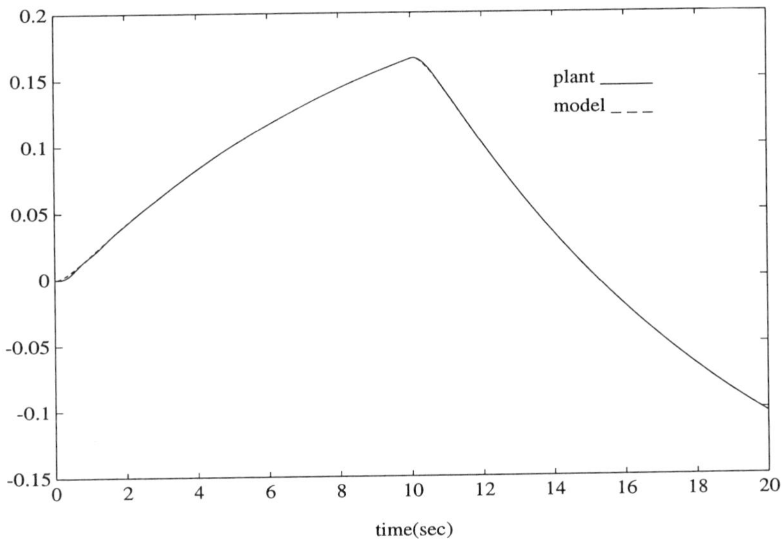

FIGURE 3.42. Plant and Model Angular Velocities $C_p = [0,\ 2]$

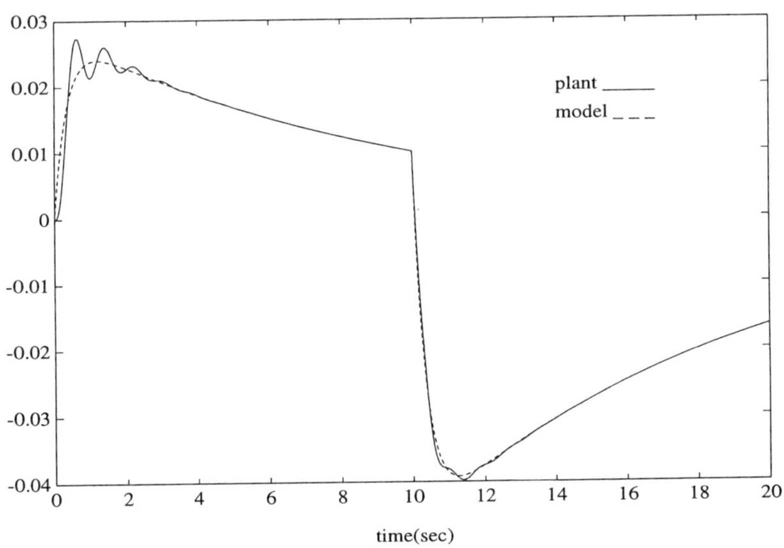

FIGURE 3.43. Plant and Model Angular Accelerations $C_p = [0,\ 1]$

3.6 Summary

This chapter has considered the use of compensation networks for alleviating the positive real conditions discussed in Chapter 2.

The simplest augmentation is the use of feedforward in the process itself. This algorithm is summarized in Section 3.2.2. However, in order to guarantee asymptotic properties of the actual output, i.e., not the augmented output, additional compensation might be needed. To this effect, an algorithm that uses a feedforward compensator about the reference model is summarized in Section 3.3.3, and a set of algorithms based upon the concept of a metasystem is summarized in Section 3.4.2.

Finally, an algorithm for a specific type of nonlinearity of known form is summarized in Section 3.5.

Appendix 3A

Proof of Positivity Lemmas

A few positivity lemmas for multivariable systems will now be presented. For the proofs of the subsequent lemmas, the following fact [98] is useful

$$G_s(s) \text{ is SPR } \Leftrightarrow G_s^{-1}(s) \text{ is SPR} \tag{3A.1}$$

Proof:

$$G_s(s) = M + jN \Rightarrow G_s^*(s) = M^T - jN^T. \tag{3A.2}$$

Then, whenever

$$G_s(s) + G_s^*(s) = M + M^T + j(N - N^T) > 0 \tag{3A.3}$$

one gets $N = N^T$ and $M > 0$ (which is positive definite but not necessarily symmetric), and whenever

$$Re\left[G_s(s)\right] = M > 0 \tag{3A.4}$$

one gets

$$G_s^{-1}(s) = (M + NM^{-1}N)^{-1} - jM^{-1}N(M + NM^{-1}N)^{-1} \tag{3A.5}$$

and then

$$Re\left[G_s^{-1}(s)\right] = (M + NM^{-1}N)^{-1} > 0. \tag{3A.6}$$

Lemma 3.1: Let $G(s)$ be any $m \times m$ strictly proper transfer matrix of arbitrary relative MacMillan degree. $G(s)$ is not necessarily stable or minimum–phase. Let K_y be a nonsingular constant output feedback matrix such that the closed–loop transfer matrix

$$G_{CL}(s) = \left[I + G(s)K_y\right]^{-1} G(s) \tag{3A.7}$$

is asymptotically stable. Then the *augmented* open–loop transfer matrix

$$G_a(s) = G(s) + K_y^{-1} \tag{3A.8}$$

is ASPR.

Proof: Let $G(s)$ have the realization

$$\dot{x}(t) = Ax(t) + Bu(t) \tag{3A.9}$$

$$y(t) = Cx(t) \tag{3A.10}$$

It is now easy to see that the following statements are true since they follow one another.

A_1 $G_a(s) = K_y^{-1} + G(s)$ is ASPR.

A_2 $G_a(s)$ is strictly minimum phase (as the constant K_y^{-1} makes the relative degree of $G_a(s)$ to be n/n).

A_3 $G_a^{-1}(s) = \left[K_y^{-1} + C(sI - A)^{-1}B\right]^{-1}$ is asymptotically stable.

But using the matrix inversion lemma, [99] $G_a^{-1}(s) = K_y - K_y C(sI - A + BK_y C)^{-1} BK_y$, $G_a^{-1}(s)$ is asymptotically stable, and thus $G_a(s)$ is ASPR.
Lemma 3.2: Let $G(s)$ be defined as in Lemma 3.1. Let

$$H(s) = K(1 + qs) \tag{3A.11}$$

be a stabilizing closed loop compensator of $G(s)$. Then the augmented plant

$$G_a(s) = G(s) + H^{-1}(s) = G(s) + \frac{K^{-1}}{1 + qs} \tag{3A.12}$$

is ASPR.
Proof: Let $G(s)$ have the realization (3A.9)–(3A.10) and let

$$u(t) = -K\left[y(t) + q\dot{y}(t)\right] + Ku_c(t) \tag{3A.13}$$

be a stabilizing controller. Then

$$u(t) = -KC\left[x(t) + qAx(t) + qBu(t)\right] + Ku_c(t) \tag{3A.14}$$

$$u(t) = -(I + qKCB)^{-1}KC(I + qA)x(t) + (I + qKCB)^{-1}Ku_c(t) \tag{3A.15}$$

$$u(t) = -(K^{-1} + qCB)^{-1}C(I + qA)x(t) + (K^{-1} + qCB)^{-1}u_c(t). \tag{3A.16}$$

Substituting (3A.16) into (3A.9) gives

$$\dot{x}(t) = \left[A - B(K^{-1} + qCB)^{-1}C(I + qA)\right]x(t) + B(K^{-1} + qCB)^{-1}u_c(t) \tag{3A.17}$$

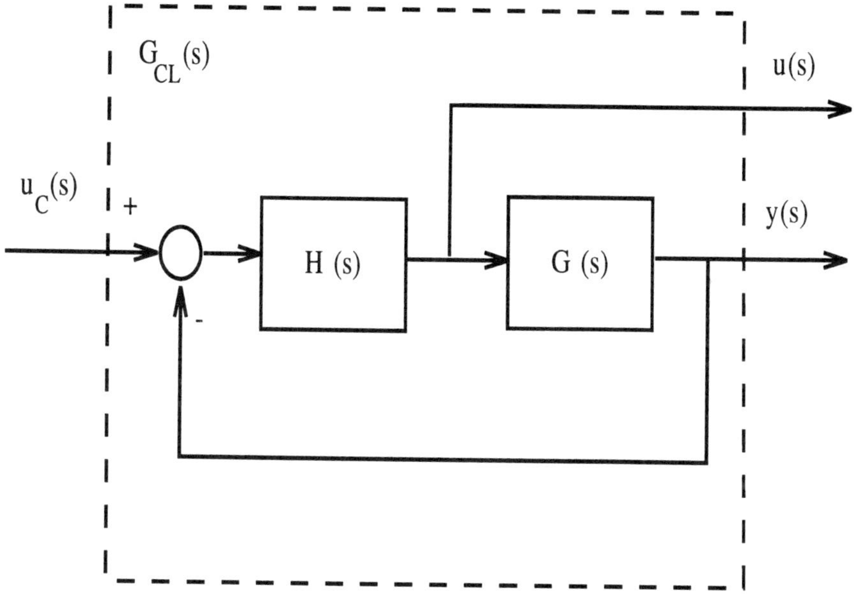

FIGURE 3A.1. The Closed–Loop System

which is, by assumption, asymptotically stable since (3A.12) is a stabilizing controller. The stability matrix of (3A.17) will be needed in the following proofs.

It is now easy to see that the following statements are equivalent as they follow one another.

A_4 $G_a(s) = G(s) + \frac{K^{-1}}{1+qs}$ is ASPR.

A_5 $G_a(s)$ is strictly minimum–phase and of relative degree $(n - m)/n$. In other words, all $n - m$ finite zeros of $G_a(s)$ are in the open left halfplane.

A_6 All $n - m$ finite poles of $G_a^{-1}(s)$ are in the open left halfplane. In other words, the improper transfer matrix $G_a^{-1}(s)$ is bounded for all finite s such that $Re\,[s] \geq 0$.

A_7 $G_a^{-1}(s) = \left[G(s) + \frac{K^{-1}}{1+qs}\right]^{-1}$ is bounded for finite s such that $Re\,[s] \geq 0$.

A_8 $G_a^{-1}(s) = \left[\frac{K^{-1}}{1+qs} + C(sI - A)^{-1}B\right]^{-1}$ is bounded for finite s such that $Re\,[s] \geq 0$.

The matrix inversion lemma $(P+MQN)^{-1} = P^{-1}M(Q^{-1}-NP^{-1}M)^{-1} NP^{-1}$ gives from A_8:

$$G_a^{-1}(s) = (1+qs)K - (1+qs)KC\left[sI - A + BK(1+qs)C\right]^{-1} BK(1+qs)$$

$$G_a^{-1}(s) = (1+qs)K -$$

$$(1+qs)KC\left[s(I+qBKC)(-A+BKC)\right]^{-1}(I+BKC)^{-1}BK(1+qs)$$

$$G_a^{-1}(s) = (1+qs)K -$$

$$(1+qs)KC\left[sI + (I+qBKC)^{-1}(-A+BKC)\right]^{-1}(I+BKC)^{-1}BK(1+qs).$$

Now again use the matrix inversion lemma $(I+qBKC)^{-1} = I - qB(K^{-1} +qCB)^{-1}C$ and

$$E \triangleq sI + \left[I - qB(K^{-1}+qCB)^{-1}C\right](-A+BKC) =$$

$$= sI - A + qB(K^{-1}+qCB)^{-1}CA + BKC - qB(K^{-1}+qCB)^{-1}CBKC =$$

$$sI - A + qB(K^{-1}+qCB)^{-1} + qCB)^{-1}CA +$$

$$B\left[I - q(K^{-1}+qCB)^{-1}CB\right]KC.$$

The last term in E is

$$B\left[I - q(K^{-1}+qCB)^{-1}CB\right]KC =$$

$$B\left[I + q(K^{-1}+qCB-qCB)^{-1}CB\right]KC =$$

$$B(I+qKCB)^{-1}KC = B(K^{-1}+qCB)^{-1}C$$

and then

$$\begin{aligned} E &= sI - A + qB(K^{-1}+qCB)^{-1}CA + B(K^{-1}+qCB)^{-1}C = \\ &= sI - A + B(K^{-1}+qCB)^{-1}C(I+qA) \end{aligned}$$

and finally

$$G_a^{-1}(s) = (1+qs)K - (1+qs)KC$$

$$\left[sI - A + B(K^{-1}+qCB)^{-1}C(I+qA)\right]^{-1} \tag{3A.18}$$

$$(I+BKC)^{-1}BK(1+qs).$$

Thus, since (3A.17) is asymptotically stable, $G_a^{-1}(s)$ *is* bounded for any finite s such that $\mathbf{Re}[s] \geq 0$. Therefore, from the statements (A4–A8), $G_a(s)$ is ASPR.

Lemma 3.3: Let $G(s)$ be defined as in Lemma 3.1. Let $H(s)$ be any dynamic stabilizing controller. Then

$$G_a(s) = G(s) + H^{-1}(s) \tag{3A.19}$$

is ASPR if the MacMillan degree of $G_a(s)$ is p/p or $(p-m)/p$, for any p.

Proof: Consider for illustration, the asymptotically stable transfer matrix $G_{CL}(s)$ from $u_c(s)$ to $u(s)$ in Figure 3A.1

$$u(s) = G_{CL}(s)u_c(s) = [I + H(s)G(s)]^{-1} H(s)u_c(s). \tag{3A.20}$$

Then the following statements are equivalent:

A_9 All finite poles of

$$G_{CL}(s) = [I + H(s)G(s)]^{-1} H(s) = \left[H^{-1}(s) + G(s)\right]^{-1}$$

are in the open left halfplane.

A_{10} $G_a(s) = G(s) + H^{-1}(s) = G_{CL}^{-1}(s)$ is strictly minimum–phase (all finite zeros of $G_a(s)$ are in the open left halfplane).

Thus, if $\deg\left[H^{-1}(s)\right] = p/p$ or $(p-m)/p$, $\deg\left[G_a(s) = G_{CL}^{-1}(s)\right] = (p+n)/(p+n)$ or $(p+n-m)/(p+n)$, and $G_a(s)$ is ASPR.

Frequency domain relations have been used to show that stabilizability of the controlled plant and parallel feedforward lead to almost positive realness of the augmented plant. Yet, the proofs of stability in this book use the time domain positive real relations.

It is beneficial, therefore, to show directly that the use of stabilizability properties and parallel feedforward guarantees satisfaction of the passivity relations (2.184)–(2.186). To this end, assume that the fictitious stabilized closed–loop plant $G_{CL}(s) = \left[H^{-1}(s) + G(s)\right]^{-1}$ has the representation

$$\dot{x}(t) = A_{CL}x(t) + B_{CL}u(t) \tag{3A.21}$$

$$y(t) = C_{CL}x(t) + D_{CL}u(t). \tag{3A.22}$$

Assume further that $G_{CL}(s)$ is asymptotically stable, and therefore, that there exist some positive definite symmetric matrices, P and Q_1 such that if one selects the Lyapunov function $V(t) = x^T(t)Px(t)$, its derivative along the trajectories of (3A.21)–(3A.22) becomes

$$\dot{V}(t) = x^T(t)\left[PA_{CL} + A_{CL}^T P\right]x(t) = -x^T(t)Q_1x(t). \tag{3A.23}$$

Because Q_1 is positive definite, it is easy to see that there exists a (unknown) positive definite gain matrix $\tilde{K}_e$ (not necessarily large but sufficiently large) such that

$$x^T(t)Q_1x(t) - x^T(t)\left[PB_{CL} - C_{CL}^T\right]\left[D_{CL} + D_{CL}^T + \tilde{K}_e + \tilde{K}_e^T\right]^{-1} \tag{3A.24}$$
$$\left[B_{CL}^T P - C_{CL}\right]x(t) = x^T(t)Qx(t)$$

for some positive definite matrix Q. Substituting $x^T(t)Q_1x(t)$ from (3A.24) into (3A.23) gives:

$$x^T(t)\left[PA_{CL} + A_{CL}^T P\right]x(t) =$$
$$= -x^T(t)\{Q + \left[PB_{CL} - C_{CL}^T\right]\left[D_{CL} + \tilde{K}_e + D_{CL}^T + \tilde{K}_e^T\right]^{-1} \tag{3A.25}$$
$$\left[B_{CL}^T P - C_{CL}\right]\}x(t)$$

which, upon comparison with (2.187), is seen to be the strict passivity relation for the system $G_1(s) = G_{CL}(s) + \tilde{K}_e$ with representation:

$$\dot{x}(t) = A_{CL}x(t) + B_{CL}u(t) \tag{3A.26}$$

$$y(t) = C_{CL}x(t) + \left[D_{CL} + \tilde{K}_e\right]u(t). \tag{3A.27}$$

The inverse system $G_s = G_1^{-1}$ then becomes:

$$G_1^{-1}(s) = \{\left[H_{(s)}^{-1} + G(s)\right]^{-1} + \tilde{K}_e\}^{-1} \tag{3A.28}$$
$$= \{I + \left[H^{-1}(s) + G(s)\right]\tilde{K}_e\}\left[H^{-1}(s) + G(s)\right],$$

as shown in Figure 3A.2.

This has the representation:

$$\dot{x}(t) = A_2x(t) + B_2u(t) \tag{3A.29}$$

$$y(t) = C_2x(t) + D_2u(t) \tag{3A.30}$$

where

$$A_2 = A_{CL} - B_{CL}\left[D_{CL} + \tilde{K}_e\right]^{-1}C_{CL} \tag{3A.31}$$

$$B_2 = B_{CL}\left[D_{CL} + \tilde{K}_e\right]^{-1} \tag{3A.32}$$

$$C_2 = \left[D_{CL} + \tilde{K}_e\right]^{-1} C_{CL} \tag{3A.33}$$

$$D_2 = \left[D_{CL} + \tilde{K}_e\right]^{-1}. \tag{3A.34}$$

Then for this system,

$$\begin{aligned}
&x^T(t)\left[PA_2 + A_2^T P\right]x(t) = x^T(t)\left[PA_{CL} + A_{CL}^T P\right]x(t) \\
&-x^T(t)\left[\{PB_{CL}\right]\left[D_{CL} + \tilde{K}_e\right]^{-1} C_{CL} \\
&+C_{CL}^T\left[D_{CL}^T + \tilde{K}_e\right]^{-1} B_{CL}^T P\}x(t) = x^T(t)\{Q + \left[PB_{CL} - C_{CL}^T\right] \\
&\left[D_{CL} + D_{CL}^T + \tilde{K}_e + \tilde{K}_e^T\right]^{-1}\left[B_{CL}^T P - C_{CL}\right] \\
&+PB_{CL}\left[D_{CL} + \tilde{K}_e\right]^{-1} C_{CL}C_{CL}^T\left[D_{CL}^T + \tilde{K}_e^T\right]^{-1} B_{CL}^T P\}x(t) \\
&= -x^T(t)\{Q + \left[PB_{CL} + C_{CL}^T\right]\left[D_{CL} + D_{CL}^T + \tilde{K}_e + \tilde{K}_e^T\right]^{-1} \\
&\left[B_{CL}^T P + C_{CL}\right]\}x(t) = \\
&-x^T(t)\{Q + \left[PB_{CL} + C_{CL}^T\right]\left[D_{CL} + \tilde{K}_e\right]^{-1}\left[D_{CL} + \tilde{K}_e\right] \\
&\left[D_{CL} + D_{CL}^T + \tilde{K}_e + \tilde{K}_e^T\right]^{-1}\left[B_{CL}^T P + C_{CL}\right]\}x(t) \\
&= -x^T(t)\{Q + \left[PB_2 + C_2^T\right]\left[D_{CL} + D_{CL}^T + \tilde{K}_e + \tilde{K}_e^T\right] \\
&\left[B_2^T P - C_2\right]\}x(t) \\
&= -x^T(t)\{Q + \left[PB_2 + C_2^T\right]\left[D_2 + D_2^T\right]^{-1}\left[B_2^T P - C_2\right]\}x(t).
\end{aligned} \tag{3A.35}$$

Finally considering only the matrices involved in the first and last lines of (3A.35), gives:

$$PA_2 + A_2^T P + \left[PB_2 - C_2^T\right]\left[D_2 + D_2^T\right]^{-1}\left[B_2^T P - C_2\right] + Q = 0 \tag{3A.36}$$

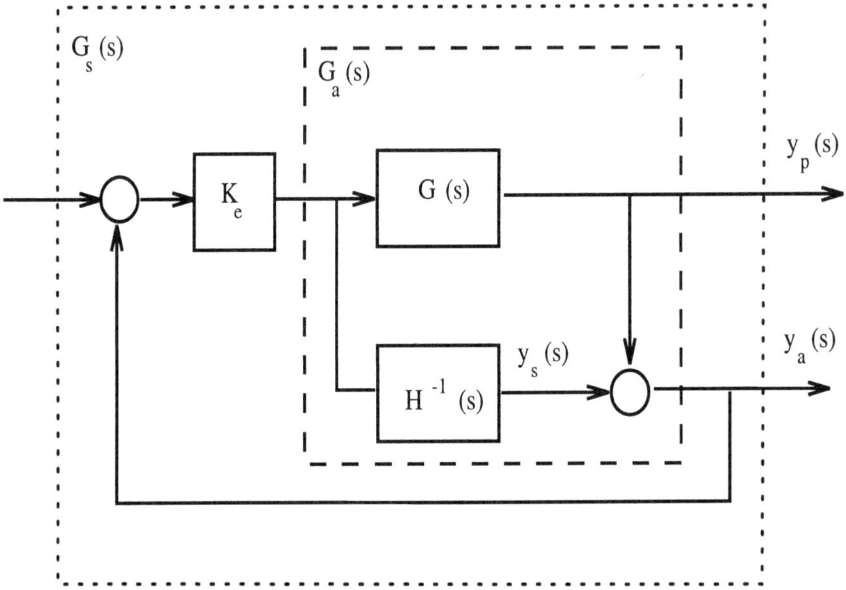

FIGURE 3A.2. The Augmented SPR System

which by comparing with (2.187) shows that the system $G_s(s)$ of Figure 3A.2 is strictly positive real (SPR), and therefore the augmented system $G_a(s)$ is almost strictly positive real (ASPR).

Appendix 3B

Proof of Theorem 3.1

As in [97], stability will be shown using the Lyapunov function

$$V(e_1 K_I) = e^T(t)Pe(t) + tr[S(K_I(t) - \tilde{K}(t))T^{-1}(K_I(t) - \tilde{K}(t))^T S^T] \quad (3B.1)$$

where "tr" denotes the trace operator, T and P are constant positive definite symmetric matrices, S is a nonsingular matrix, and $\tilde{K}$ is partitioned in the same manner as K, i.e.

$$\tilde{K}(t) = [\tilde{K}_e, \tilde{K}_q(t), \tilde{K}_u(t)] \quad (3B.2)$$

where $\tilde{K}_q = [\tilde{K}_x : 0]$.

Note that P and $\tilde{K}_e$ are constant since F_p and G_p are not time varying. Manipulations similar to those in [97] and [31] result in

$$\begin{aligned}
\dot{V} &= e^T(t)(PF_p + F_p^T P)e(t) - 2e^T(t)PG_pH_pe(t)r^T(t)\bar{T}r(t) - \\
&\quad 2e^T PG_pK_I r(t) \\
&+ 2e^T(t)PG_pS_{21}(t)q_m(t) + 2e^T(t)PG_pS_{22}(t)u_m(t) \\
&+ 2tr(S(K_I(t) - \tilde{K}(t))T^{-1}\dot{K}_I^T(t)S^T) \\
&- 2tr(S(K_I(t) - \tilde{K}(t))T^{-1}\dot{\tilde{K}}^T(t)S^T).
\end{aligned} \quad (3B.3)$$

Because $R_p(s)$ was chosen such that the system is ASPR, P, H, and $\tilde{K}_e$ can be selected such that

$$H_p = (S^T S)^{-1}\, G_p^T P \quad (3B.4)$$

and

$$P(F_p - G_p\tilde{K}_eH_p) + (F_p - G_p\tilde{K}_eH_p)^T P < 0. \quad (3B.5)$$

Letting $T = Diag(T_e, T_q, T_u)$, and using the previously defined partitioning of K_I and $\tilde{K}$, results in:

$$
\begin{aligned}
\dot{V} &= e^T(t)[P(F_p - G_p\tilde{K}_e H_p) + (F_p - G_p\tilde{K}_e H_p)^T\ P]e(t) \\
&- 2e^T(t)\ PG_p(S^TS)^{-1}\ G_p^T\ Pe(t)\ r^T(t)\bar{t}r(t) \\
&+ 2e^T(t)\ PG_p(S_{21_x}(t) - \tilde{K}_x)\ x_m(t) - 2tr(t)\ \{S(K_{I_x}(t) \\
&\quad -\tilde{K}_x(t))T_x^{-1}\dot{\tilde{K}}_x^T(t)S^T\} \qquad \text{(3B.6)} \\
&+ 2e^T(t)\ PG_p(S_{22}(t) - \tilde{K}_u)\ u_m(t) - 2Tr(t)\ \{S(K_{Iu}(t) \\
&\quad -\tilde{K}_u(t))T_u^{-1}\dot{\tilde{K}}_u^T(t)S^T\}.
\end{aligned}
$$

Now consider for illustrative purposes, the first trace term on the right side of (3B.6); namely,

$$
-2t_r\{S[K_{I_x}(t) - \tilde{K}_x(t)]T_x^{-1}\dot{\tilde{K}}_x^T(t)S^T\}. \qquad \text{(3B.7)}
$$

Define:

$$
\dot{\tilde{K}}_x(t) = (S^TS)^{-1}G_p^TPe(t)x_m^T(t)\tilde{T}_x(t) \qquad \text{(3B.8)}
$$

where $\tilde{T}_x(t)$ is a weighting matrix to be determined.

With this definition (3B.7) can be rewritten as

$$
-2e^T(t)PG_p[K_{px}(t) - \tilde{K}_x(t)]T_x^{-1}\tilde{T}_x(t)x_m(t). \qquad \text{(3B.9)}
$$

Combining (3B.9) with the third term on the right side of (3B.7) gives

$$
\begin{aligned}
\dot{V}_1 = {} & 2e^T(t)PG_p(S_{21_x}(t) - \tilde{K}_x(t))x_m(t) \\
& -2e^T(t)PG_p(K_{I_x}(t) - \tilde{K}_x(t))T_x^{-1}\tilde{T}_x(t)x_m(t).
\end{aligned} \qquad \text{(3B.10)}
$$

In a similar manner, if

$$
\dot{\tilde{K}}_u = (S^TS)^{-1}G_p^TPe(t)u_m(t)\tilde{T}_u(t), \qquad \text{(3B.11)}
$$

then the last two terms of (3B.6) become

$$
\begin{aligned}
\dot{V}_2 = {} & 2e^T(t)PG_p(S_{22}(t) - \tilde{K}_u(t))u_m(t) \\
& -2e^T(t)PG_p(K_{I_u}(t) - \tilde{K}_u(t))T_u^{-1}\tilde{T}_u(t)u_m(t).
\end{aligned} \qquad \text{(3B.12)}
$$

Thus, if $\tilde{T}_x(t)$ and $\tilde{T}_u(t)$ can be selected so that

$$\dot{V}_1 = \dot{V}_2 = 0 \tag{3B.13}$$

then $\dot{V}$ will be negative definite with respect to $e(t)$. This may be accomplished by two possible procedures, namely:

1) Find $\tilde{T}_x(t)$ and $\tilde{T}_u$ such that the following (linear) equations are satisfied.

$$(S_{21_z}(t) - \tilde{K}_x(t)) - (K_{I_z}(t) - \tilde{K}_x(t))T_x^{-1}\tilde{T}_x(t) = 0 \tag{3B.14}$$

and

$$(S_{22}(t) - \tilde{K}_u(t)) - (K_{I_u}(t) - \tilde{K}_u(t))T_u^{-1}\tilde{T}_u(t) = 0. \tag{3B.15}$$

Of course, the solution of $\tilde{T}_x(t)$ and $\tilde{T}_u(t)$ is dependent upon the solvability of the above equations.

2) If (B14) and/or (B15) are not solvable, then find $\tilde{T}_x(t)$ and/or $\tilde{T}_u(t)$ such that

$$\dot{V}_1 = 0 \tag{3B.16}$$

$$\dot{V}_2 = 0. \tag{3B.17}$$

Each of these relations is a single equation with several unknowns; namely the elements of $\tilde{T}_x(t)$ and $\tilde{T}_u(t)$. Thus, since T_x and T_u are positive definite, a solution should exist as long as

$$(K_{I_x}(t) - \tilde{K}_x(t)) \neq 0 \tag{3B.18}$$

and

$$(K_{I_u}(t) - \tilde{K}_u(t)) \neq 0. \tag{3B.19}$$

Therefore, assuming that

$$\dot{V}_1 = \dot{V}_2 = 0, \tag{3B.20}$$

$$\begin{aligned} \dot{V} &= e^T(t)[P(F_p - G_p\tilde{K}_eH_p) + (F_p - G_p\tilde{K}_eH_p)^T P]e(t) \\ &- 2e^T(t)PG_p(S^TS)^{-1}G_p^TPe(t)r^T(t)\bar{T}r(t). \end{aligned} \tag{3B.21}$$

In the singular case, when the integral adaptive gains equal the ideal gains, i.e.,

$$K_{I_x}(t) - \tilde{K}_x(t) = 0 \tag{3B.22}$$

and/or

$$K_{I_u}(t) - \tilde{K}_u(t) = 0 \tag{3B.23}$$

the trace terms in (3B.6) vanish. To illustrate the resulting effects of both (3B.23) and (3B.24) being satisfied, note that since

$$S_{21_x}(t) = K_x(t) \tag{3B.24}$$

and

$$S_{22}(t) = K_u(t), \tag{3B.25}$$

the remaining last two terms become:

$$\begin{aligned}
&2e^T(t)PG_p(S_{21_x}(t) - \tilde{K}_x(t))x_m(t)\\
&+2e^T(t)\ PG_p(S_{22}(t) - \tilde{K}_u(t))u_m(t) =\\
&2e^T(t)PG_p(K_{p_x}(t))x_m(t) + 2e^T(t)PG_p(K_{p_u})u_m(t) =\\
&2e^T(t)PG_pe_z(t)x_m^T(t)\bar{T}_xx_m(t) + 2e^T(t)PG_pe_z(t)u_m^T\bar{T}_uu_m(t) =\\
&2e^T(t)PG_p(S^TS)^{-1}G_p^TPe(t)x_m^T(t)\bar{T}_xx_m(t)+\\
&2e^T(t)PG_p(S^TS)^{-1}G_p^TPe(t)u_m^T(t)\bar{T}_uu_m(t).
\end{aligned} \tag{3B.26}$$

However, note that the second term on the right side of $\dot{V}$ can be rewritten based upon the defined partitioning as:

$$\begin{aligned}
&-2e^T(t)PG_p(S^TS)^{-1}G_p^T\ Pe(t)r^T(t)\bar{T}r(t) =\\
&-2e^T(t)PF_p(S^TS)^{-1}G_p^T\ Pe(t)[e_z^T(t)\bar{T}_ee_z(t)\\
&+x_m^T(t)\bar{T}_xx_m(t) + u_m^T(t)\bar{T}_uu_m(t)].
\end{aligned} \tag{3B.27}$$

Combining (3B.21) with (3B.27) gives:

$$\begin{aligned}
\dot{V} &= e^T(t)[P(F_p - G_p\tilde{K}_eH_p) + (F_p - G_p\tilde{K}_eH_p)^TP]e(t)\\
&- 2e(t)^TPG_p(S^TS)^{-1}G_p^TPe(t)e_z^T(t)\bar{T}_ee_z(t).
\end{aligned} \tag{3B.28}$$

Thus, since $\dot{V} \leq 0, V$ is bounded. This implies boundedness of both $e(t)$ and of $(K_{I_e}(t) - \tilde{K}_e)$ or equivalently boundedness of $K_{I_e}(t)$. However since

$$K_{I_e}(t) = \int_0^t e_z(\tau)e_z^T(\tau)T_e d\tau, \tag{3B.29}$$

boundedness of $K_{I_e}(t)$ along with uniform continuity of $\dot{e}_z(t)$ imply that both $e_z(t)$ and $\dot{e}_z(t)$ approach zero (the desired result).

To show uniform continuity of $\dot{e}_z(t)$, it is sufficient to show boundedness of $\ddot{e}_z(t)$ [14]. To demonstrate this boundedness property, substitute into (3.101) the following relations from (3.85) and (3.86):

$$\begin{aligned} S_{21_x}(t) &= K_x(t) \\ S_{22}(t) &= K_u(t). \end{aligned} \tag{3B.30}$$

Then the error equation becomes:

$$\dot{e}(t) = (F_p - G_p K_e(t))e(t). \tag{3B.31}$$

Now $\dot{e}(t)$ will be bounded, since $e(t)$ is bounded and $K_e(t) = K_{I_e}(t) + K_{p_e}(t)$ is bounded.

Differentiating (3B.31) gives:

$$\ddot{e}(t) = (F_p - G_p K_e(t))\dot{e}(t) - G_p \dot{K}_e(t)e(t). \tag{3B.32}$$

Now the first term on the right side of (3B.32) is bounded since both K_e and $\dot{e}(t)$ are bounded. The last term is bounded since

$$\begin{aligned} \dot{K}_e(t) &= \dot{K}_{I_e}(t) + K_{\dot{p}_e}(t) = \\ & e_z(t)e_z^T(t)T + \dot{e}_z(t)e_z(t)\bar{T} + e_z(t)\dot{e}_z(t)\bar{T}, \end{aligned} \tag{3B.33}$$

and both $e_z(t)$ and $\dot{e}_z(t)$ are bounded.

Thus, in summary, if $R_p(s)$ of (3.8) is chosen such that $[R_p(s) + G_p(s)]$ is ASPR and if a solution exists to (3.65)–(3.68), then the controller defined by (3.44)–(3.50) will result in asymptotic output tracking.

Appendix 3C

Proof of Theorem 3.2

Stability is analyzed using a Lyapunov approach by forming a quadratic function which is positive definite in the state variables of the adaptive system, $e_x(t)$, and $K_I(t)$. Then for, $T > 0$, $P > 0$, the Lyapunov function, which is positive definite, is chosen as

$$V(e_x, K_I) \;=\; e_x^T(t)Pe_x(t) + tr\left[(K_I(t) - \tilde{K})T^{-1}(K_I(t) - \tilde{K})^T\right] \quad (3C.1)$$

where $\tilde{K}$ is a constant gain matrix which does not appear in the adaptive control algorithm.

Let $V(e_x, K_I) \;=\; V_1(e_x, K_I) + V_2(e_x, K_I)$
where

$$V_1(e_x(t)) = e_x^T(t)Pe_x(t) \quad (3C.2)$$

$$V_2(e_x, K_I) = tr\left[(K_I(t) - \tilde{K})T^{-1}(K_I(t) - \tilde{K})^T\right]. \quad (3C.3)$$

For convenience, suppress the dependence on time. Then

$$\dot{V}_1 \;=\; \dot{e}_x^T Pe_x \;+\; e_x^T P\dot{e}_x. \quad (3C.4)$$

Substitute $\dot{e}_x$ from (3.174) to obtain

$$\dot{V}_1 \;=\; e_x^T(PA_c + A_c^TP)e_x \;-\; z^TB^TPe_x \;-\; e_x^TPBz \;-\; F_1^TPe_x \;-\; e_x^TPF_1 \quad (3C.5)$$

where F_1 is defined by (3.175).

Next, substitute (3.176) into (3C.5) to obtain

$$\dot{V}_1 \;=\; -e_x^TRe_x - e_x^TLL^Te_x \;-\; 2e_x^TPBz \;-\; 2e_x^TPF_1. \quad (3C.6)$$

Then, add and subtract $-2e_x^TLWz \;+\; z^TW^TWz$ in (3C.6) to obtain

$$\begin{aligned}\dot{V}_1 &= -e_x^TRe_x - (Le_x^T - Wz)^T(Le_x^T - Wz) + \\ &2e_x^T(-LW - PB)z + z^TW^TWz - 2e_x^TPF_1.\end{aligned} \quad (3C.7)$$

Next, use (3C.3) to obtain

$$\dot{V}_2 = tr\left[\dot{K}_I T^{-1}(K_I - \tilde{K})^T\right] + tr\left[(K_I - \tilde{K})T^{-1}(\dot{K}_I)^T\right]. \quad (3C.8)$$

Substitute $\dot{K}_I$ from (3.155) into (3C.8) to obtain

$$\dot{V}_2 = tr\left[vr^T(K_I - \tilde{K})^T\right] + tr\left[(K_I - \tilde{K})(vr^T)^T)\right]. \quad (3C.9)$$

Substitute $K_I = K - K_p = K - vr^T\bar{T}$ into (3C.9), and use the definition $z = (K - \tilde{K})r$ to obtain

$$\dot{V}_2 = tr\left[vz^T\right] + tr\left[zv^T\right] - 2v^T vr^T\bar{T}r. \quad (3C.10)$$

Substitute $v = Qe_{yv} + GKr = QCe_v + GKr$ into (3C.10) to obtain

$$\begin{aligned}
\dot{V}_2 &= tr\left[(QCe_v + GKr)z^T\right] + tr\left[z(QCe_v + GKr)^T\right] - 2v^T vr^T\bar{T}r \\
&= 2e_v^T C^T Q^T z + tr\left[G(Kr - \tilde{K}r + \tilde{K}r)z^T\right] + \\
&\quad tr\left[z(Kr - \tilde{K}r + \tilde{K}r)^T G^T\right] - 2v^T vr^T\bar{T}r \\
&= 2e_v^T C^T Q^T z + z^T(G + G^T)z + 2z^T G(\tilde{K}_a Ce_v + \tilde{K}_x x_m + \tilde{K}_u u_m) \\
&\quad -2v^T vr^T\bar{T}r \\
&= 2e_v^T C^T(Q^T + \tilde{K}_a G^T)z + z^T(G + G^T)z + 2z^T G(\tilde{K}_x x_m + \tilde{K}_u u_m) \\
&\quad -2v^T vr^T\bar{T}r.
\end{aligned} \quad (3C.11)$$

Substitute (3.173) for $e_v(t)$ into (3C.11) to obtain

$$\begin{aligned}
\dot{V}_2 &= 2(e_x^T - x_0^{*T})C^T(Q^T + \tilde{K}_a^T G^T)z + z^T(G + G^T)z + \\
&\quad 2z^T G(\tilde{K}_x x_m + \tilde{K}_u u_m) \\
&\quad -2v^T vr^T\bar{T}r = 2e_x^T C^T(Q^T + \tilde{K}_a^T G^T)z + z^T(G + G^T)z \\
&\quad -2v^T vr^T\bar{T}r + 2z^T\left[G(\tilde{K}_x x_m + \tilde{K}_u u_m) - (Q + G\tilde{K}_a)Cx_0^*\right] \\
&= 2e_x^T C^T(Q^T + \tilde{K}_a^T G^T)z + z^T(G + G^T)z - \\
&\quad 2v^T vr^T\bar{T}r - 2z^T F_2
\end{aligned} \quad (3C.12)$$

where

$$F_2 = -G(\tilde{K}_x x_m + \tilde{K}_u u_m) + (Q + G\tilde{K}_a)Cx_0^*. \tag{3C.13}$$

Combine (3C.7) and (3C.12) to obtain

$$\begin{aligned}\dot{V} &= -e_x^T R e_x - (L^T e_x - Wz)^T (L^T e_x - Wz) + \\ &2e_x^T \left[-LW - PB + C^T(Q^T + \tilde{K}_a^T G^T)\right] z + \\ &z^T(W^T W + G + G^T)z - 2v^T v r^T \bar{T} r - \\ &2e_x^T P F_1 - 2z^T F_2.\end{aligned} \tag{3C.14}$$

Finally, substitute (3.177) and (3.178) into (3C.14) and use $e_x^T R e_x = e_x^T(R + R^T)e_x/2$ to obtain

$$\begin{aligned}\dot{V}(e_x, K_I) \;=\; &-e_x^T(t)(R + R^T)e_x/2 - \left[L^T e_x(t) - Wz(t)\right]^T \\ &\left[L^T e_x(t) - Wz(t)\right] + z^T(t)(J + J^T + G + G^T)z(t) \\ &-2v^T(t)v(t)r^T(t)\bar{T}r(t) - 2e_x^T P F_1 - 2z^T F_2.\end{aligned} \tag{3C.15}$$

Observe that there exist some positive constants $\alpha_1, \alpha_2, \ldots, \alpha_5$ such that

$$\begin{aligned}\dot{V}(e_x, K_I) \;\leq\; &-\alpha_1 \|e_x(t)\|^2 - \alpha_2 \| \left[K(t) - \tilde{K}\right] r(t)\|^2 \\ &-\alpha_3 \|v(t)\|^2 \|r(t)\|^2 \\ &+\alpha_4 \|e_x(t)\| \;+\; \alpha_5 \| \left[K(t) - \tilde{K}\right] r(t)\|.\end{aligned} \tag{3C.16}$$

If either $\|e_x(t)\|$ or $\| \left[K(t) - \tilde{K}\right] r(t)\|$ increase beyond some bound, then the negative quadratic terms in (3C.16) will become dominant, and thus $\dot{V}$ becomes negative. The quadratic form of the Lyapunov function $V(e_x, K_I)$ then guarantees that $e_x(t)$ and $e_y(t)$ are bounded.

Appendix 3D

Proof of Theorem 3.3

The Lyapunov derivative is given by (3C.15) where

$$F_1(t) = [-\dot{x}^*(t) + Ax^*(t)] - B\tilde{K}_a C x_0^*(t) + B\left[\tilde{K}_x x_m(t) + \tilde{K}_u u_m\right] \quad (3D.1)$$

$$F_2(t) = (Q + G\tilde{K}_a)Cx_0^*(t) - G\left[\tilde{K}_x x_m(t) + \tilde{K}_u u_m(t)\right]. \quad (3D.2)$$

Next, conditions will be obtained such that $F_1(t) = 0$ and $F_2(t) \to 0$. Use (3.201) and $\tilde{K}_a C x_0^*(t) = \tilde{K}_f C_f x_f^*(t)$ to obtain

$$F_1(t) = -Bu_p^*(t) - B\tilde{K}_f C_f x_f^*(t) + B\left[\tilde{K}_x x_m + \tilde{K}_u u_m\right] \quad (3D.3)$$

$$F_2(t) = (Q_f + G\tilde{K}_f)C_f x_f^*(t) - G\left[\tilde{K}_x x_m(t) + \tilde{K}_u u_m\right]. \quad (3D.4)$$

Then, use (3.206) and (3.207) to yield

$$\begin{aligned}
&F_1(t) = -B\left[S_{31}x_m(t) + S_{32}u_m\right] \\
&-B\tilde{K}_f C_f\left[S_{21}x_m(t) + S_{22}u_m\right] \\
&+B\left[\tilde{K}_x x_m(t) + \tilde{K}_u u_m\right] \\
&= -B\left[(\tilde{K}_x - S_{31} - \tilde{K}_f C_f S_{21})x_m(t)\right. \\
&\left.+(\tilde{K}_u - S_{32} - \tilde{K}_f C_f S_{22})u_m\right]
\end{aligned} \quad (3D.5)$$

Now consider $F_2(t)$:

$$\begin{aligned}
F_2(t) &= (Q_f + G\tilde{K}_f)C_f\left[S_{21}x_m(t) + S_{22}u_m(t)\right] \\
&\quad -G\left[\tilde{K}_x x_m(t) + \tilde{K}_u u_m\right] \\
&= \left[(Q_f + G\tilde{K}_f)C_f S_{21} - G\tilde{K}_x\right] x_m(t) \\
&\quad +\left[(Q_f + G\tilde{K}_f)C_f S_{22} - G\tilde{K}_u\right] u_m \\
&= \left[(Q_f C_f S_{21} + G(\tilde{K}_f C_f S_{21} - \tilde{K}_x)\right] x_m(t) \\
&\quad +\left[Q_f C_f S_{22} + G(\tilde{K}_f C_f S_{22} - \tilde{K}_u)\right] u_m.
\end{aligned} \tag{3D.6}$$

Next, choose $G = -EB_f$ and $Q_f C_f = EA_f$ to obtain

$$\begin{aligned}
F_2(t) &= \left[EA_f S_{21} - EB_f(\tilde{K}_f C_{pf} S_{21} - \tilde{K}_x)\right] x_m(t) \\
&\quad +\left[EA_f S_{22} - EB_f(\tilde{K}_f C_f S_{22} - \tilde{K}_u)\right] u_m \\
&= EA_f S_{21} x_m(t) - EB_f(\tilde{K}_f C_f S_{21} + \tilde{K}_x - S_{31} + S_{31})x_m(t) \\
&\quad +EA_f S_{22} u_m - EB_f(\tilde{K}_f C_f S_{22} + \tilde{K}_u + S_{32} - S_{32})u_m \\
&= EA_f\left[S_{21}x_m(t) + S_{22}u_m\right] + EB_f\left[S_{31}x_m(t) + S_{32}u_m\right] \\
&\quad +EB_f(\tilde{K}_x - \tilde{K}_f C_f S_{21} - S_{31})x_m(t) \\
&\quad +EB_f(\tilde{K}_u - \tilde{K}_f C_f S_{22} - S_{32})u_m \\
&= E\left[A_f x_f^*(t) + B_f u_p^*(t)\right] \\
&\quad +EB_f(\tilde{K}_x - \tilde{K}_f C_f S_{21} - S_{31})x_m \\
&\quad +EB_f(\tilde{K}_u - \tilde{K}_f C_f S_{22} - S_{32})u_m \\
&= ES_{21}\dot{x}_m(t) + EB_f(\tilde{K}_x - \tilde{K}_f C_f S_{21} - S_{31})x_m(t) \\
&\quad +EB_f(\tilde{K}_u - \tilde{K}_f C_f S_{22} - S_{32})u_m.
\end{aligned} \tag{3D.7}$$

Choose

$$\tilde{K}_x = \tilde{K}_f C_f S_{21} + S_{31} \tag{3D.8}$$

$$\tilde{K}_u = \tilde{K}_f C_f S_{22} + S_{32}. \tag{3D.9}$$

Then $F_1(t) = 0$ and $F_2(t) = ES_{21}\dot{x}_m(t)$, which vanishes asymptotically because u_m is a constant for $t \geq t_1$.

Finally, the Lyapunov derivative becomes

$$\begin{aligned} &\dot{V}(e_x, K_I) = -e_x^T(t)Re_x(t) - \left[L^T e_x(t) - Wz(t)\right]^T \\ &\left[L^T e_x(t) - Wz(t)\right] \\ &+z^T(t)(J + J^T + G + G^T)z(t) - 2v^T(t)v(t)r^T(t)\bar{T}r(t) \\ &+2z^T(t)ES_{21}\dot{x}_m(t) \end{aligned} \tag{3D.10}$$

where $R > 0$, $\bar{T} \geq 0$, and $J + J^T + G + G^T < 0$.

Note that $\dot{V}(e_x, K_I)$ is not negative definite or semi–definite due to the last term in (3D.10). Define

$$\dot{V}(e_x, K_I) \;=\; W_1(e_x, K_I) \;+\; W_2(e_x, K_I) \tag{3D.11}$$

such that

$$W_1(e_x, K_I) \leq \tilde{W}_1(e_x, K_I) \;\leq\; 0 \tag{3D.12}$$

where

$$\begin{aligned} &W_1(e_x, K_I) = -e_x^T(t)Re_x(t) - \left[L^T e_x(t) - Wz(t)\right]^T \\ &\left[L^T e_x(t) - Wz(t)\right] \\ &+z(t)^T(J + J^T + G + G^T)z(t) - 2v^T(t)v(t)r^T(t)\bar{T}(t), \\ &W_2(e_x, K_I) \;=\; 2z^T(t)ES_{21}\dot{x}_m(t) \\ &\tilde{W}_1(e_x, K_I) = -\alpha_1\|e_x(t)\|^2 - \alpha_2\| \left[K(t) - \tilde{K}\right] r(t)\|^2 \\ &-\alpha_3\|v(t)\|^2\|r(t)\|^2 \end{aligned} \tag{3D.13}$$

where α_1, α_2, and α_3 are some positive constants.

Then,

$$\lim_{t\to\infty} W_2(e_x, K_I) = 0 \tag{3D.14}$$

which yields

$$\dot{V}(e_x, K_I) \to W_1(e_x, K_I) \text{ as } t \to \infty \tag{3D.15}$$

where $W_1(e_x, K_I)$ is called the "limiting derivative of the Lyapunov function."

Consider Bar–Kana's [51] extension of LaSalle's Invariance Set Principle [100], [101] for nonautonomous differential equations when $\dot{V}(t)$ is not necessarily negative semi–definite.
Let

$$\dot{x} = f(x,t) \tag{3D.16}$$

be a general nonlinear nonautonomous differential equation and assume that

$$\left| \int_{\alpha}^{\beta} f(x,t)\, d\tau \right| < \mu(\beta - \alpha) \tag{3D.17}$$

where the function $\mu(t)$ is a modulus of continuity for the integral and $\mu(\beta - \alpha)$ is bounded for any finite interval $\beta - \alpha$ (where α and β are the limits of the integral). Let $V(x)$ be a differentiable function bounded from below. Assume that the derivative $\dot{V}(x,t)$ "along the trajectories" of (3D.16) is

$$\dot{V}(x,t) = W_1(x,t) + W_2(x,t) \tag{3D.18}$$

such that

$$W_1(x,t) \le \tilde{W}_1(x) \le 0 \tag{3D.19}$$

where $\tilde{W}_1(x)$ is a continuous function of x and $W_2(x,t)$ is a continuous function of x and piecewise continuous in t satisfying

$$\lim_{t\to\infty} W_2(x,t) = 0. \tag{3D.20}$$

Thus, $V(x,t) \to W_1(x,t)$ as $t \to \infty$ and $W_1(x,t)$ is called the "limiting derivative of the Lyapunov function".

Theorem: Under the assumptions described by (3D.16)–(3D.20), all bounded solutions of (3D.15) asymptotically approach the set

$$\Omega \triangleq \{x | W_1(x) \equiv 0\}. \tag{3D.21}$$

Theorem 3.2 has previously shown that all states, gains, and errors in the adaptive system are bounded. Therefore, $V(e_x,K_I)$ is bounded which is required by LaSalle's invariance set principle. In addition, $\dot{V}(e_x,K_I)$ will approach $W_1(e_x,K_I)$ as $t \to \infty$. Furthermore, $W_1(e_x,K_I)$ is bounded by $\tilde{W}_1(e_x,K_I) \leq 0$. Thus, $\dot{V}(e_x, K_I)$ will approach $W_1(e_x,K_I)$ which is negative semi–definite. Thus, the modified version of LaSalle's invariance set principle implies that all signals in the adaptive control system are bounded and that the output tracking error vanishes asymptotically.

Appendix 3E

Proof of Theorem 3.4

Stability is analyzed using a Lyapunov approach by forming a quadratic function which is positive definite in the state variables of the adaptive system, $e_{xp}(t)$, and $K_I(t)$. For positive definite symmetric matrices P and T, the Lyapunov function is chosen as the positive definite quadratic form:

$$V(e_{xp}, K_I) = e_{xp}^T(t)Pe_{xp}(t) + tr\left[(K_I(t) - \tilde{K})T^{-1}(K_I(t) - \tilde{K})^T\right] \quad (3E.1)$$

where $\tilde{K}$ is a constant gain matrix which does not appear in the adaptive control algorithm.

Let $V(e_{xp}, K_I) = V_1(e_{xp}, K_I) + V_2(e_{xp}, K_I)$ where

$$V_1(e_{xp}(t)) = e_{xp}^T(t)Pe_{xp}(t) \quad (3E.2)$$

$$V_2(e_x, K_I) = tr\left[(K_I(t) - \tilde{K})T^{-1}(K_I(t) - \tilde{K})^T\right]. \quad (3E.3)$$

The dependence on time is suppressed for convenience. Then

$$\dot{V}_1 = \dot{e}_{xP}^T Pe_{xp} + e_{xp}^T P\dot{e}_{xp}. \quad (3E.4)$$

Substitute $\dot{e}_{xp}$ from (3.351) to obtain

$$\dot{V}_1 = e_{xp}^T(PA_c + A_c^T P)e_{xp} - z^T B^T Pe_{xp} - e_{xp}^T PBz - F_1^T Pe_{xp} - e_{xp}^T PF_1 \quad (3E.5)$$

where F_1 is defined by (3.354).

Next, substitute (3.355) into (3E.5) to obtain

$$\dot{V}_1 = -e_{xp}^T LL^T e_{xp} - 2e_{xp}^T PB_p z - 2e_{xp}^T PF_1. \quad (3E.6)$$

Next, use (3E.3) to obtain

$$\dot{V}_2 = tr\left[\dot{K}_I T^{-1}(K_I - \tilde{K})^T\right] + tr\left[(K_I - \tilde{K})T^{-1}(\dot{K}_I)^T\right]. \quad (3E.7)$$

Substitute $\dot{K}_I$ from (3.334) into (3E.7) to obtain

$$\dot{V}_2 = tr\left[vr^T(K_I - \tilde{K})^T\right] + tr\left[(K_I - \tilde{K})(vr^T)^T\right]. \tag{3E.8}$$

Substitute $K_I = K - K_p = K - vr^T\bar{T}$ into (3E.8) and use the definition

$$z = (K - \tilde{K})r \tag{3E.9}$$

to obtain

$$\dot{V}_2 = tr\left[vz^T\right] + tr\left[zv^T\right] - 2v^T vr^T\bar{T}r \tag{3E.10}$$

Substitute $v = e_{yp} = C_p e_{xp}$ into (3E.10) to obtain

$$\begin{aligned}\dot{V}_2 &= tr\left[C_p e_{xp} z^T\right] + tr\left[z(C_p e_{xp})^T\right] - 2v^T vr^T\bar{T}r \\ &= 2e_{xp}^T C_p^T z - 2v^T vr^T\bar{T}r.\end{aligned} \tag{3E.11}$$

Combine (3E.6) and (3E.11) to obtain

$$\dot{V} = -e_{xp}^T LL^T e_{xp} + 2e_{xp}^T(C_p^T - PB_p)z - 2v^T vr^T\bar{T}r - 2e_{xp}^T PF_1. \tag{3E.12}$$

Finally substitute (3.356) into (3E.12) to obtain

$$\dot{V}(e_{xp}, K_I) = -e_{xp}^T(t)LL^T e_{xp}(t) - 2v^T(t)v(t)r^T(t)\bar{T}r - 2e_x^T PF_1 \tag{3E.13}$$

where

$$F_1(t) = -\dot{x}_p^*(t) + \left[A_p S_{11} + B_p \tilde{K}_x\right] x_m(t) +$$

$$\left[A_p S_{12} + B_p \tilde{K}_u\right] u_m(t) + \left[A_\gamma + B_p \tilde{K}_\gamma\right] \gamma(C_p x_p, u_p, t)$$

$$= -S_{11}\dot{x}_m(t) + \left[A_p S_{11} + B_p \tilde{K}_x\right] x_m(t)$$

$$+ \left[A_p S_{12} + B_p \tilde{K}_u\right] u_m + \left[A_\gamma + B_p \tilde{K}_\gamma\right] \gamma(C_p x_p, u_p, t)$$

$$= -S_{11}\left[A_m x_m(t) + B_m u_m(t)\right] + \left[A_p S_{11} + B_p \tilde{K}_x\right] x_m(t) \tag{3E.14}$$

$$+ \left[A_p S_{12} + B_p \tilde{K}_u\right] u_m(t) + \left[A_\gamma + B_p \tilde{K}_\gamma\right] \gamma(C_p x_p, u_p, t)$$

$$= \left[A_p S_{11} + B_p \tilde{K}_x - S_{11} A_m\right] x_m(t) +$$

$$\left[A_p S_{12} + B_p \tilde{K}_u - S_{11} B_m\right] u_m(t)$$

$$+ \left[A_\gamma + B_p \tilde{K}_\gamma\right] \gamma(C_p x_p, u_p, t)$$

Let $\tilde{K}_\gamma = S_\gamma$, $\tilde{K}_x = S_{21}$, and $\tilde{K}_u = S_{22}$ to obtain

$$\begin{aligned} F_1(t) &= \left[A_p S_{11} + B_p S_{21} - S_{11} A_m\right] x_m(t) \\ &+ \left[A_p S_{12} + B_p S_{22} - S_{11} B_m\right] u_m(t) + \left[A_\gamma + B_p S_\gamma\right] \gamma(C_p x_p, u_p t). \end{aligned} \tag{3E.15}$$

Finally, use (3.344) and rank $[B_p]$ = rank $[B_p | A_\gamma]$ to obtain $F_1(t) = 0$. Then,

$$\dot{V}(e_{xp}, K_I) = -e_{xp}^T(t) L L^T e_{xp}(t) - 2v^T(t)v(t)r^T(t)\bar{T}r(t) \tag{3E.16}$$

which is negative definite in $e_{xp}(t)$ and $v(t)$. Therefore, $e_{xp}(t)$ is asymptotically vanishing, which together with (3.350) implies that the output error $e_{yp}(t)$ is asymptotically vanishing. Furthermore, the adaptive gain $K_I(t)$ is bounded because $V(e_{xp},\ K_I)$ cannot increase beyond its initial value.

Chapter 4

Robust Design Procedures

This chapter discusses a simple modification to be made to the integral adjustment gain $K_I(t)$ in order to account for the effects of disturbances and noise that lead to a persistent nonzero error. These modifications are presented and illustrated in Sections 4.2.4.3, and 4.4 respectively for the basic algorithm, the algorithm with reference model feedforward, and for the algorithms based upon supplementary dynamics.

4.1 Introduction

The techniques developed in Chapter 3 have shown how vague knowledge about basic stabilizability properties can be used in order to design adaptive controllers without requiring precise knowledge of the plant parameters. For tutorial clarity, the controlled plants were assumed to function in an ideal environment, with no input or output disturbances. It is easy to see that these adaptive algorithms must be adjusted in order to maintain robustness under realistic operational conditions.

To illustrate, consider a SISO plant. The integral component of the error gain is from (2.269):

$$K_{Ie}(t) = (T_e) \int_0^t e^2(\tau) d\tau + K_{Ie}(0). \tag{4.1}$$

It is useful to recall that in ideal conditions, the integral gain increases as long as there is an error. When it reaches some value, it has a stabilizing effect on the plant, the error starts decreasing and, under the assumptions of the previous chapter, the error ultimately vanishes and the adaptive gain stops increasing, maintaining some (stabilizing) constant value.

In realistic environments (i.e., in the presence of complex commands or noise and disturbances) the error can never be zero. Even if it is acceptably small, the adaptive gain is continually increasing. Furthermore, every sudden command change implies higher and higher values for the integral adaptive gain. Although ASPR configurations are high gain stable and theoretically maintain stability even for infinite gains, the adaptive gains become very large even when not needed, and in extreme cases, they may lead to divergence of the adaptive control system, and/or digital instabilities.

To compensate for disturbances this section adopts the leaky integrator concept of Ioannou and Kokotovic [44] that makes the simple adaptive algorithm very powerful and robust in very difficult realistic environments, without having to give up on the simplicity of implementation [49], [102], [69], [20].

The proof of stability is given in this chapter to show that the adaptive control algorithm maintains robust control of large systems in realistic operational environments. It should be mentioned that, in order to guarantee robust stability, perfect tracking is not obtained in general, but the adaptive controller maintains a small tracking error over large ranges of non–ideal conditions and uncertainties.

4.2 Robust Redesign of the Basic Adaptive Algorithm

4.2.1 Algorithm Description

In realistic environments, the multivariable controlled plant is represented by

$$\dot{x}_p(t) = A_p x_p(t) + B_p u_p(t) + d_i'(t) \tag{4.2}$$

$$y_p(t) = C_p x_p(t) + d_0(t) \tag{4.3}$$

where $x_p(t)$ is the large plant n–state vector, $u_p(t)$ is the m–input control vector, $y_p(t)$ is the disturbed plant m–output vector, and the terms $d_i'(t)$ and $d_0(t)$ represent bounded, unknown, and unmeasurable plant and output disturbances respectively.

As in Chapter 2, the plant output $y_p(t)$ is required to follow the output of the reference model:

$$\dot{x}_m(t) = A_m x_m(t) + B_m u_m(t) \tag{4.4}$$

$$y_m(t) = C_m x_m(t) \tag{4.5}$$

where $x_m(t)$ is the n_m–state vector, $u_m(t)$ is the m–input command vector, and $y_m(t)$ is the m–output vector of the model. Again, the reference model is only used to represent the desired input–output behavior of the plant, but it is free otherwise, and in general

$$n \;=\; \dim\left[x_p\right] >> \dim\left[x_m\right] = n_m \tag{4.6}$$

For example, a two–input two–output 100 state plant with unknown or poorly defined parameters may be required to follow the input–output behavior of a well–behaving two–input two–output two–state model. These demanding assumptions are made, because this method can be used when it is not realistic to assume that one can model or identify all parameters of the plant, even if the plant order *were* known.

Note that one can only monitor the disturbed output tracking error

$$e_y(t) \quad = \quad y_m(t) \quad - \quad y_p(t) \tag{4.7}$$

and use it for generating the adaptive control gains. This disturbed error will finally be minimized by the adaptive system. Robustness with disturbances is maintained, but one must differentiate between two cases. In the first case, if $d_o(t)$ represents output disturbance (such as wind gusts affecting the position of the fins that control an airplane), then $y_p(t)$ is indeed the output to be controlled in spite of the disturbance. On the other hand, if $d_o(t)$ represents measurement noise and if it is considerably large, the designer may want to use a filtered [52] version of (2.269) for implementation of the adaptive algorithm.

The simple structure for the controller is maintained as:

$$u_p(t) \quad = \quad K_{e_y}(t)e_y(t) + K_{x_m}(t)x_m(t) + K_{u_m}(t)u_m(t) = K(t)r(t) \tag{4.8}$$

where

$$K(t) \quad = \quad \begin{bmatrix} K_{e_y}(t) & K_{x_m}(t) & K_{u_m}(t) \end{bmatrix} \tag{4.9}$$

$$r^T(t) \quad = \quad \begin{bmatrix} e_y^T(t) & x_m^T(t) & u_m^T(t) \end{bmatrix}. \tag{4.10}$$

The adaptive controller uses all values that can be measured, namely, $e_y(t), x_m(t), u_m(t)$. As in Chapter 3 the adaptive gains, $K(t)$, are calculated as a combination of "proportional" and "integral" terms of the form:

$$K(t) \quad = \quad K_p(t) + K_I(t) \tag{4.11}$$

$$K_p(t) = \begin{bmatrix} e_y(t)e_y^T(t)\bar{T}_{e_y} & e_y(t)x_m^T(t)\bar{T}_{x_m} & e_y(t)u_m^T(t)\bar{T}_{u_m} \end{bmatrix} = e_y(t)r^T(t)\bar{T} \tag{4.12}$$

$$\dot{K}_I(t) \quad = \quad e_y(t)r^T(t)T \quad - \quad \sigma K_I(t) \tag{4.13}$$

where T and $\bar{T}$ are (selected) positive definite scaling matrices, but where the "integral" gain (4.13) differs from the basic adaptive algorithm.

While the integral gain given in (4.13) is used to guarantee convergence, the proportional term $K_p(t)$ adds immediate penalty for large errors and leads the system very quickly towards small tracking errors. The σ–term in (4.13) is influenced by an idea of Ioannou [44] and is introduced in order to avoid divergence of the integral gains in the presence of disturbances. Without the σ–term, $K_I(t)$ is a perfect integrator and may steadily increase whenever *perfect* following ($e_y(t) = 0$) is not possible, and may thus reach unnecessarily large values, or may even diverge. With the σ–term, $K_I(t)$ is obtained from a first order filtering of $e_y(t)r^T(t)T$ and therefore cannot diverge, unless $e_y(t)$ diverges (Appendix 4A). However, in this context, the σ–term does more for the concept of "adaptive" control. It is worth mentioning that there is no "optimal" gain value any more, that the adaptive controller wants to achieve and no external persistent excitation is ever used here. On the contrary, the gains increase only if high gains are needed (if the errors attempt to increase) and decrease if they are not needed any more. They are also allowed to change at any rate without affecting stability, such that the designer can adjust this rate to fit the specific needs of the particular plant.

To establish stability of the modified adaptive algorithm, assume either that the system itself satisfies the positive real constraint, defined in Theorem 2.1, or else suitable feedforward augmentation is incorporated. To illustrate this latter situation, for the plant feedforward augmentation defined by (3.9)–(3.16), the composite dynamics become:

$$\dot{x}(t) = Ax(t) + Bu_p(t) + d_i(t) \tag{4.14}$$

$$y_a(t) = Cx(t) + d_o(t) \tag{4.15}$$

where

$$A = \begin{pmatrix} A_p & 0 \\ 0 & A_s \end{pmatrix} \tag{4.16}$$

$$B = \begin{pmatrix} B_p \\ B_s \end{pmatrix} \tag{4.17}$$

$$C = \begin{pmatrix} C_p & D_p \end{pmatrix} \tag{4.18}$$

$$d_i = \begin{pmatrix} d_i' \\ 0 \end{pmatrix} \tag{4.19}$$

and where the transfer function

$$C(sI - A + B\tilde{K}_e C)^{-1}B \tag{4.20}$$

is positive real for some $\tilde{K}_e$.

It is useful to recall that various "ideal trajectories" were defined for reference, as those trajectories that allow perfect tracking in ideal conditions, without disturbances. Although various ideal trajectories have been defined in Chapter 2 as a function of the various tracking tasks they all satisfy the perfect tracking condition $y_a^*(t) = Cx^*(t) = C_m x_m(t) = y_m(t)$.

The adaptive controller attempts to move the plant towards those "ideal trajectories" that would guarantee perfect tracking.

a) When the input command is a constant, the ideal trajectories (2.19) have the form

$$x^*(t) = S_{11}x_m(t) + S_{12}u_m(t) \tag{4.21}$$

and the ideal control signal is

$$u^*(t) = S_{21}x_m(t) + S_{22}u_m(t). \tag{4.22}$$

For perfect following, the various gains in (4.21)–(4.22) must satisfy the conditions

$$\begin{bmatrix} A & B \\ C & 0 \end{bmatrix} \begin{bmatrix} S_{11} & S_{12} \\ S_{21} & S_{22} \end{bmatrix} = \begin{bmatrix} S_{11}A_m & S_{11}B_m \\ C_m & 0 \end{bmatrix}. \tag{4.23}$$

b) For perfect tracking with general input commands that can be represented as the output of an unknown generator of the form

$$\dot{v}_m(t) = A_v v_m(t) \tag{4.24}$$

$$u_m(t) = C_v v_m(t) \tag{4.25}$$

the ideal trajectories (2.57) have been defined as

$$x^*(t) = X_{11}x_m(t) + X_{12}u_m(t) \tag{4.26}$$

and the ideal control signal as

$$u^*(t) = \tilde{K}_x x_m(t) + \tilde{K}_u u_m(t). \tag{4.27}$$

For perfect following, the various gains in (4.26)–(4.27) must satisfy the conditions

$$\begin{bmatrix} A & B \\ C & 0 \end{bmatrix} \begin{bmatrix} X \\ K \end{bmatrix} = \begin{bmatrix} XA_v \\ C_m E \end{bmatrix} \tag{4.28}$$

$$X = X_{11}E + X_{12} \tag{4.29}$$

$$K = \tilde{K}_x E + \tilde{K}_u C_v \tag{4.30}$$

$$A_m E - E A_v + B_m C_v = 0. \tag{4.31}$$

These equations are satisfied if

$$\dim\,(v_m) \;\leq\; \dim\,(u_m) \;+\; \dim\,(x_m). \tag{4.32}$$

c) In a more general case, the definitions (4.24)–(4.27) are maintained, but the input command is allowed to be very rich, so condition (4.32) is not satisfied. In this case, the controlled plant may not be able to achieve perfect following. Yet, to measure the performance of the system, one defines those special trajectories that would guarantee perfect tracking if the plant could reach and move along them. These bounded "ideal target trajectories" defined in Section 2.2.3 exist generically, because the corresponding equations

$$CX_{11} \;=\; C_m \tag{4.33}$$

$$CX_{12}C_v = 0 \tag{4.34}$$

have solutions for the matrices X_{11} and X_{12}, in general. As shown in Section 2.2.3, perfect tracking would be obtained if the "ideal" gains of the "ideal control" $u_p^*(t)$

$$u_p^*(t) \;=\; \tilde{K}_x x_m(t) \;+\; \tilde{K}_u u_m(t) \tag{4.35}$$

were such that the 'bias' term

$$\begin{aligned} E(t) = (AX_{11} - X_{11}A_m + B\tilde{K}_x)x_m + (AX_{12}C_v + \\ B\tilde{K}_u C_v - X_{11}B_m C_v - X_{12}C_v A_v)v_m \end{aligned} \tag{4.36}$$

vanished. However, since perfect tracking is not necessarily assumed to be obtainable, the ideal gains $\tilde{K}_x$ and $\tilde{K}_u$ (unknown and not needed for implementation) are only assumed to be those gains that minimize some norm of (4.36). Again, $E(t)$ is not assumed to be zero and in general remains as a source for bounded tracking errors.

d) For the most general input commands, when the input commands are not even assumed to be differentiable, the ideal target trajectories are defined as

$$x^*(t) = Xx_m(t) \tag{4.37}$$

and such bounded ideal target trajectories exist if the equation

$$CX = C_m \tag{4.38}$$

has a solution for the matrix X. Perfect tracking would be obtainable if the "ideal" gains of the "ideal control" $u^*(t)$

$$u^*(t) = \tilde{K}_x x_m(t) + \tilde{K}_u u_m(t) \tag{4.39}$$

were such that the 'bias' term

$$E(t) = (AX - XA_m + B\tilde{K}_x)x_m + (B\tilde{K}_u - XB_m)u_m \tag{4.40}$$

vanished. When perfect tracking can not be obtained, the ideal gains $\tilde{K}_x$ and $\tilde{K}_u$ (unknown and not needed for implementation) are only assumed to be those gains that minimize some norm of (4.40). $E(t)$ is not zero, in general, and again is a source for nonzero tracking errors.

Because robust stability rather than perfect tracking is sought when input and output disturbances are present, the most general cases will be treated in detail, and the implications of the other conditions in particular situations will be mentioned when it is appropriate.

Define the state error as the difference between the ideal target state and the actual state:

$$e_x(t) = x^*(t) - x(t). \tag{4.41}$$

Recalling that $y_a^*(t) = y_m(t)$, the output error is defined as:

$$e_y(t) = y_m(t) - y_a(t) = y_a^*(t) - y_a(t) = Ce_x(t) - d_o(t). \tag{4.42}$$

Differentiate $e_x(t)$ in (4.41), and add and subtract $Ax^*(t)$ to get

$$\begin{aligned} \dot{e}_x(t) = \dot{x}^*(t) - \dot{x}(t) = \dot{x}^*(t) - Ax^*(t) + Ax^*(t) - \dot{x}(t) = \\ X\dot{x}_m(t) - Ax^*(t) + Ax^*(t) - \dot{x}(t). \end{aligned} \tag{4.43}$$

In the most general case, after substituting the appropriate expressions (4.37)–(4.40) in (4.43) correspondingly, one obtains

$$\dot{e}_x(t) = XA_m x_m(t) + XB_m u_m(t) - AXx_m(t) + Ax^*(t) - \\ [Ax(t) + Bu_p(t) + d_i(t)]\,. \tag{4.44}$$

Rearrange the terms in (4.44) and add and subtract $B[\tilde{K}_e e_y(t) + \tilde{K}_x x_m(t) + \tilde{K}_u u_m(t)]$ to get

$$\begin{aligned} \dot{e}_x(t) = {} & XA_m x_m(t) + XB_m u_m(t) - AXx_m(t) + A\left[x^*(t) - x(t)\right] \\ & -BK(t)r(t) - d_i(t) \\ + {} & B[\tilde{K}_e e_y(t) + \tilde{K}_x x_m(t) + \tilde{K}_u u_m(t)] \\ - {} & B[\tilde{K}_e e_y(t) + \tilde{K}_x x_m(t) + \tilde{K}_u u_m(t)]. \end{aligned} \tag{4.45}$$

Application of (4.42) and some algebraic manipulations give the following differential equation for the state error

$$\dot{e}_x(t) = A_c e_x(t) - B\left[K(t) - \tilde{K}\right] r(t) - F(t) \tag{4.46}$$

where

$$A_c = A - B\tilde{K}_e C \tag{4.47}$$

$$\tilde{K} = \begin{bmatrix} \tilde{K}_e & \tilde{K}_{x_m} & \tilde{K}_{u_m} \end{bmatrix} \tag{4.48}$$

$$\begin{aligned} F(t) = {} & E(t) - B\tilde{K}_e d_o(t) + d_i(t) \\ = {} & (AX - XA_m + B\tilde{K}_x)x_m(t) - (XB_m - B\tilde{K}_u)u_m(t) - \\ & B\tilde{K}_e d_o(t) + d_i(t). \end{aligned} \tag{4.49}$$

Note that the bounded bias term $F(t)$ does not vanish, in general, even if the disturbances vanish, as it was assumed that $E(t)$ in (4.40) is not necessarily zero.

It can be shown that the equation (4.46) holds also in the particular cases a), b), and c). Only the 'bias' term $F(t)$ changes in the different situations. In cases a) and b) the bias term $F(t)$ becomes

$$F(t) = -B\tilde{K}_e d_o(t) + d_i(t) \tag{4.50}$$

when either conditions (4.23) or conditions (4.28)–(4.31) are satisfied. In case c), the bias term is

$$\begin{aligned} F(t) &= (A_p X_{11} - X_{11} A_m + B_p \tilde{K}_x) x_m + (A_p X_{12} C_v \\ &+ B_p \tilde{K}_u C_v - X_{11} B_m C_v - X_{12} C_v A_v) v_m + B\tilde{K}_e d_o(t) + d_i(t) \end{aligned} \tag{4.51}$$

One can prove now the following theorem of stability for the proposed adaptive control procedure.

Theorem 1: Assume that the controlled plant (4.14)–(4.15) is ASPR and that the input and output disturbances are bounded. Then the adaptive control system is globally stable with respect to boundedness. In other words, all values (states, gains, and errors) involved in the control of the augmented process (4.14)–(4.15) via the adaptive procedure (4.9)–(4.13), are bounded. Furthermore, the output tracking error $e_y(t)$ can be directly controlled and thus reduced via the adaptation coefficient $\bar{T}$.

Proof: The detailed proof of stability is presented in Appendix 4A. As shown in Appendix 4A, one selects the positive definite quadratic Lyapunov function

$$V(e_x, K_I) = e_x^T(t) P e_x(t) + tr\left[(K_I(t) - \tilde{K})T^{-1}(K_I(t) - \tilde{K})^T\right] \tag{4.52}$$

for the proof of stability of the dynamical system described by (4.13) and (4.46). The derivative of the Lyapunov function along the trajectories of the system is

$$\begin{aligned} \dot{V}(e_x, K_I) = \; & -e_x^T(t) Q e_x(t) - 2\sigma tr\left[(K_I(t) - \tilde{K})T^{-1}(K_I(t) - \tilde{K})^T\right] \\ & -2e_y^T(t) e_y(t) e_y^T(t) \bar{T}_{e_y} e_y(t) \\ & -2e_y^T(t) e_y(t)\left[x_m^T(t) \bar{T}_{x_m} x_m(t) + u_m^T(t) \bar{T}_{u_m} u_m(t)\right] \\ & -2\sigma tr\left[(K_I(t) - \tilde{K})T^{-1}\tilde{K}^T\right] \\ & -2e_x^T(t) P F(t) - 2d_o^T(t)(K_I(t) - \tilde{K}) r(t) \\ & -2d_o^T(t) e_y(t) e_y^T(t) \bar{T}_{e_y} e_y(t) \\ & -2d_o^T(t) e_y(t)\left[x_m^T(t) \bar{T}_{x_m} x_m(t) + u_m^T(t) \bar{T}_{u_m} u_m(t)\right]. \end{aligned} \tag{4.53}$$

The detailed analysis in Appendix 4A shows that whenever the errors or the gains increase beyond some bound, the first four negative definite or semidefinite terms in $\dot{V}(e_x, K_I)$ become dominant over the last four not necessarily negative terms, and $\dot{V}(e_x, K_I)$ then becomes negative. Therefore, all values involved in the adaptive control system, states, gains, tracking errors are bounded.

Of particular importance is the term $-2e_y^T(t)e_y(t)e_y^T(t)\bar{T}_{e_y}e_y(t)$ in (4.53). This term is the result of using the proportional adaptive gain $K_p(t)$; it is negative definite quartic in terms of $e_y(t)$, and weighted by the adaptation coefficient $\bar{T}_{e_y}$. Large values of $\bar{T}_{e_y}$ increase the weighting of this negative term in (4.53) whenever the tracking errors are nonzero, thus leading to ultimate tracking with small errors. The coefficients $\bar{T}_{x_m}$ and $\bar{T}_{u_m}$, have a similar effort on the final tracking errors whenever $x_m(t)$ are $u_m(t)$ are not zero.

It is not as easy to define the influence of the integral adaptation coefficient T on the tracking error, although it is remarkable in applications. A qualitative argument may help understanding this effect. Assume that close to the steady state, the integral adaptive gain is "slowly varying" and thus, $\dot{K}_I \cong 0$, and from (4.13) one gets $K_I = e_y r^T T/\sigma$. Thus, the larger T and the smaller σ, the larger $K_I(t)$ for the same error $e_y(t)$. Consequently, the larger is the weighting of the negative quadratic term $-2\sigma tr\left[(K_I - \tilde{K})T^{-1}(K_I - \tilde{K})\right]$ in the balance between negative terms (that lead to convergence) and disturbing factors, thus pushing the tracking errors to smaller values.

It is worth mentioning that by adopting the σ-term in order to prevent divergence in nonideal conditions, the designer gives up perfect tracking even in ideal situations. In those cases, defined as a) and b) above, when perfect tracking is possible, and $F(t) = 0$ in (4.55), steady state is obtained at the balance between the negative definite and the nonnegative definite terms. Even in the absence of any external disturbances, $\dot{V}$ as defined in (4.55) will contain the term, $2\sigma tr\left[K_I - \tilde{K}\right]T^{-1}\tilde{K}^T$, which is not necessarily negative. Steady state is obtained at the balance between this term and the negative definite terms, so that the steady state tracking errors are not necessarily zero. It is clear again, however, that the higher the adaptation coefficients, the smaller the final tracking errors. It is worth remembering that ASPR configurations maintain stability with arbitrarily high gains. Finally, the designer must decide on the trade–off between possibly larger tracking errors or larger control gains.

4.2.2 Illustrative Examples

In order to evaluate the effectiveness of the algorithm defined in the previous section, in an environment with disturbances, the two systems defined in (3.23) and (3.31) were again considered.

These were the stable system:

$$P_1(s) = \frac{y_p(s)}{u_p(s)} = \frac{2}{(s+1)}\frac{229}{(s^2+30s+229)} \tag{4.54}$$

and the unstable system:

$$P_2(s) = \frac{y_p(s)}{u_p(s)} = \frac{200}{(s-1)(s^2+8s+100)}. \tag{4.55}$$

The output of each plant was required to follow the output of the reference model:

$$\frac{y_m(s)}{u_m(s)} = \frac{1}{1+s/3}. \tag{4.56}$$

For each of these systems two different reference commands were considered; namely the step defined as:

$$u_m = 1.0, \; t \geq 0. \tag{4.57}$$

and the modulated square wave defined by:

$$\begin{aligned} u_m &= 1.0 + .5\sin(.5t), \quad 0 \leq t \leq 20. \\ & \quad -1.0 + .5\sin(.5t), \quad t \geq 20. \end{aligned} \tag{4.58}$$

An output disturbance $d(t)$ was added to the systems state x_p so that

$$y_p = x_p + d(t) \tag{4.59}$$

where

$$\begin{aligned} d = 0 \qquad & t \leq 18. \\ .2\dot{d} + d = .25\sin(6t) \quad & t \geq 18. \end{aligned} \tag{4.60}$$

A disturbance

$$d_1(t) = .5\cos(2t) \tag{4.61}$$

was also added to the computed adaptive control of (3.103) giving:

$$u_p = K(t)r + d_1(t). \tag{4.62}$$

With this representation, the disturbance denoted as $d_i(t)$ in (4.14) will in actuality be $Bd_i(t)$ which will of course still be bounded.

The feedforward filter of (3.11) and (3.12) was

$$R_p(s)/U_p(s) = .05/(.2s+1). \tag{4.63}$$

The corresponding weighting matrices (3.19, 3.20) were

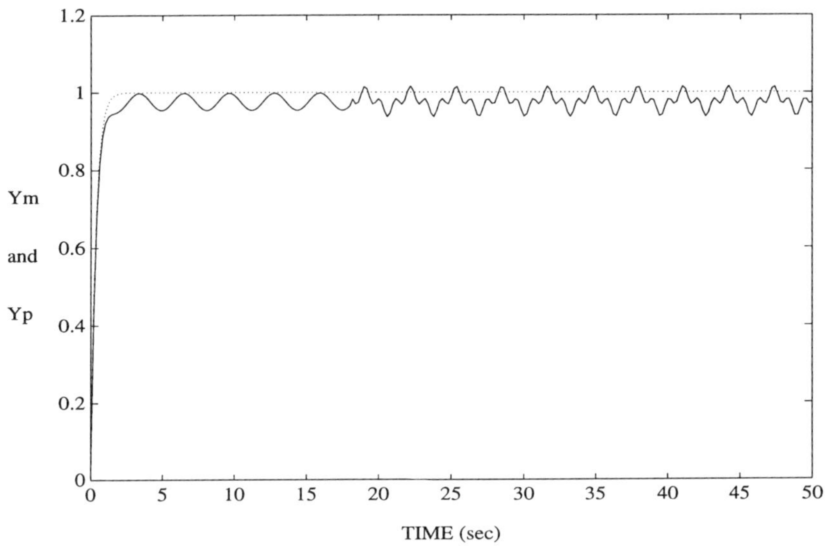

FIGURE 4.1. Feedforward Around the Plant; Stable Example, Outputs $y_p(t)$ and $y_m(t)$)

$$
\begin{aligned}
T &= 10,000 * I_3 \\
\text{and} \qquad \bar{T} &= 1,000 * I_3
\end{aligned}
$$

where I_3 is a 3 x 3 identity matrix, and σ was selected as 0.0 or 0.1.

For the stable plant, the output of the plant $y_p(t)$ and the output of the model $y_m(t)$ are given in Figure 4.1, which shows that the plant follows the model quite well, and that the effect of the disturbances is largely reduced. The bias between $y_m(t)$ and $y_p(t)$ is introduced by the parallel feedforward, and is the price this algorithm pays for robust stability of all states and gains. Of course, as the adaptive controller fights the disturbances in order to maintain the output $y_p(t) = x_p(t) + d(t)$ close to $y_m(t)$, the internal value $x_p(t)$ is strongly affected by the disturbances. It is worth mentioning that this example has been tested with both $\sigma = 0.$ and $\sigma = 0.1$. While the performance is not practically affected, the behavior of the adaptive gains changes considerably. Figure 4.2 shows that the error integral gain K_{I_e} for $\sigma = 0.1.$, both increases and decreases in accordance with the specific tracking situation. On the other hand, as shown in Figure 4.3, when $\sigma = 0.$, this adaptive gain only increases whenever the error is not zero, and may ultimately diverge.

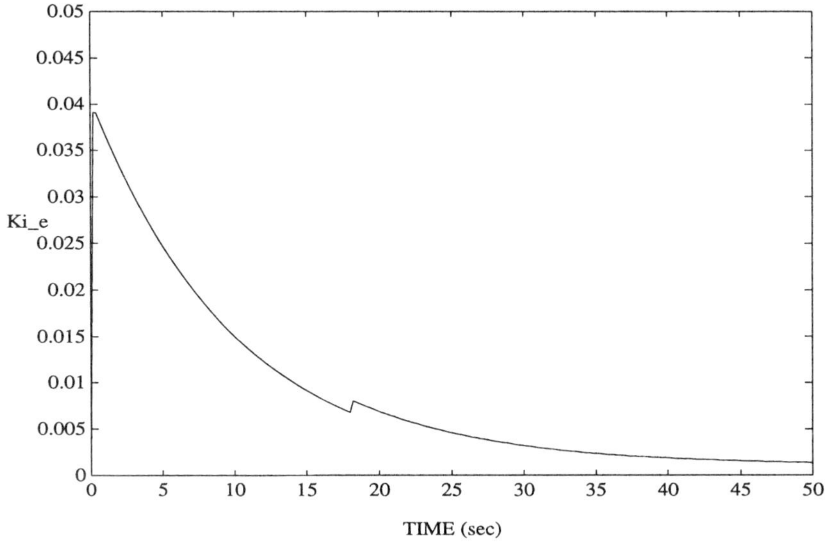

FIGURE 4.2. Feedforward Around the Plant; Stable Example, Adaptive Gain with $\sigma = 0.1$

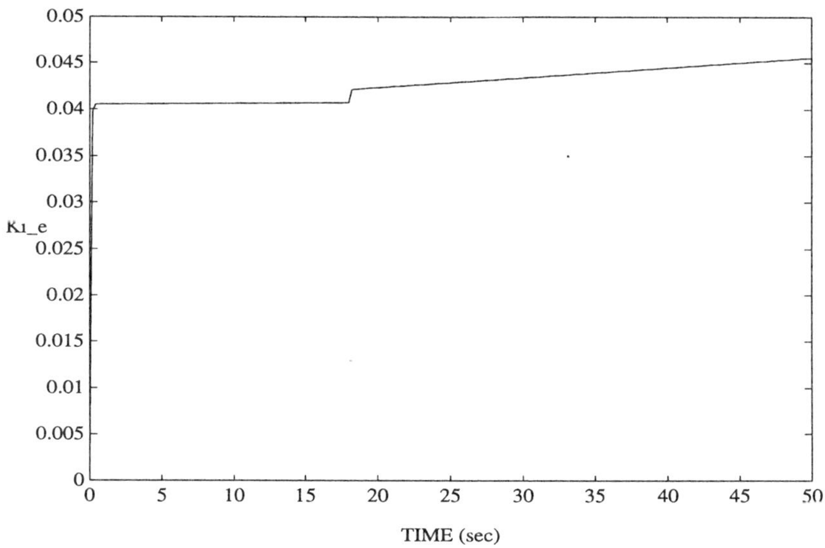

FIGURE 4.3. Feedforward Around the Plant; Stable Example, Adaptive Gain with $\sigma = 0.0$

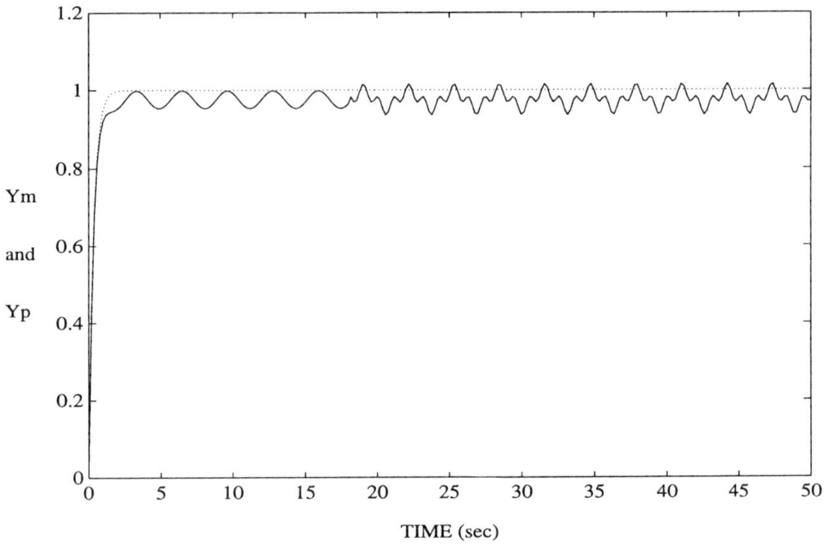

FIGURE 4.4. Feedforward Around the Plant; Unstable Example with Step Input, Outputs $y_p(t)$ and $y_m(t)$

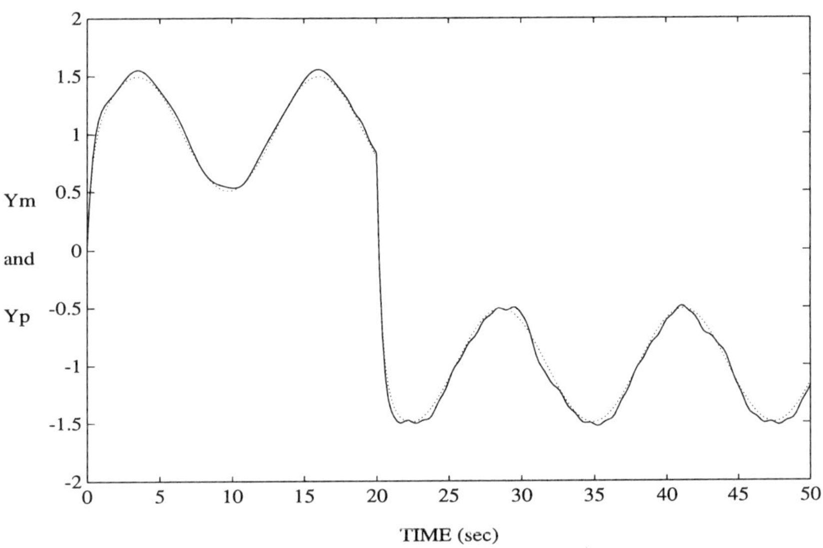

FIGURE 4.5. Feedforward Around the Plant, Unstable Example with Modulated Square–Wave Input, Outputs $y_p(t)$ and $y_m(t)$

To test the applicability of the adaptive control procedure in a more challenging situation, the unstable process of (4.55) was considered.

The reader is encouraged to sketch the Root-locus for this plant for a better understanding of the difficulty that such a plant presents to the designer. Figure 4.4 shows again that the plant tracks the model in spite of the disturbances. The larger bias shows that the parallel feedforward gain that is needed to guarantee the stability for this specific plant, may be too large, as far as good tracking is concerned.

The second example was also run with the more complex modulated square wave input command defined in (4.58). The outputs given in Figure 4.5 again show satisfactory tracking with bounded errors, in spite of the large input and output disturbances.

4.3 Robustness Considerations with Feedforward in the Reference Model

4.3.1 Algorithm Description

When the previous basic algorithm for the disturbed system of (4.2) and (4.3), is modified to include feedforward in both plant and model as discussed in Section 3.3, the statement of Theorem 1 from the previous section will still be valid provided that the integral gain update of (3.122) is again as in (4.13) replaced by

$$\dot{K}_I(t) = (z_m(t) - z_p(t))r^T(t)T - \sigma K_I(t). \tag{4.64}$$

This can be shown in a manner similar to that of the previous section by noting that the augmented system of (3.53) and (3.54) in the presence of the bounded input and output disturbances, $d_i(t)$ and $d_o(t)$ defined in (4.14) and (4.15), becomes

$$\dot{q}_p(t) = F_p q_p(t) + G_p u_p(t) + d_i(t) \tag{4.65}$$

$$z_p(t) = H_p q_p(t) + d_0(t). \tag{4.66}$$

The error equation of (3.101) then becomes:

$$\begin{aligned}\dot{e}(t) = F_p e(t) \quad &+ \quad G_p[S_{21}(t)q_m(t) + S_{22}(t)u_m \\ &- \quad K_p(t)r(t) - K_I(t)r(t)] - d_i(t) = \\ (F_p - G_p\tilde{K}_e H_p)e(t) \quad &+ \quad G_p[\tilde{K}eH_p e(t) + S_{21}(t)q_m(t) + S_{22}(t)u_m\end{aligned}$$

$$- \quad K_p(t)r(t) - K_I(t)r(t)] - d_i(t). \tag{4.67}$$

However since

$$\begin{aligned} e_z(t) &= z_m(t) - z_p(t) = H_p(q_p^*(t) - q_p(t)) - d_0 \\ &= H_p e(t) - d_0 \end{aligned} \tag{4.68}$$

the above error equation (4.67) becomes

$$\begin{aligned} \dot{e}(t) &= (F_p - G_p\tilde{K}eH_p)e(t) + G_p\tilde{K}_e e_z(t) + S_{21}(t)q_m(t) \\ &+ S_{22}(t)u_m - K_p(t)r(t) - K_I(t)r(t)] \\ &- d_i(t) + G_p\tilde{K}_e d_0(t). \end{aligned} \tag{4.69}$$

The resulting stability analysis follows as in Appendix 4B of Chapter 3 if (4.64) and (4.69) are used in place of (3.99) and (3.101) respectively.

The resulting Lyapunov function derivative $\dot{V}(t)$ as shown in (4B.21) or (4B.28) will then contain the additional component

$$\begin{aligned} \dot{V}_s(t) &= -2\sigma tr(s(K_T - \tilde{K})T^{-1}K_I^T S^T \\ &+ 2e^T(t)P(G_p\tilde{K}_e d_o(t) - d_i(t)) - 2d_o^T(S^T S)(K_I - \tilde{K})r \\ &+ 2e^T PG_p d_0(t)r^T\bar{T}r = -2\sigma tr(S(K_I - \tilde{K})T^{-1}(K_I - \tilde{K})^T S^T) \\ &- 2\sigma tr(S(K_I - \tilde{K})T^{-1}(\tilde{K}^T S^T) \\ &+ 2e^T(t)P[G_p\tilde{K}_e d_0(t) - d_i(t)] - 2d_o^T(S^T S)(K_I - \tilde{K})r \\ &+ 2e^T PG_p d_0(t)r^T\bar{T}r. \end{aligned} \tag{4.70}$$

In a manner similar to the approach used in Appendix 4A, boundedness of e and $(K_I - \tilde{K})$ can be shown since the terms involving the (bounded) disturbances will, for large values of e and $(K_I - \tilde{K})$, be dominated by higher order terms.

One possible exception might be a situation in which $\tilde{K}(t)$ increases faster than does $(K_I - \tilde{K})$. Such a possibility, can be avoided by employing the revised integral update (4.64) on only the error gain so that

$$\dot{K}_{I_e}(t) = (z_m(t) - z_p(t))(z_m(t) - z_p(t))^T T - \sigma K_{I_e}(t). \qquad (4.71)$$

In this case the first two terms of $\dot{V}_s$ will be

$$\begin{aligned} &- \quad 2\sigma tr(S(K_{I_e}(t) - \tilde{K}_e)T_e^{-1}(K_{I_e} - \tilde{K}_e)S^T) \\ &- \quad 2\sigma tr(S(K_{I_e}(t) - \tilde{K}_e)\tilde{K}_e)\tilde{K}_e^T S^T). \qquad (4.72) \end{aligned}$$

Since $\tilde{K}_e$ is constant, there will be no problem in having the second term exceed the first in magnitude for large values of $(K_{I_e} - \tilde{K}_e)$.

The net result will as in Appendix 4A be the boundedness of $e_z(t)\dot{e}_z(t)$ and $K_{I_e}(t)$. Of course as expected this in turn will imply only the boundedness of $e_y(t)$.

4.3.2 Illustrative Examples

For the stable plant as defined by (4.54), with a step command, Figures 4.6 and 4.7 show respectively the plant and model outputs for σ (4.64) equal to 0.0 and to 0.1. Corresponding gains are shown in Figures 4.8 and 4.9. Tuning parameter values were

$$T = 1000 * I_3 \qquad (4.73)$$

$$\bar{T} = 500 * I_3, \qquad (4.74)$$

where I_3 denotes a (3 x 3) identity matrix.

The feedforward filter (3.126) was:

$$R_p(s)/U_p(s) = 250/(0.3s + 1) \qquad (4.75)$$

Of interest is the fact that the response is not significantly affected by using the σ factor in the integral gain update formula, while as expected, the integral gain on the error is reduced and bounded.

Similar behavior was also observed for the modulated square wave command as seen in Figures 4.10, 4.11, 4.12, and 4.13.

Although the preceding results showed that the adaptive algorithm of Section 3.3 could control very satisfactorily in the presence of disturbances, the more interesting test is the unstable system defined in (4.55). For this case the tuning parameters were:

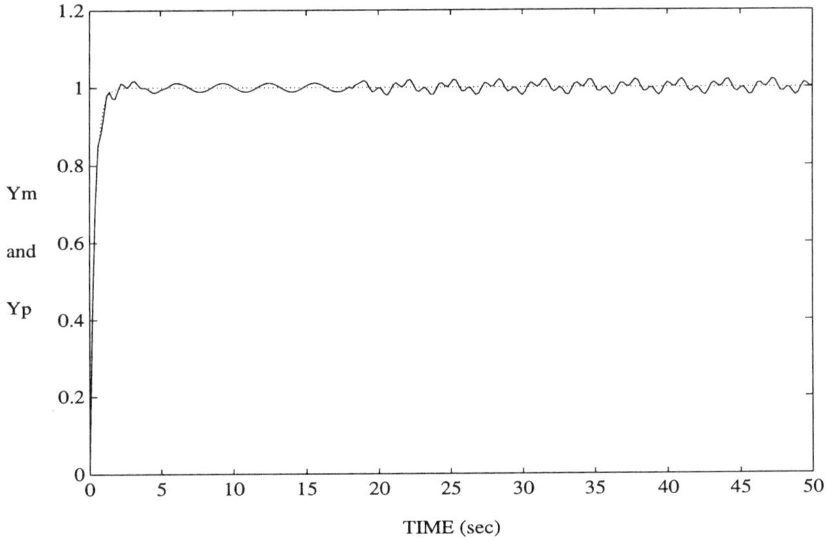

FIGURE 4.6. Plant and Model Feedforward, Stable Example, Plant and Model Outputs, $\sigma = 0.0$

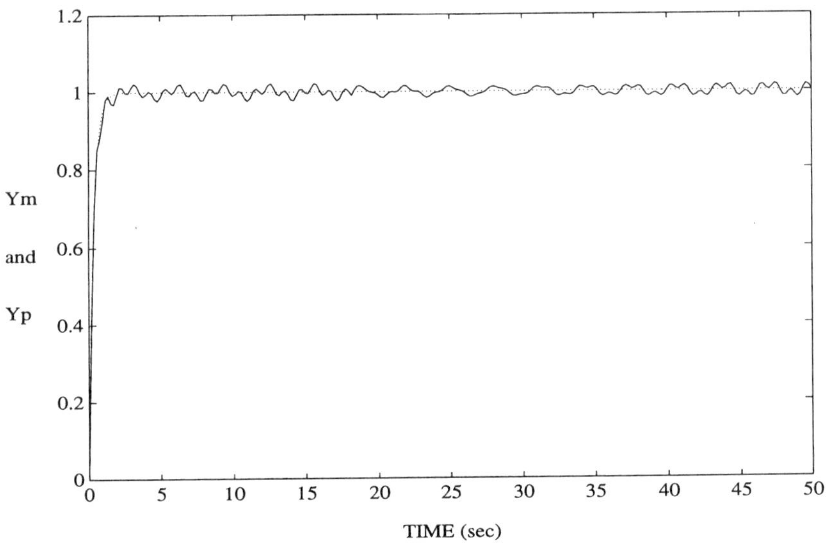

FIGURE 4.7. Plant and Model Feedforward, Stable Example, Plant and Model Outputs, $\sigma = 0.1$

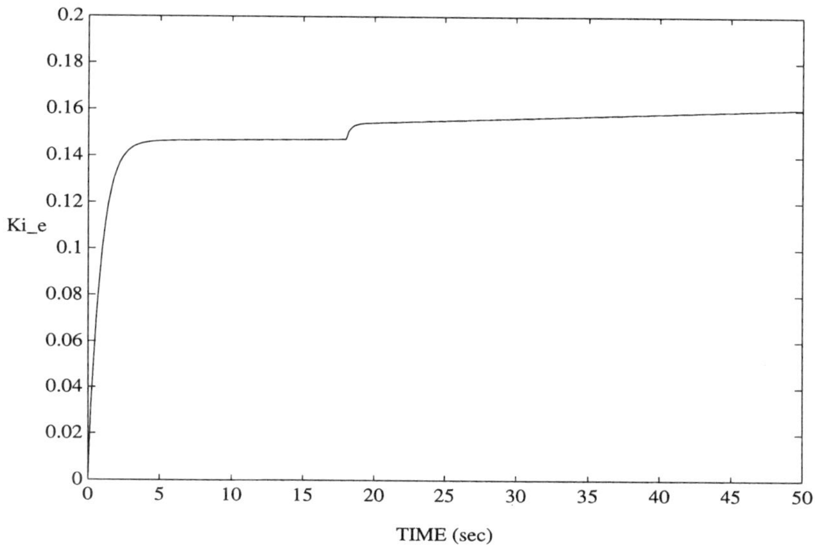

FIGURE 4.8. Plant and Model Feedforward, Stable Example, Control Gains, $\sigma = 0.0$

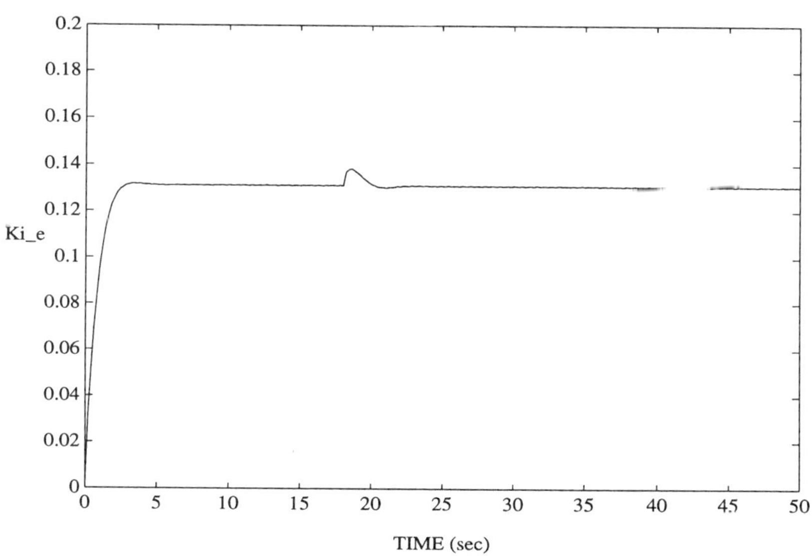

FIGURE 4.9. Plant and Model Feedforward, Stable Example, Control Gains, $\sigma = 0.1$

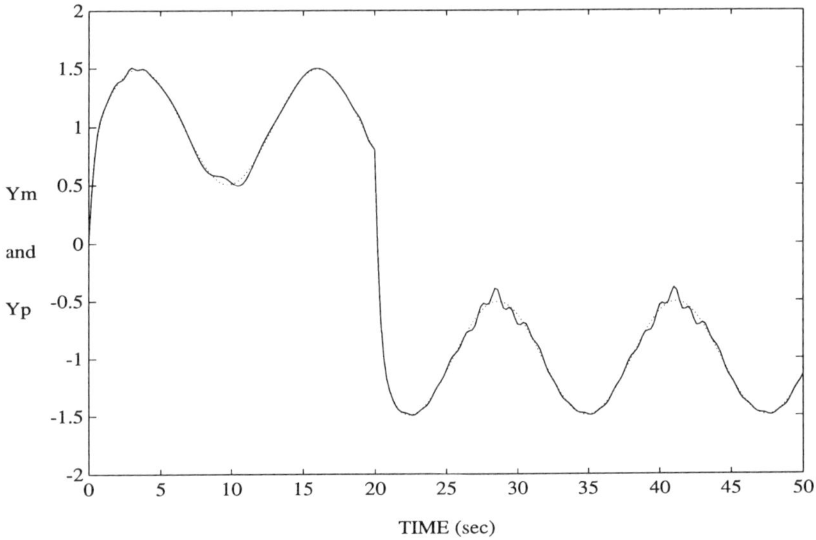

FIGURE 4.10. Plant and Model Feedforward, Stable Example, Modulated Square Wave Input, Plant and Model Outputs, $\sigma = 0.0$

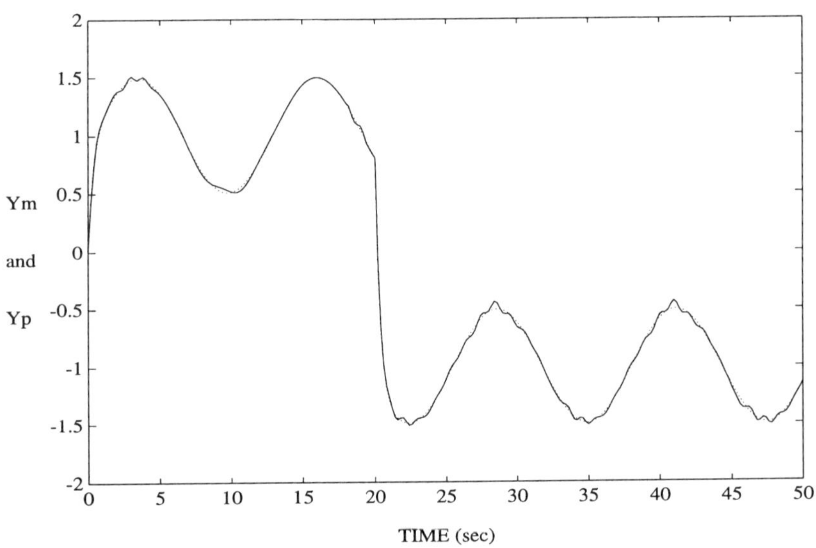

FIGURE 4.11. Plant and Model Feedforward, Stable Example, Modulated Square Wave Input, Plant and Model Outputs, $\sigma = 0.1$

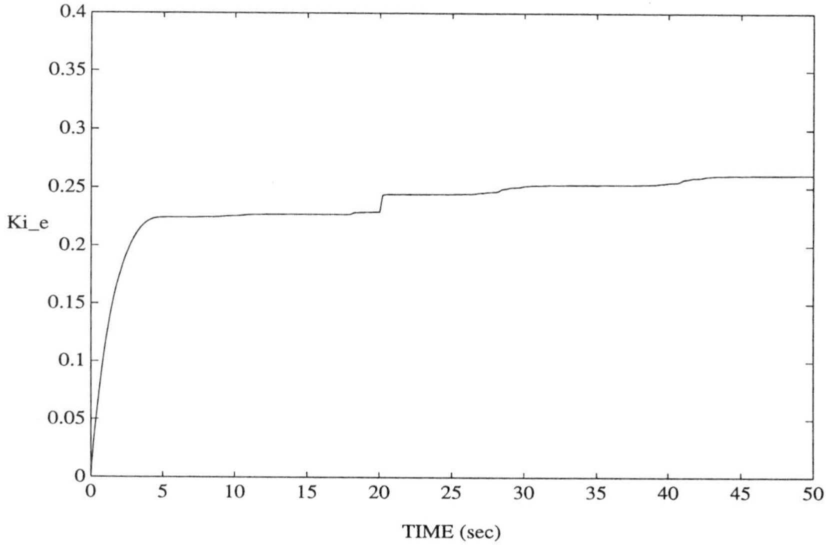

FIGURE 4.12. Plant and Model Feedforward, Stable Example, Modulated Square Wave Input, Control Gains, $\sigma = 0.0$

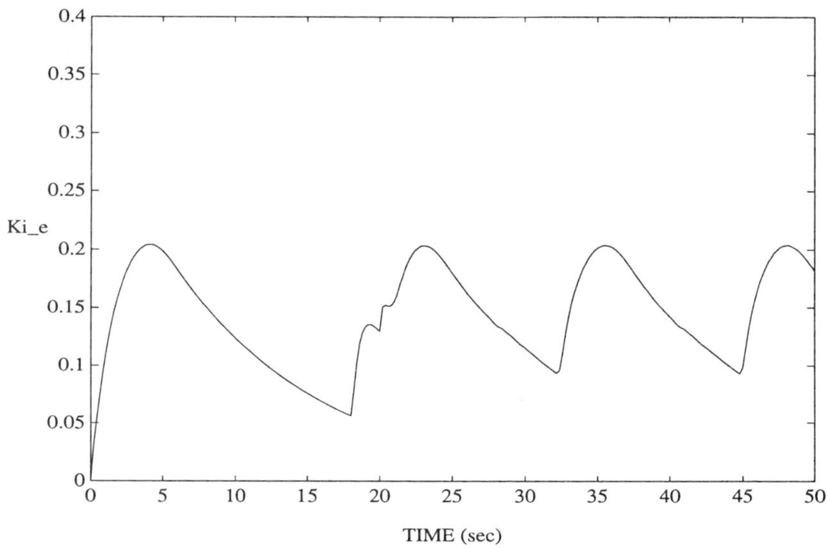

FIGURE 4.13. Plant and Model Feedforward, Stable Example, Modulated Square Wave Input, Control Gains, $\sigma = 0.1$

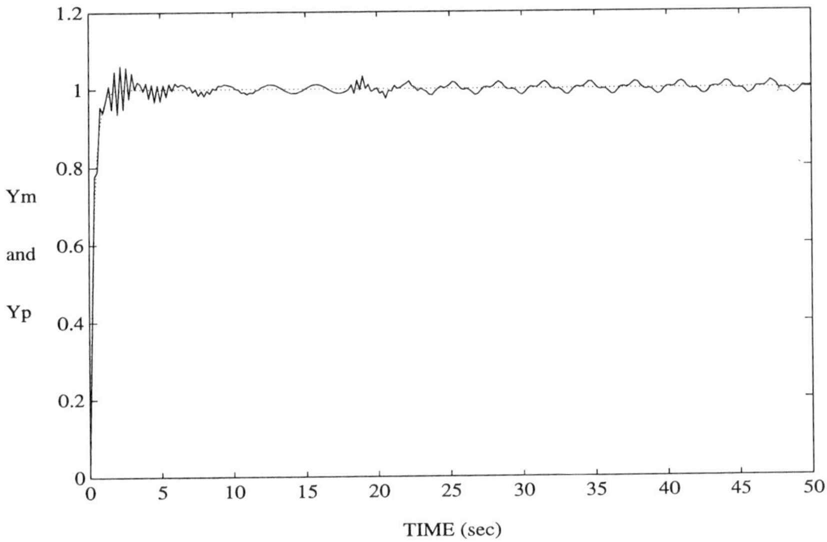

FIGURE 4.14. Plant and Model Feedforward, Unstable Example, Plant and Model Outputs, $\sigma = 0.0$

$$T = 5000 * I_3 \tag{4.76}$$
$$\bar{T} = 5000 * I_3 \tag{4.77}$$

and the feedforward filter (3.126) was:

$$R_p(s)/U_p(s) = 2500/(0.3s + 1). \tag{4.78}$$

Figures 4.14 and 4.15 show respectively the plant and model outputs for σ (4.64) equal to 0.0 and to 0.1. Corresponding gains are shown in Figures 4.16 and 4.17. Of interest is the fact that the model tracking is now somewhat affected by using the σ factor in the integral gain update formula, although as expected, the integral gain on the error is again reduced and bounded.

Model following and gain responses for the modulated square wave command are presented in Figures 4.18, 4.19, 4.20, and 4.21. Although the tracking error differences are not as large as for the step command, it is evident that the use of the σ factor in an unstable system can result in some (possibly insignificant) compromise in the tracking, but again with a gain that will not grow without bound in the presence of persistent error.

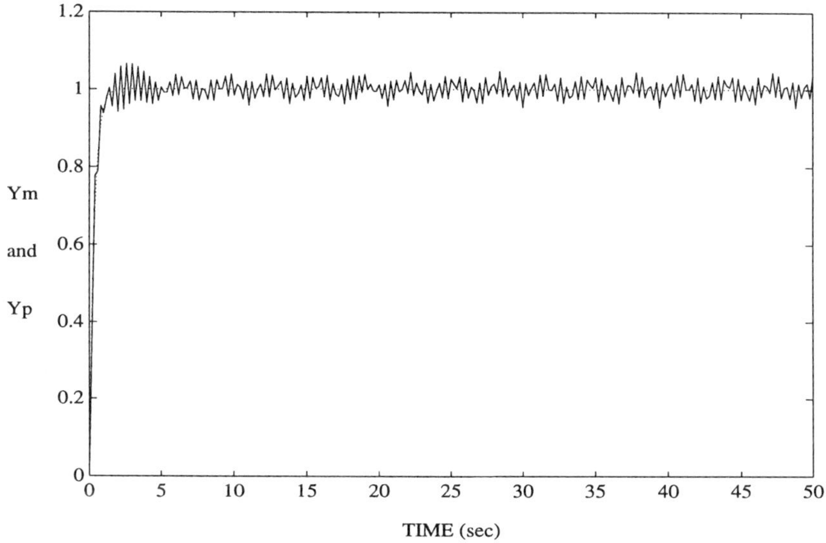

FIGURE 4.15. Plant and Model Feedforward, Unstable Example, Plant and Model Outputs, $\sigma = 0.1$

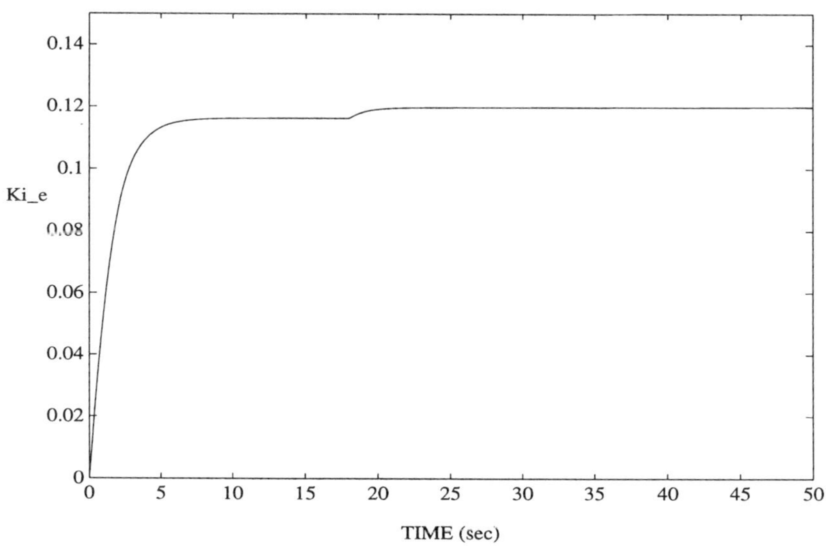

FIGURE 4.16. Plant and Model Feedforward, Unstable Example, Control Gains, $\sigma = 0.0$

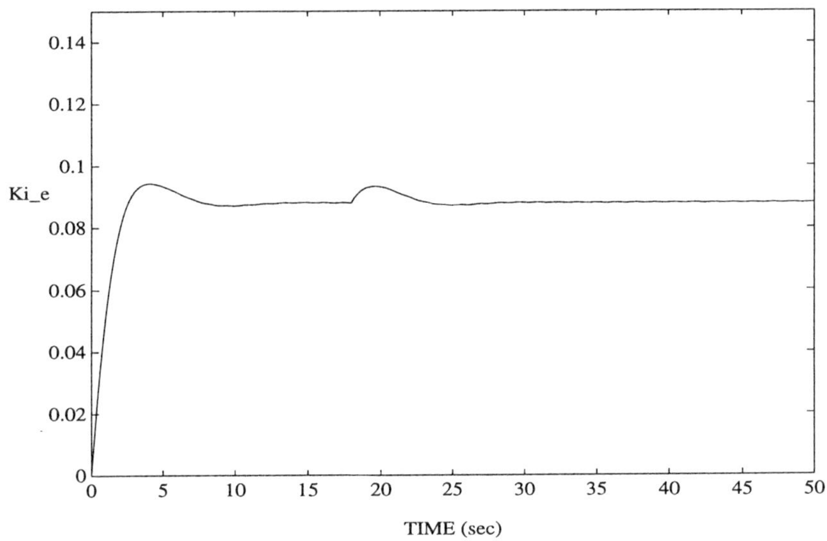

FIGURE 4.17. Plant and Model Feedforward, Unstable Example, Control Gains, $\sigma = 0.1$

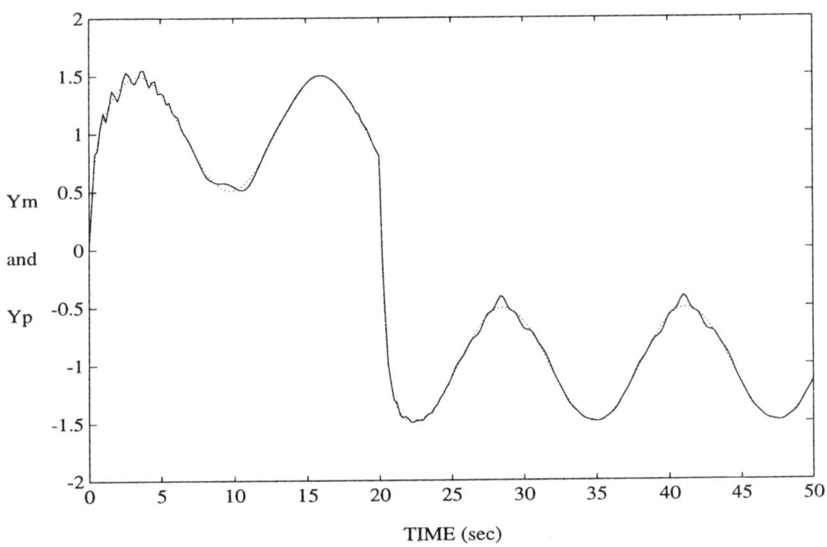

FIGURE 4.18. Plant and Model Feedforward, Unstable Example, Modulated Square Wave Input, Plant and Model Outputs, $\sigma = 0.0$

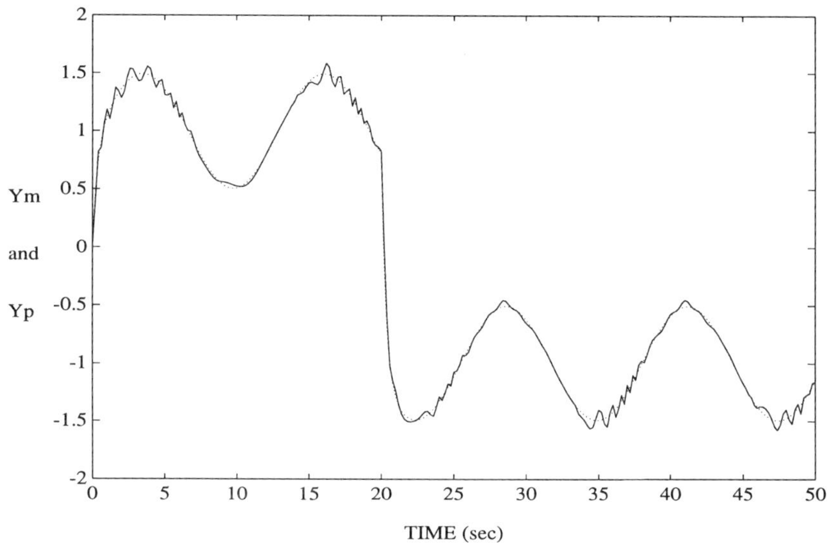

FIGURE 4.19. Plant and Model Feedforward, Unstable Example, Modulated Square Wave Input, Plant and Model Outputs, $\sigma = 0.1$

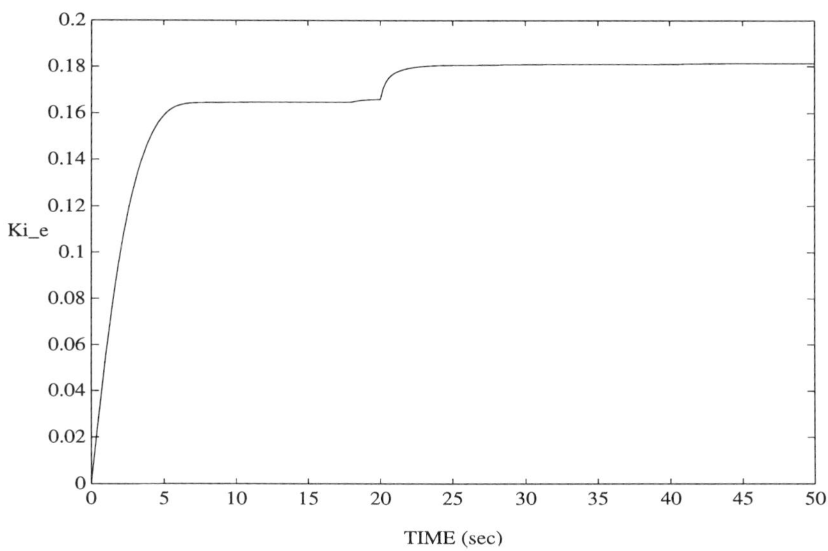

FIGURE 4.20. Plant and Model Feedforward, Unstable Example, Modulated Square Wave Input, Control Gains, $\sigma = 0.0$

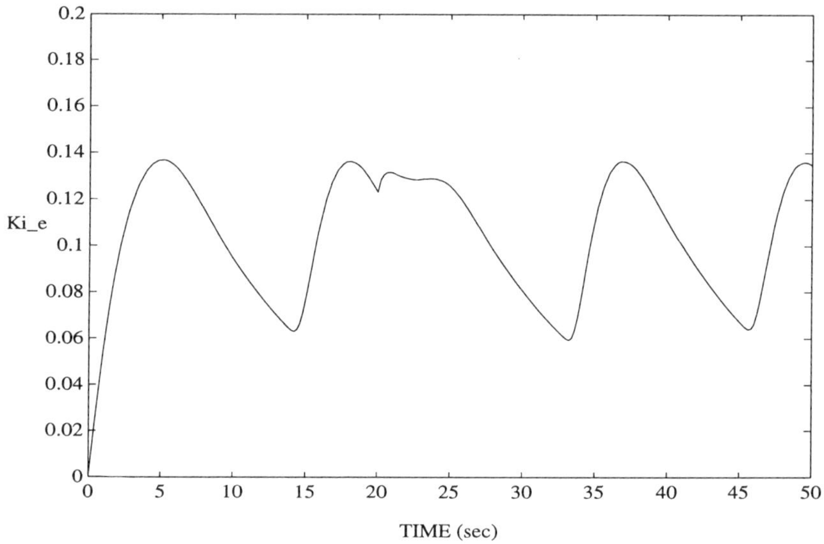

FIGURE 4.21. Plant and Model Feedforward, Unstable Example, Modulated Square Wave Input, Control Gains, $\sigma = 0.1$

4.4 Robust Redesign for Supplementary Dynamics

4.4.1 Algorithm Description

The algorithms of Section 3.4, which utilize supplementary dynamics, are extended to incorporate the leaky integrator concept of Ioannou [44]. A stability proof is presented which shows that the error between the plant and model outputs is bounded in the presence of bounded input and output disturbances.

The plant is described by (4.2)–(4.3) and the reference model is described by (4.4) and (4.5). The supplementary dynamics , which are repeated below for convenience, are described by (3.130) and (3.131).

$$\dot{x}_f(t) = A_f x_f(t) + B_f u_f(t) \tag{4.79}$$

$$y_f(t) = C_f x_f(t). \tag{4.80}$$

As before, the form for $u_f(t)$ will depend upon whether the supplementary dynamics are inserted in a parallel path, a feedback path, or a cascade path with the plant. We shall assume that the target trajectories of (4.37) exist. That is, we assume that

$$\text{rank } [C_p \; C_m] = \text{rank } [C_p]. \tag{4.81}$$

The different expressions for the plant input $u_p(t)$ and the supplementary input $u_f(t)$ are substituted into the plant and supplementary dynamics given by (4.2) and (4.79), respectively. Then, the plant and supplementary dynamics are concatenated to form an $(n + n_f)$ order metasystem.

In Algorithm 1, supplementary dynamics are inserted into the output feedback path such that the control law is of the form

$$u_p(t) = K_e(t)e_{yp}(t) - K_f(t)y_f(t) + K_x(t)x_m(t) + K_u u_m(t) \tag{4.82}$$

$$u_f(t) = y_m(t) - y_p(t) = y_m(t) - C_p x_p(t) - d_0(t). \tag{4.83}$$

The plant and supplementary dynamics are concatenated to form the following metasystem:

$$\begin{aligned}\begin{bmatrix} \dot{x}_p(t) \\ \dot{x}_f(t) \end{bmatrix} &= \begin{bmatrix} A_p & 0 \\ -B_f C_p & A_f \end{bmatrix} \begin{bmatrix} x_p(t) \\ x_f(t) \end{bmatrix} \\ &+ \begin{bmatrix} B_p \\ 0 \end{bmatrix} [K_e(t), K_f(t), K_x(t), K_u(t)] \begin{bmatrix} e_{yp}(t) \\ -y_f(t) \\ x_m(t) \\ u_m(t) \end{bmatrix} \\ &+ \begin{bmatrix} d_i'(t) \\ B_f[y_m(t) - d_0(t)] \end{bmatrix}. \end{aligned} \tag{4.84}$$

Algorithm 2 is characterized by supplementary dynamics which are in parallel with the plant. The control law is of the form:

$$u_p(t) = u_f(t) = K_e(t)e_{yp}(t) - K_f(t)y_f(t) + K_x(t)x_m(t) + K_u u_m(t). \tag{4.85}$$

The plant and supplementary dynamics are concatenated to form the following metasystem:

$$\begin{aligned}\begin{bmatrix} \dot{x}_p(t) \\ \dot{x}_f(t) \end{bmatrix} &= \begin{bmatrix} A_p & 0 \\ 0 & A_f \end{bmatrix} \begin{bmatrix} x_p(t) \\ x_f(t) \end{bmatrix} \\ &+ \begin{bmatrix} B_p \\ B_f \end{bmatrix} [K_e(t), K_f(t), K_x(t), K_u(t)] + \begin{bmatrix} d_i'(t) \\ 0 \end{bmatrix}. \end{aligned} \tag{4.86}$$

Algorithm 3 is characterized by supplementary dynamics which are in cascade with the plant. The control law is of the form:

$$u_p(t) = D_f u_f(t) - y_f(t) = D_f u_f(t) - C_f x_f(t) \tag{4.87}$$

$$u_f(t) = K_e(t) e_{yp}(t) - K_f(t) y_f(t) + K_x(t) x_m(t) + K_u u_m(t). \tag{4.88}$$

The metasystem for Algorithm 3 is of the form:

$$\begin{bmatrix} \dot{x}_p(t) \\ \dot{x}_f(t) \end{bmatrix} = \begin{bmatrix} A_p & -B_p C_f \\ 0 & A_f \end{bmatrix} \begin{bmatrix} x_p(t) \\ x_f(t) \end{bmatrix} + \begin{bmatrix} B_p D_f \\ B_f \end{bmatrix}$$

$$[K_e(t), K_f(t), K_x(t), K_u(t)] \times \begin{bmatrix} e_{yp}(t) \\ -y_f(t) \\ x_m(t) \\ u_m(t) \end{bmatrix} + \begin{bmatrix} d_i'(t) \\ 0 \end{bmatrix}. \tag{4.89}$$

The metasystem output equation is identical for all three algorithms and is given by

$$\begin{bmatrix} y_p(t) \\ y_f(t) \end{bmatrix} = \begin{bmatrix} C_p & 0 \\ 0 & C_f \end{bmatrix} \begin{bmatrix} x_p(t) \\ x_f(t) \end{bmatrix} + \begin{bmatrix} d_0(t) \\ 0 \end{bmatrix}. \tag{4.90}$$

We may write the metasystem equations more compactly as

$$x(t) = Ax(t) + BK(t)r(t) + \tilde{d}_i(t) \tag{4.91}$$

$$y(t) = Cx(t) + \tilde{d}_0(t). \tag{4.92}$$

The above description of the metasystem is to be compared with the disturbance free case given by (3.143)–(3.144). The definitions for $x(t)$, $y(t)$, A, B, and C are identical to the disturbance free case given by (3.145)–(3.148). However, it is important to note the new definitions for $\tilde{d}_i(t)$ and $\tilde{d}_0(t)$ which are given as follows:
For Algorithm 1:

$$\tilde{d}_i(t) = \begin{bmatrix} d_i'(t) \\ B_f[y_m(t) - d_0(t)] \end{bmatrix} \qquad \tilde{d}_o(t) = \begin{bmatrix} d_0(t) \\ 0 \end{bmatrix} \tag{4.93}$$

for Algorithm 2:

$$d_i(t) = \begin{bmatrix} d_i'(t) \\ 0 \end{bmatrix} \qquad \tilde{d}_0(t) = \begin{bmatrix} d_0(t) \\ 0 \end{bmatrix} \tag{4.94}$$

for Algorithm 3:

$$\tilde{d}_i(t) = \begin{bmatrix} d_i'(t) \\ 0 \end{bmatrix} \qquad \tilde{d}_0(t) = \begin{bmatrix} d_0(t) \\ 0 \end{bmatrix}. \tag{4.95}$$

The adaptive controller is described by (3.149)–(3.158) with the exception of the integral gain equation which is modified to include Ioannou's [44] leaky integrator concept. The integral gain adaptation in the presence of disturbances is given by:

$$\dot{K}_I(t) = v(t)r^T(t)T - \sigma K_I(t). \tag{4.96}$$

4.4.2 Error System Equations

The plant state error is defined as

$$e_{xp}(t) = x_p^*(t) - x_p(t) \tag{4.97}$$

and the plant output error as

$$e_{yp}(t) = y_m(t) - y_p(t) = y_p^*(t) - y_p(t) = C_p e_{xp}(t) - d_0(t). \tag{4.98}$$

The supplementary dynamics state error is defined as

$$e_{xf}(t) = x_f^*(t) - x_f(t). \tag{4.99}$$

The metastate error is of the form

$$e_x(t) = \begin{bmatrix} e_{xp}(t) \\ e_{xf}(t) \end{bmatrix}. \tag{4.100}$$

Then, the error derivative will be

$$\dot{e}_x(t) = x^*(t) - \dot{x}(t) = x^*(t) - Ax(t) - BK(t)r(t) - \tilde{d}_i(t). \tag{4.101}$$

Furthermore, suppose there exists a constant matrix K given by

$$\tilde{K} = [\tilde{K}_a, \tilde{K}_x, \tilde{K}_u] \tag{4.102}$$

such that

$$A_c = A - B\tilde{K}_a C \tag{4.103}$$

is a stability matrix.

Next, use (4.101) and (4.102) to obtain

$$\dot{e}_x(t) = \dot{x}^*(t) - A[x^*(t) - e_x(t)] - B[(K(t) - \tilde{K})r(t) + \tilde{K}r(t)] \\ -\tilde{d}_i(t). \tag{4.104}$$

Define $z(t) = [K(t) - \tilde{K}]r(t)$. Then, (4.104) becomes

$$\dot{e}_x(t) = \dot{x}^*(t) - Ax^*(t) + Ae_x(t) - Bz(t) - B\tilde{K}r(t) - \tilde{d}_i(t). \tag{4.105}$$

Substitute $\tilde{K}r(t) = \tilde{K}_a e_{yv}(t) + \tilde{K}_x x_m(t) + \tilde{K}_u u_m(t)$ into (4.105) to obtain

$$\dot{e}_x(t) = \dot{x}^*(t) - Ax^*(t) + Ae_x(t) - Bz(t) - B[\tilde{K}_e e_{yv}(t) + \tilde{K}_x x_m(t) \\ +\tilde{K}_u u_m(t)] - \tilde{d}_i(t). \tag{4.106}$$

Define

$$e_v(t) = \begin{bmatrix} e_{xp}(t) \\ -x_f(t) \end{bmatrix} \tag{4.107}$$

and

$$x_0^*(t) = \begin{bmatrix} 0 \\ x_f^*(t) \end{bmatrix} \tag{4.108}$$

such that

$$e_{yv}(t) = Ce_v(t) - \tilde{d}_0 \tag{4.109}$$

and

$$e_v(t) = e_x(t) - x_0^*(t). \tag{4.110}$$

Substitute (4.103), (4.109) and (4.110) into (4.106) to obtain

$$\dot{e}_x(t) = A_c e_x(t) - Bz(t) - F_1(t) \tag{4.111}$$

where

$$F_1(t) = -\dot{x}^*(t) + Ax^*(t) - B\tilde{K}_a C x_0^*(t) \\ +B[\tilde{K}_x x_m(t) + \tilde{K}_u u_m(t) - \tilde{K}_a \tilde{d}_0(t)] + \tilde{d}_i(t). \tag{4.112}$$

4.4.3 Stability Analysis

Theorem 2: Suppose that the adaptive control algorithm is given by (3.149)–(3.158) with the integral gain given by (4.96). Further, suppose that there exists a real symmetric positive definite matrix P and real matrices L, W, K, and $R, (R + R^T) > 0$ such that

$$P(A - B\tilde{K}_a C) + (A - B\tilde{K}_a C)^T P = -LL^T - R < 0 \tag{4.113}$$

$$PB = C^T(Q^T + \tilde{K}_a^T G^T) - LW \tag{4.114}$$

$$W^T W = J + J^T \tag{4.115}$$

$$J + J^T + G + G^T < 0 \tag{4.116}$$

where the matrices T and $\bar{T}$ are positive definite symmetric and positive semi-definite symmetric, respectively.

Then, all states and errors in the adaptive system are bounded.
Proof: See Appendix 4C.

4.4.4 Illustrative Examples

In order to evaluate the effectiveness of the algorithms described in Section 4.2.1 in an environment with disturbances, the two systems defined in (3.285) and (3.315) are revisited. For each of these systems, two different reference commands are considered. The first command is a step defined by

$$u_m = 1.0; \; t \geq 0. \tag{4.117}$$

The second command is a modulated square wave which is defined by

$$u_m = \left\{ \begin{array}{ll} 1.0 + 0.5\sin(0.5t); & 0 \leq t < 20 \\ -1.0 + 0.5\sin(0.5); & t \geq 20 \end{array} \right. . \tag{4.118}$$

An output disturbance $d_0(t)$ was added to the plant output so that

$$y_p(t) = C_p x_p(t) + d_0(t) \tag{4.119}$$

where

$$d_0(t) = \left\{ \begin{array}{ll} 0; & 0 \leq t < 18 \\ 0.2\dot{d}_0 + d_0 = 0.25\sin(6t); & t \geq 18 \end{array} \right. . \tag{4.120}$$

An input disturbance $d_i'(t)$ is defined by

$$d_i'(t) = 0.5\cos(2t) \tag{4.121}$$

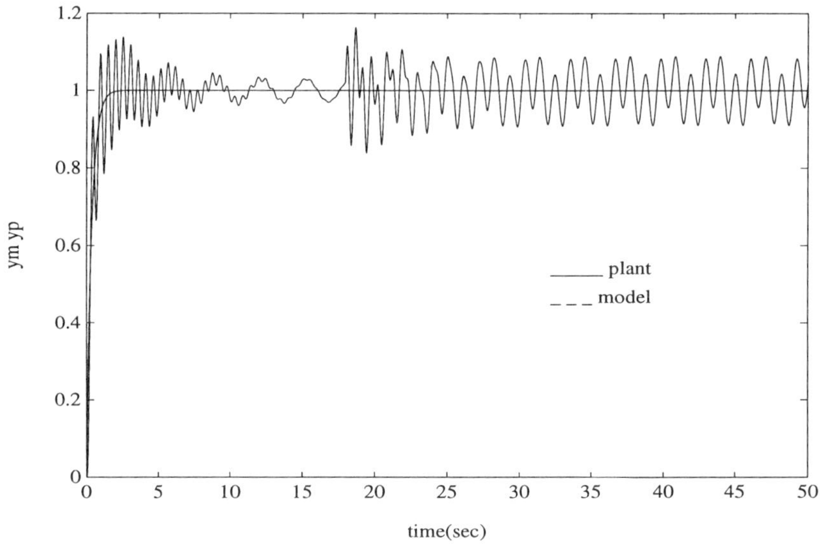

FIGURE 4.22. Stable Example: Plant and Model Outputs, $\sigma = 0$

is added to the adaptive control signal $u_p(t)$ of (4.82), (4.85), and (4.87) for algorithms 1, 2, and 3, respectively.

As in Section 3.4.3, the reference model is defined by

$$y_m(s)/u_m(s) = 1/(s/3 + 1). \tag{4.122}$$

For the stable plant as defined by (3.285), with a step command, Figures 4.22 and 4.27 show the plant and model outputs for sigma in (4.96) equal to 0.0 and 1.0, respectively. The corresponding integral gains K_{eI}, K_{fI}, K_{xI}, and K_{uI} are shown in Figures 4.23, 4.24, 4.25, and 4.26 respectively for sigma equal to 0.0 and in Figures 4.28, 4.29, 4.31, and 4.30, respectively for sigma equal to 1.0. The stabilizing compensator is given by (3.287), Algorithm 2 is used, and T and $\bar{T}$ are identity matrices. Of interest is the observation that after an initial transient, the average value of the output error is zero. Furthermore, the sigma modification has little effect on the response. The integral gains K_{eI} and K_{fI} diverge when sigma is equal to zero, but are clearly bounded when sigma is equal to 1.0. For sigma equal to 1.0, the gain K_{eI} is large at $t = 0$, then becomes small, increases again at $t = 18$ when the output disturbance is applied, and then decreases in value as the average value of the error becomes zcro.

Similar behavior of the output error is observed for the modulated square wave command. Figure 4.32 indicates that after an initial transient, the average error is again zero.

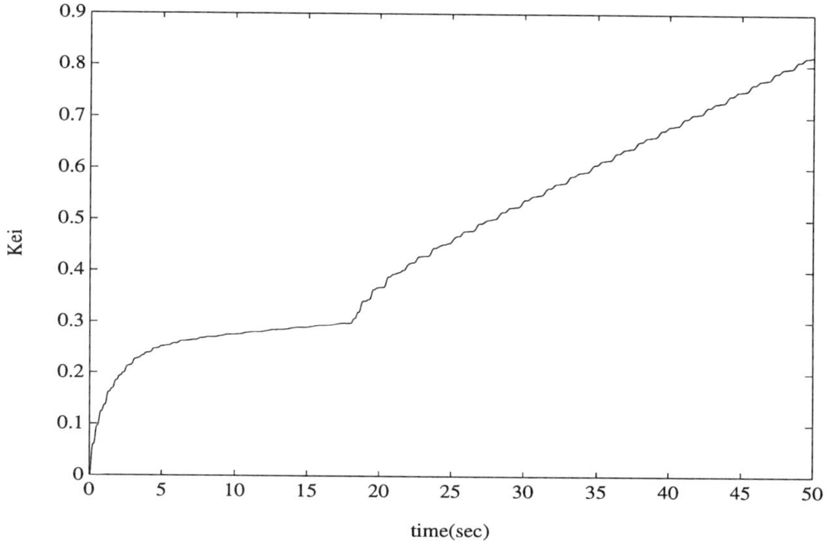

FIGURE 4.23. Stable Example: Integral Adaptive Gain $K_{Ie}, \sigma = 0$

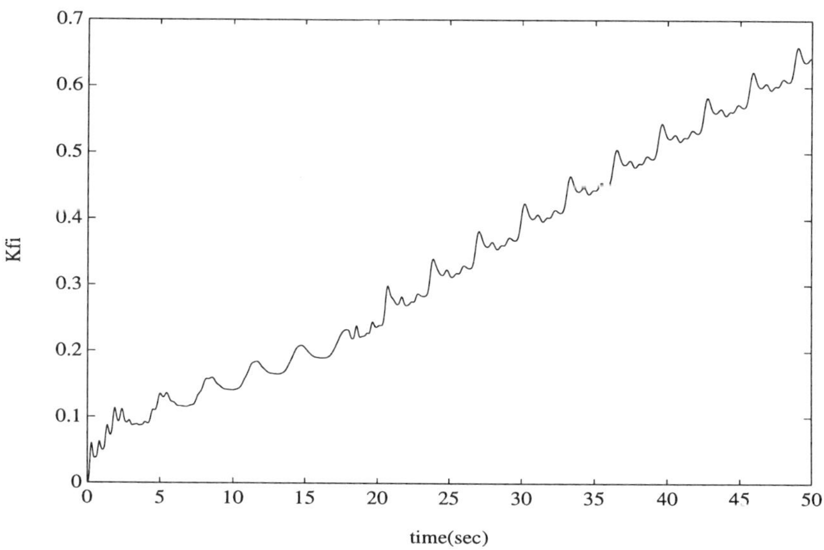

FIGURE 4.24. Stable Example: Integral Adaptive Gain $K_{If}, \sigma = 0$

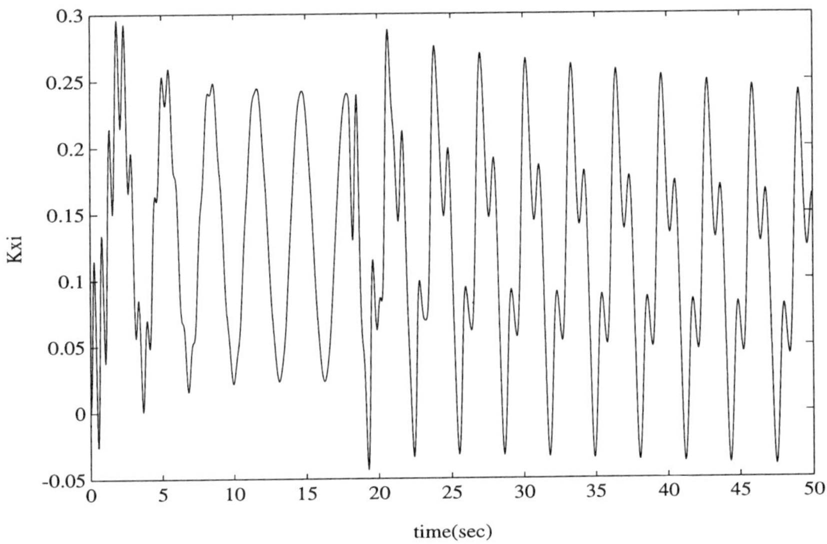

FIGURE 4.25. Stable Example: Integral Adaptive Gain $K_{Ix}, \sigma = 0$

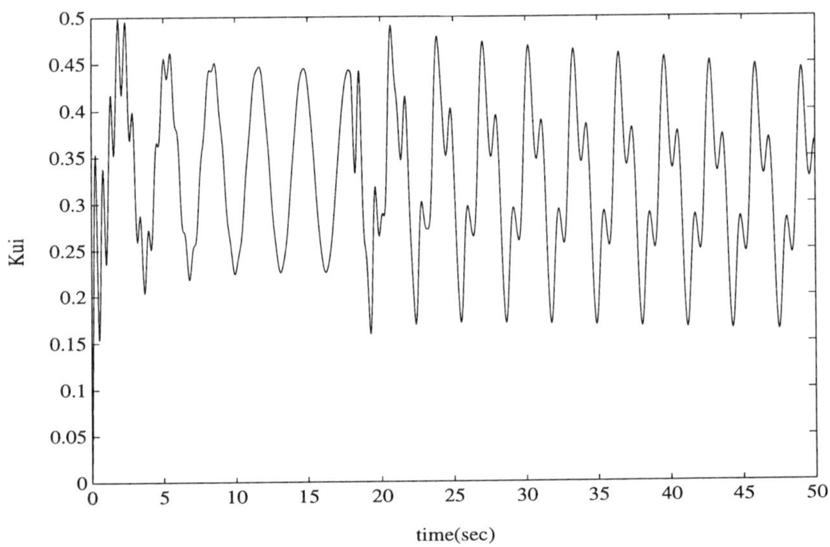

FIGURE 4.26. Stable Example: Integral Adaptive Gain $K_{Iu}, \sigma = 0$

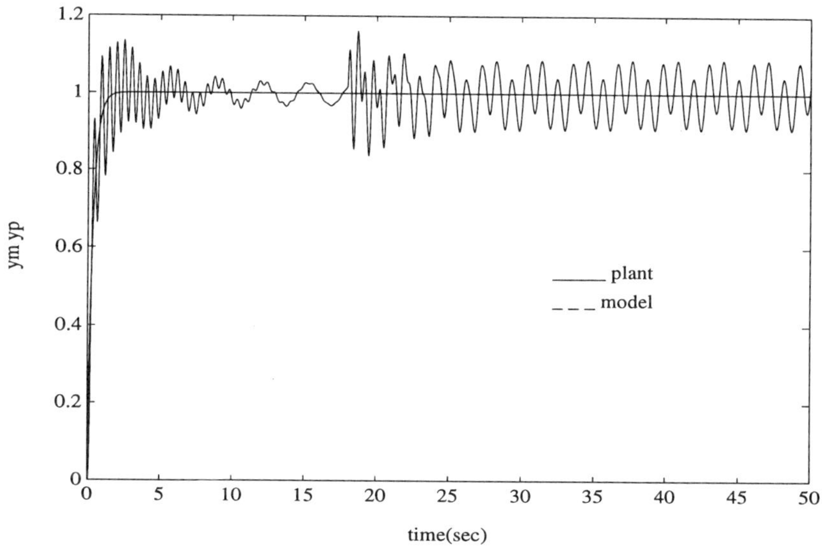

FIGURE 4.27. Stable Example: Plant and Model Outputs, $\sigma = 1$

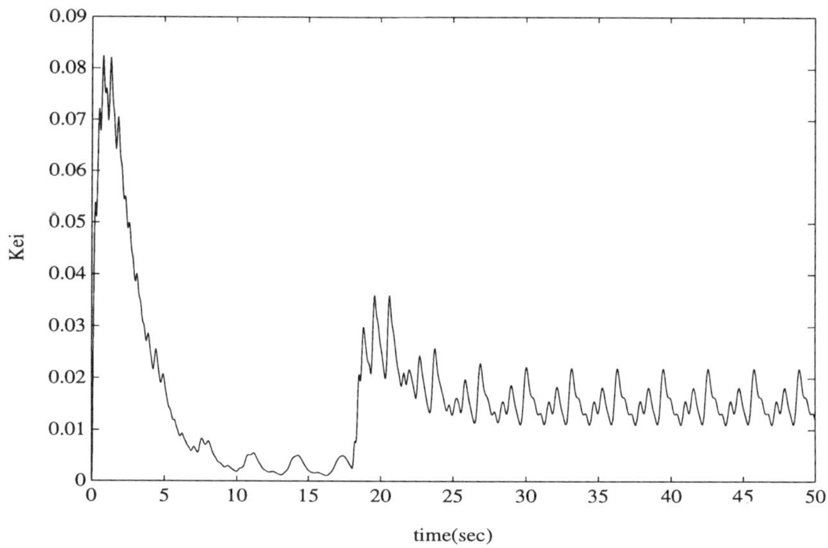

FIGURE 4.28. Stable Example: Integral Adaptive Gain $K_{Ie}, \sigma = 1$

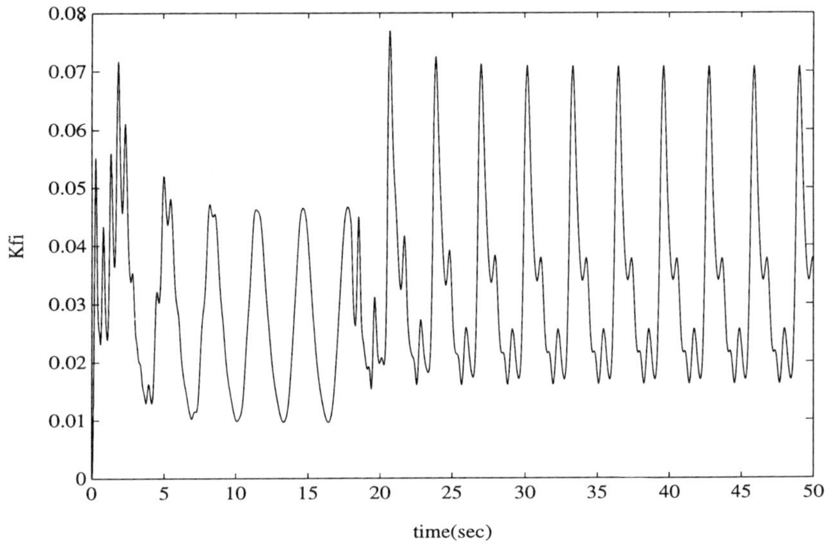

FIGURE 4.29. Stable Example: Integral Adaptive Gain $K_{If}, \sigma = 1$

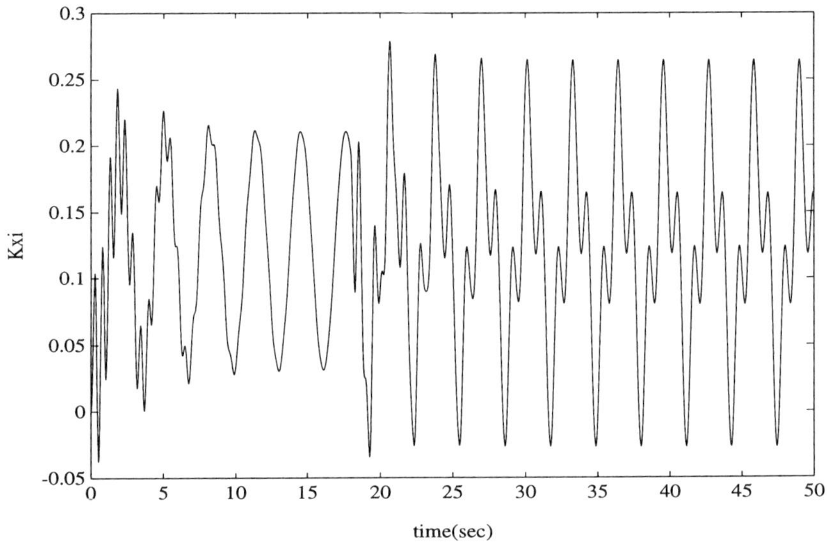

FIGURE 4.30. Stable Example: Integral Adaptive Gain $K_{Ix}, \sigma = 1$

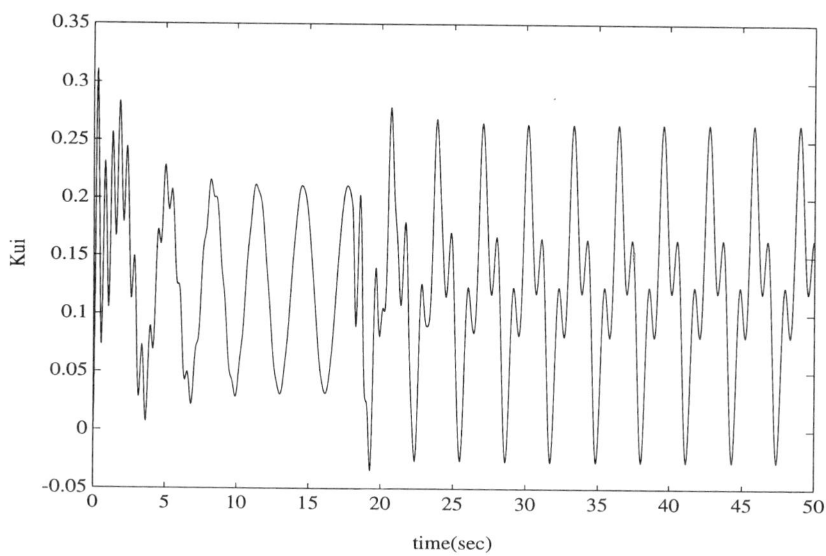

FIGURE 4.31. Stable Example: Integral Adaptive Gain $K_{Iu}, \sigma = 1$

The plant and model outputs for a step command are shown in Figures 4.33 and 4.34 for algorithms 1 and 3, respectively. Upon comparing these responses with Figure 4.27, it is observed that Algorithm 2 yields the best response.

Finally, Algorithm 2 is repeated with the stabilizing compensator defined by

$$G_c(s) = \frac{-(10s + 3.5)}{s}. \tag{4.123}$$

The plant and model outputs for a step command and a modulated square wave command are shown in Figures 4.35 and 4.36, respectively. Upon comparing these responses with Figures 4.27 and 4.32, respectively, it is observed that the compensator given by (4.123) yields a significantly improved response relative to the compensator given by (3.287).

For the unstable plant as defined by (3.315), with a step command, Figures 4.37 and 4.40 show the plant and model outputs for sigma in (4.96) equal to 0.0 and 1.0, respectively. The corresponding integral gains K_{eI}, K_{fI}, K_{xI}, and K_{uI} are shown in Figures 4.42, 4.38, 4.39, and 4.41, respectively for sigma equal to 0.0 and in Figures 4.43, 4.44, 4.46, and 4.45, respectively for sigma equal to 1.0. The stabilizing compensator is given by (3.316), Algorithm 2 is used, and T and $\bar{T}$ are identity matrices. Of interest is the observation that after an initial transient, the average value

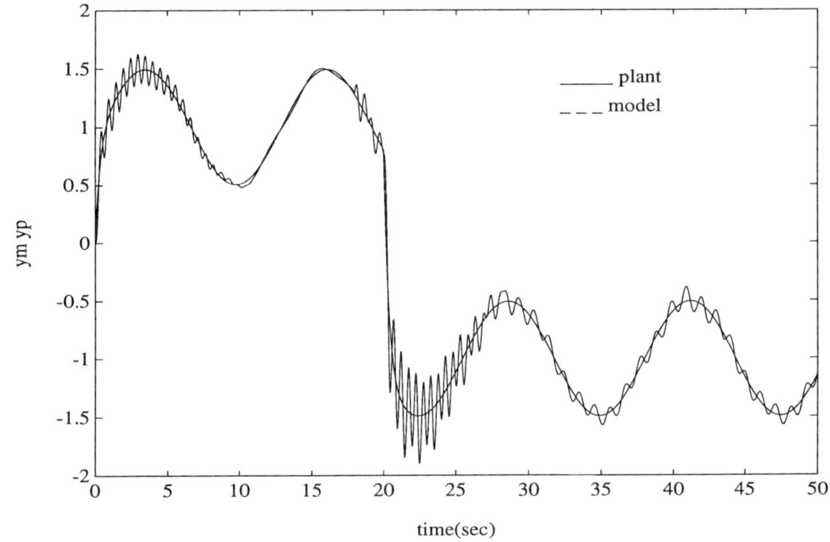

FIGURE 4.32. Stable Example: Plant and Model Outputs, Modulated Square Wave Command

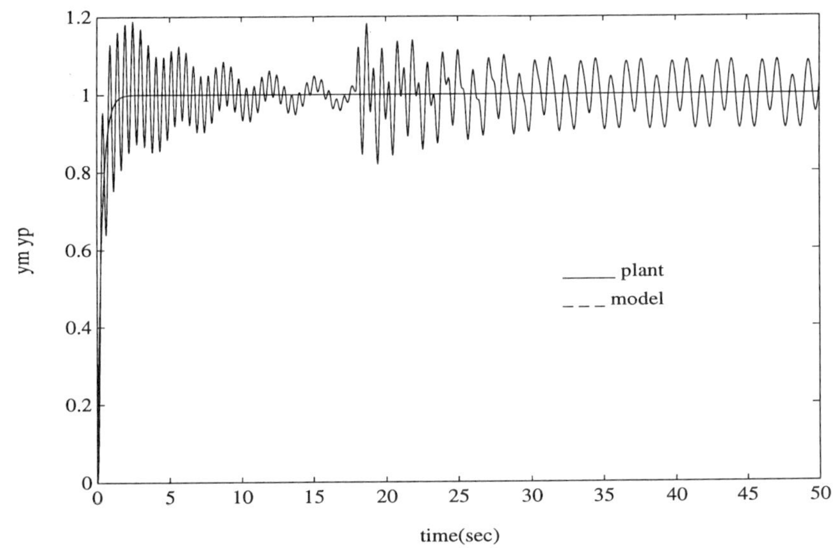

FIGURE 4.33. Stable Example: Plant and Model Outputs, Algorithm 1

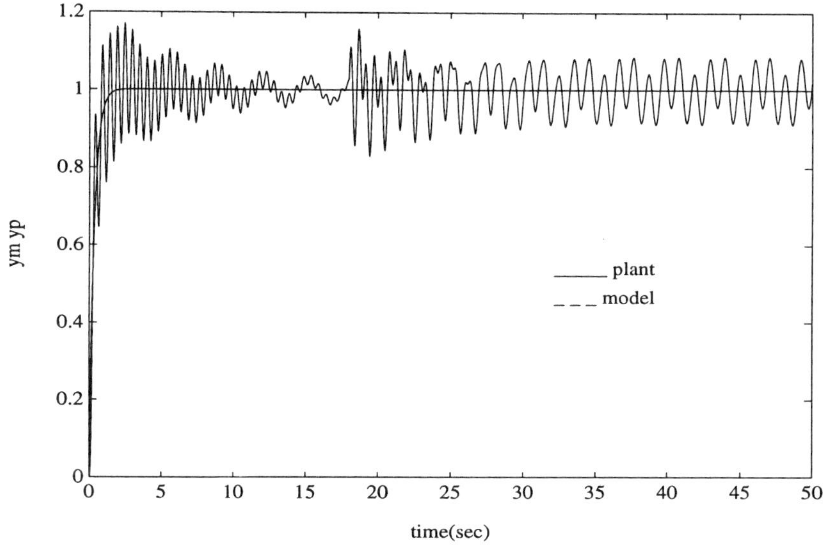

FIGURE 4.34. Stable Example: Plant and Model Outputs, Algorithm 3

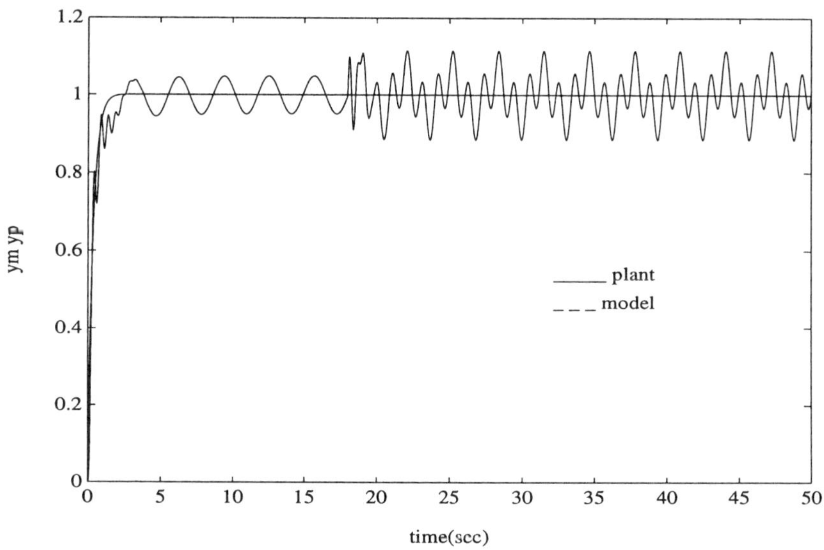

FIGURE 4.35. Stable Example: Plant and Model Outputs, New Compensator, Step Command

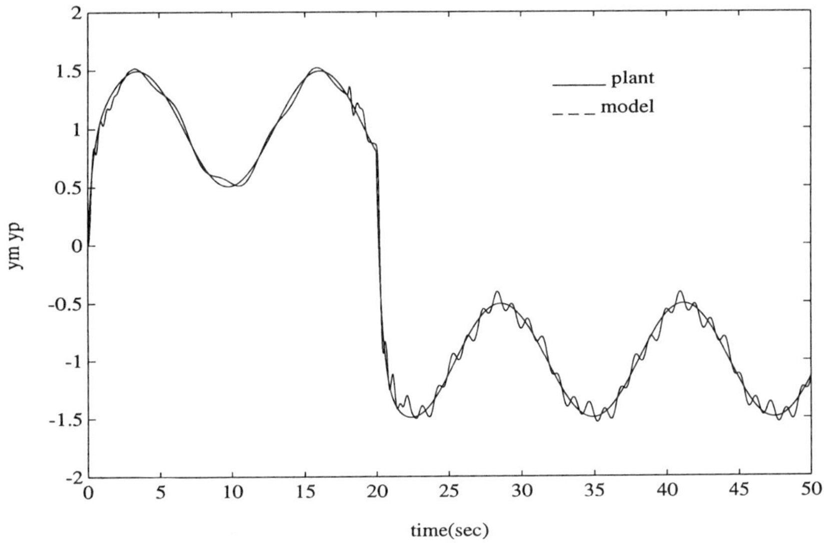

FIGURE 4.36. Stable Example: Plant and Model Outputs, New Compensator, Modulated Square Wave Command

of the output error is zero. Furthermore, the sigma modification has little effect on the response. The integral gains K_{eI} and K_{fI} diverge when sigma is equal to zero, but are clearly bounded when sigma is equal to 1.0. For sigma equal to 1.0, the gain K_{eI} is large at $t = 0$, then becomes small, increases again at t=18 when the output disturbance is applied, and then reaches a new steady state value as the average value of the error becomes zero.

Similar behavior of the output error is observed for the modulated square wave command. Figure 4.47 indicates that after an initial transient, the average error is again zero.

Finally, the simulation is repeated with the stabilizing compensator defined by

$$G_c(s) = \frac{-(3s+1)}{s}. \tag{4.124}$$

The plant and model outputs for a step command and a modulated square wave command are shown in Figures 4.48 and 4.49, respectively. Upon comparing these responses with Figures 4.40 and 4.47, respectively, it is observed that the compensator given by (4.124) yields improved disturbance attenuation relative to the compensator given by (3.316).

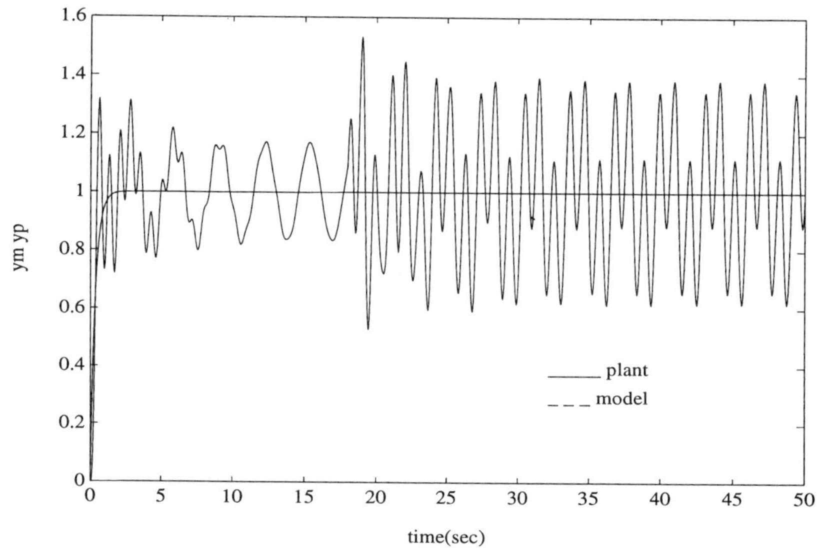

FIGURE 4.37. Unstable Example: Plant and Model Outputs, $\sigma = 0$

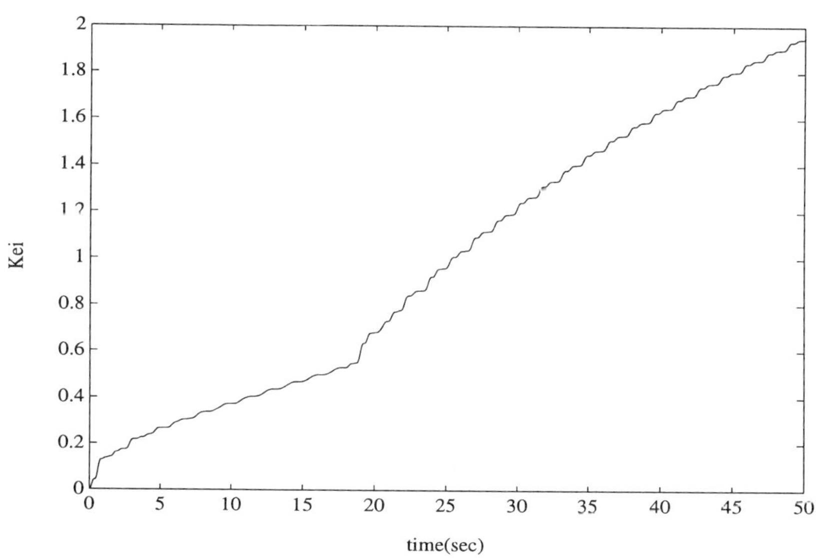

FIGURE 4.38. Unstable Example: Integral Adaptive Gain $K_{Ie}, \sigma = 0$

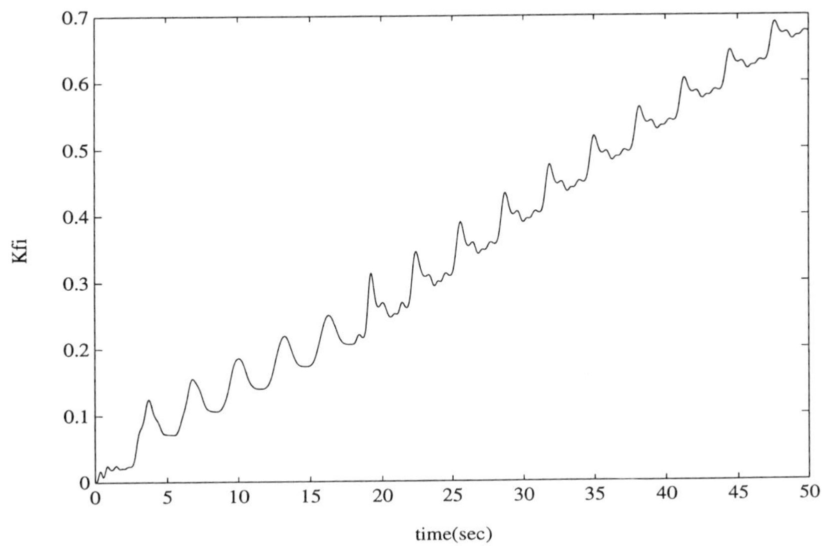

FIGURE 4.39. Unstable Example: Integral Adaptive Gain $K_{If}, \sigma = 0$

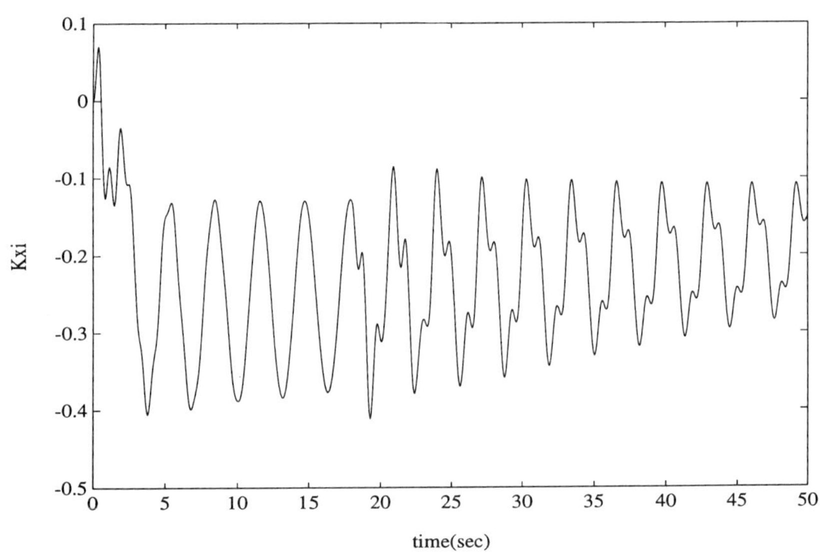

FIGURE 4.40. Unstable Example: Integral Adaptive Gain $K_{Ix}, \sigma = 0$

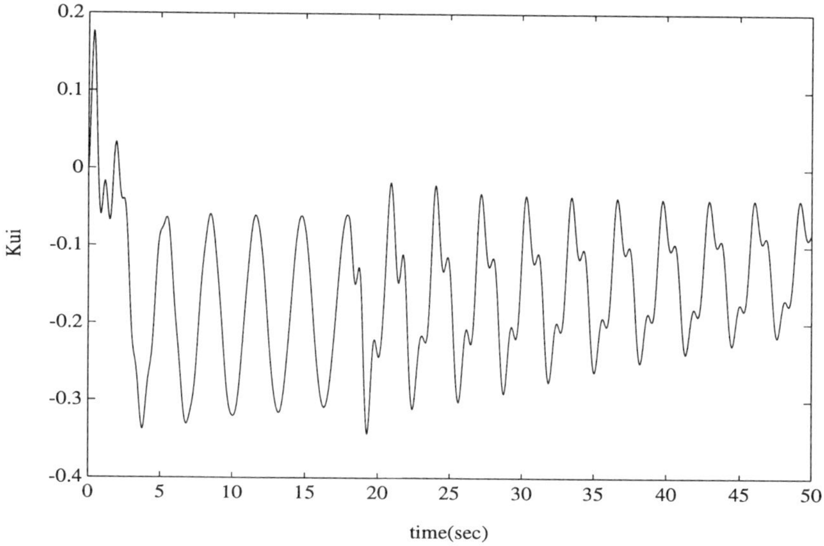

FIGURE 4.41. Unstable Example: Integral Adaptive Gain $K_{Iu}, \sigma = 0$

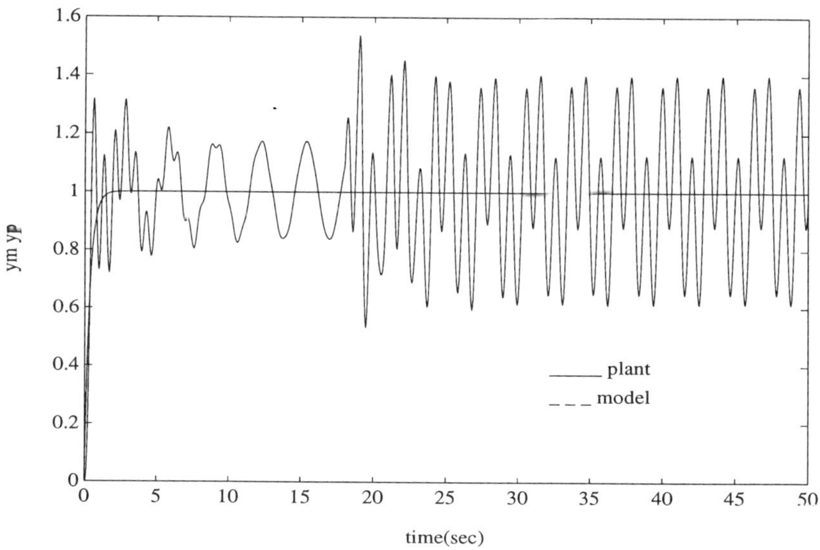

FIGURE 4.42. Unstable Example: Plant and Model Outputs, $\sigma = 1$

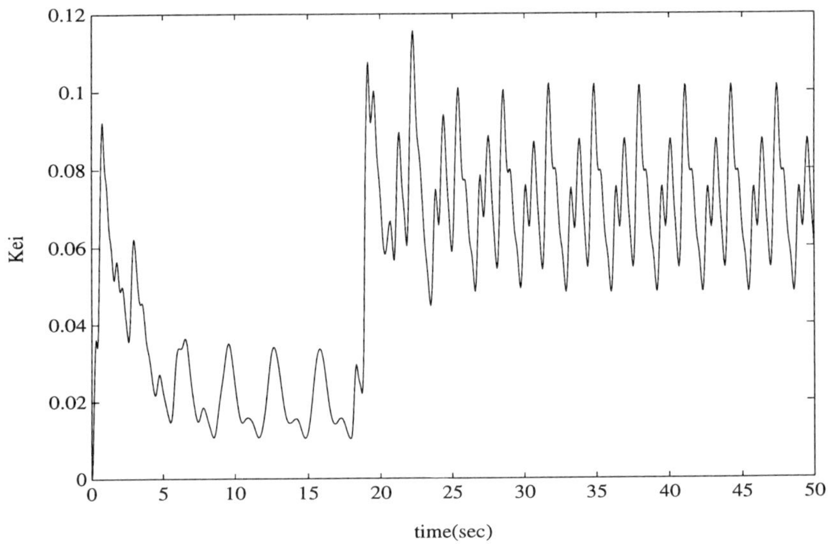

FIGURE 4.43. Unstable Example: Integral Adaptive Gain $K_{Ie}, \sigma = 1$

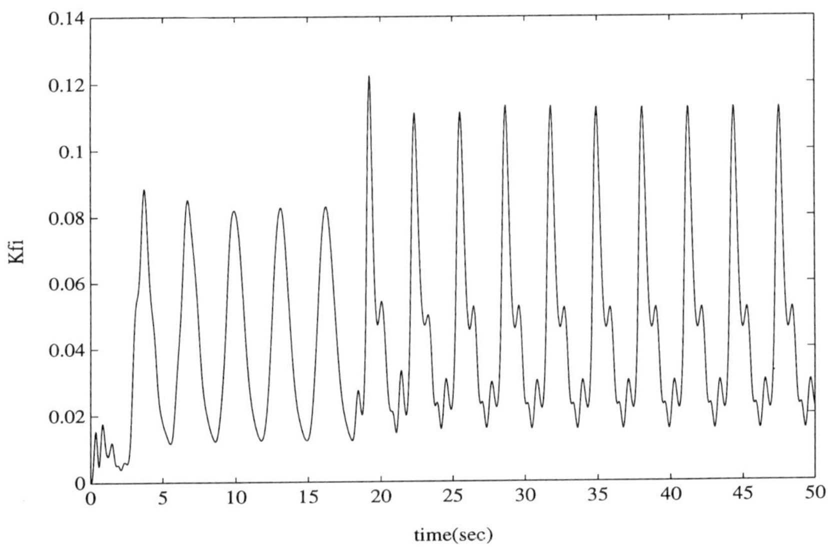

FIGURE 4.44. Unstable Example: Integral Adaptive Gain $K_{If}, \sigma = 1$

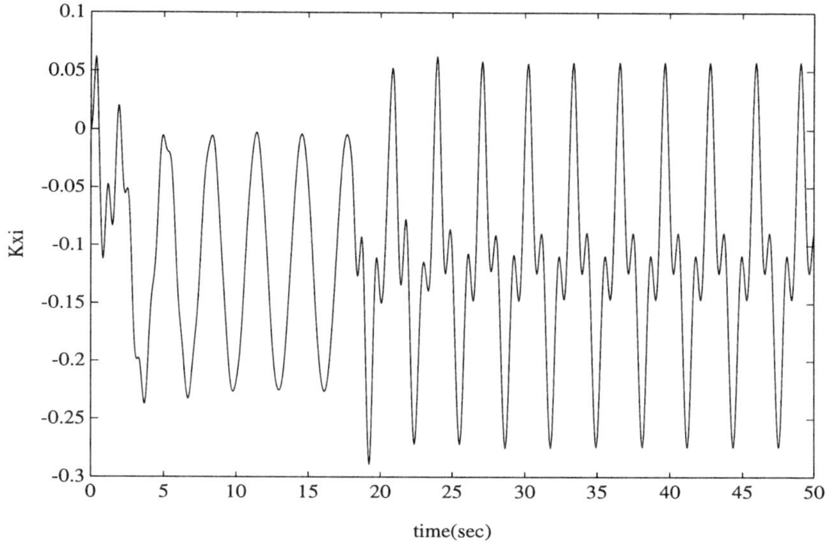

FIGURE 4.45. Unstable Example: Integral Adaptive Gain $K_{Ix}, \sigma = 1$

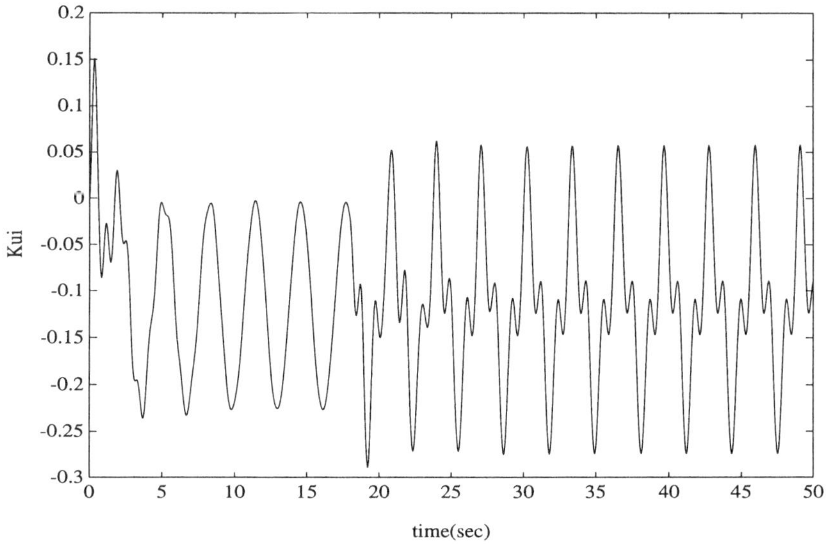

FIGURE 4.46. Unstable Example: Integral Adaptive Gain $K_{Iu}, \sigma = 1$

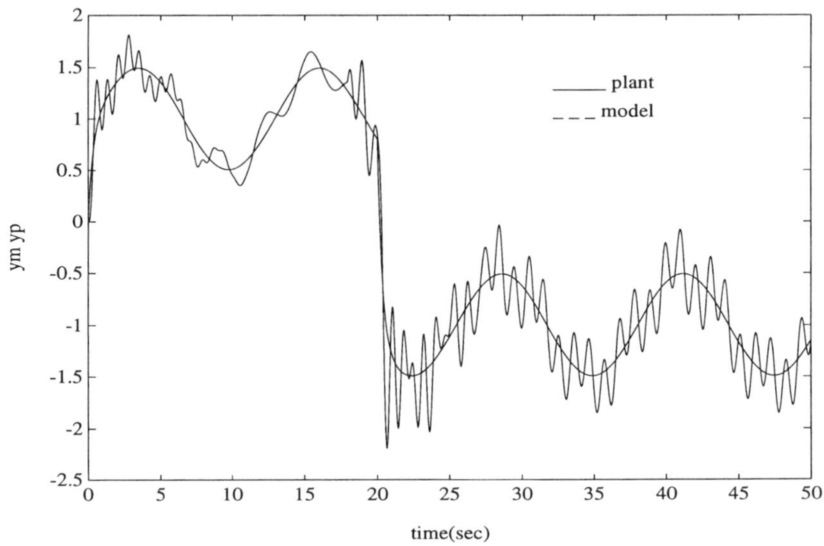

FIGURE 4.47. Unstable Example: Plant and Model Outputs, Modulated Square Wave Command

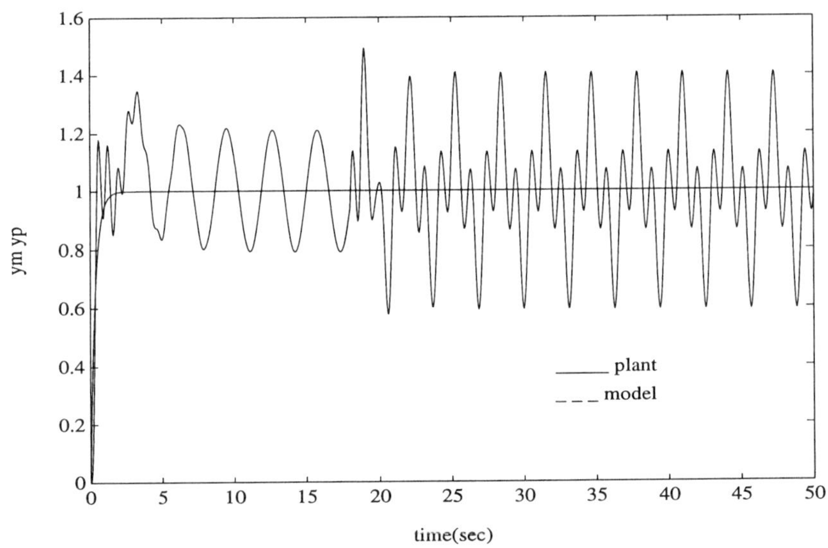

FIGURE 4.48. Unstable Example: Plant and Model Outputs, New Compensator, Step Command

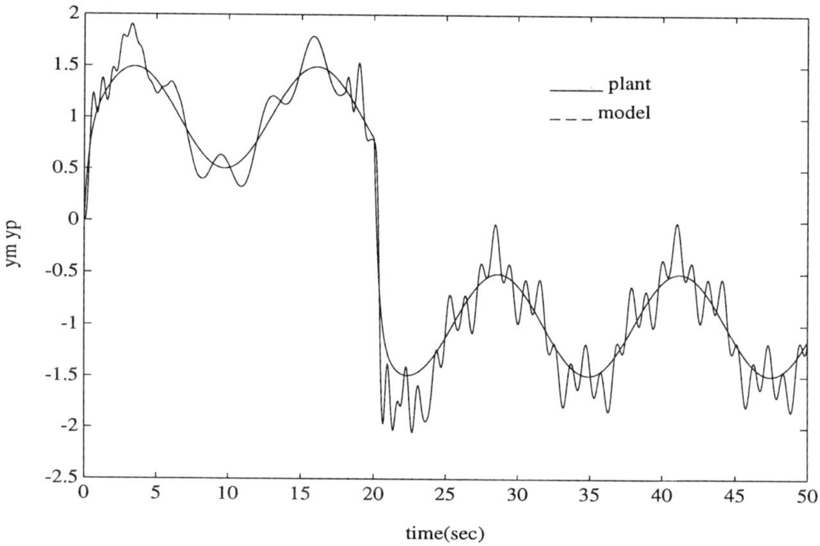

FIGURE 4.49. Unstable Example: Plant and Model Outputs, New Compensator, Modulated Square Wave Command

4.5 Bursting Phenomena and Their Elimination

This chapter has shown how the adaptive control algorithms can be adjusted to guarantee stability with respect to boundedness of all dynamic values involved in the adaptation process.

While these adjustments do indeed robustify the adaptive controllers in the presence of disturbances, some interesting, though not necessarily desired, phenomena have been observed when the robust adaptive algorithms are used with no external excitation [103], [18], [104], [105], [106], [107], [108], [109]. These effects are apparent in particular if the original controlled plant is unstable [104], [108].

For a intuitive presentation of the phenomenon, it is useful to follow the behavior of robust adaptive controller that controls a simple plant, such as [104], [108]

$$G_p(s) = \frac{y_p(s)}{u_p(s)} = \frac{s + 2}{s^2 - s - 2}. \tag{4.125}$$

The plant is unstable, but it is easy to see that the plant is ASPR (minimum-phase and relative degree 1). Therefore, a simple adaptive control algorithm with σ-term (4.13) guarantees stability with respect to boundedness, in the presence of any bounded input or output disturbances.

One may expect the tracking errors to increase with the disturbances, and to vanish if the disturbances vanish.

Assume that no input commands or external disturbances are present and that all adaptive gains are zero, initially. Thus, $u_m(t) = 0, x_m(t) = 0, y_m(t) = 0$, and therefore $K_x(t) = 0$ and $K_u(t) = 0$, and the only possible nonzero adaptive gain is the main gain $K_e(t)$. The relevant terms of the adaptive algorithm in this case are

$$e_y(t) = y_m(t) - y_p(t) \tag{4.126}$$

$$\dot{K}_{e_I}(t) = \gamma_I e_y^2(t) - \sigma K_{e_I}(t) \tag{4.127}$$

$$K_{e_P}(t) = \gamma_p e_y^2(t) \tag{4.128}$$

$$K(t) = K_e(t) = K_{e_I}(t) + K_{e_p}(t) \tag{4.129}$$

$$u_p(t) = K_e(t) e_y(t). \tag{4.130}$$

Select for illustration $\gamma_I = 1$, $\gamma_p = 0, \sigma = .03$. Since the initial gains are zero, the (unstable) plant has the normal tendency to diverge. The tracking error $e_y(t)$ starts increasing, and the adaptive gain $K(t)$ also starts increasing, until it reaches some stabilizing value [108]. As a result, the plant has now the tendency to converge. The state and output errors move towards their zero value, and from (4.127) and (4.128), the decrease of the output error leads to the decrease of the adaptive gains towards zero (Figure 4.50).

The simulation in Figure 4.50 runs for 40 sec., and the reader may be (almost) convinced that the adaptive system is indeed stable. The error $e_y(t)$ then vanishes, and because the adaptive gains (4.127) and (4.128) include $e_y^2(t)$, they start to decrease and the system apparently moves toward the equilibrium point ($e_y(t) = 0$, $K_e(t) = 0$). However, it is useful to recall here that the equilibrium point $(e_y(t) = 0, K_e(t) = 0)$ cannot be stable because the original plant is unstable.

Figure 4.51 shows the same case for 400 seconds. It can be seen that after some time the "stabilized" plant exhibits a change in behavior.

To understand the phenomenon, observe that the plant (4.125) can be stabilized by any constant output feedback gain K that satisfies $K \geq K_{\min} = 1$. Since higher values of the adaptive gain have a stabilizing effect on the plant, the state and output errors move quickly to zero and then the adaptive gain (4.129) also decreases. When the gain becomes too small, the plant is again unstable. Yet, as all errors are very small or practically zero, the effect of instability is felt only after the gain is well in the unstable region. Then, a sudden "burst" occurs, because the error increases quickly,

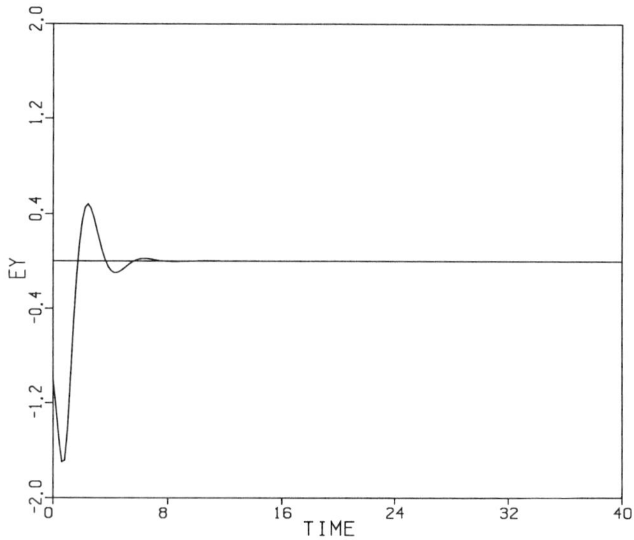

a. Output error

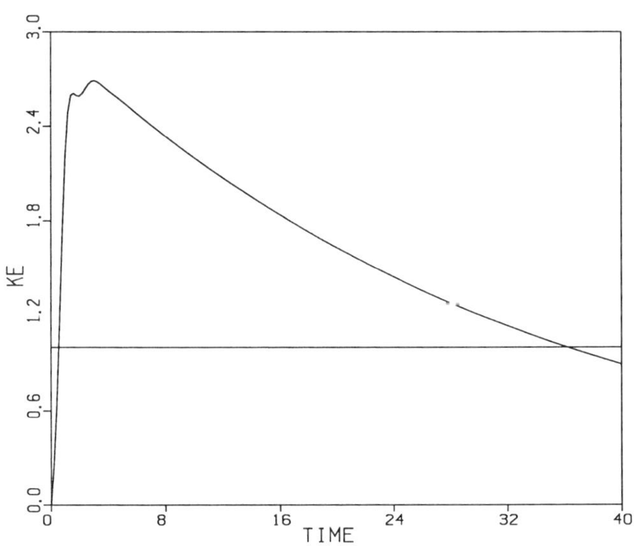

b. Adaptive error gain

FIGURE 4.50. The Tracking Error and the Adaptive Gain.

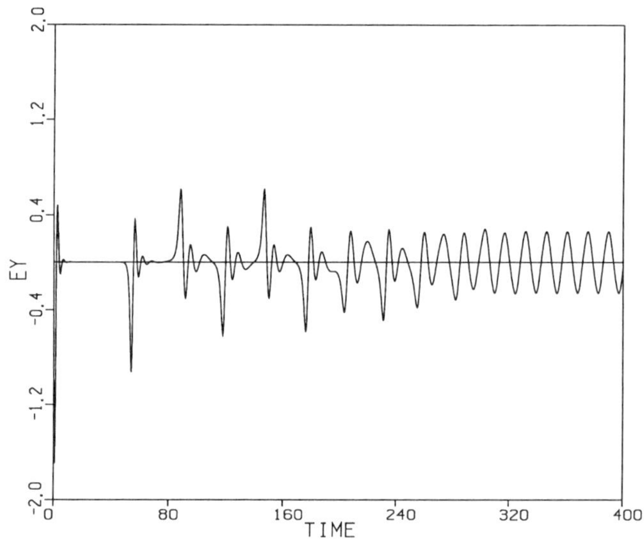

a. Output error

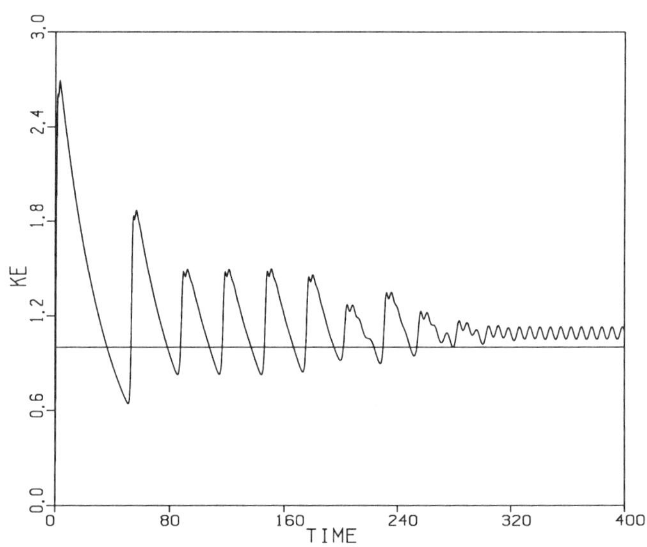

b. Adaptive error gain

FIGURE 4.51. Bursting Phenomena

followed immediately by the sudden increase in the adaptive gain. This brings the gains again into the stable region, and so on [106], [107], [108].

The particular behavior depends on the plant and upon the specific adaptation coefficients. In some cases the bursts may go on for ever. In other cases, such as in the case presented here, the amplitude of the bursts gradually decreases and the system ends with oscillations of the errors and of the adaptive gain. This results in an interesting nonlinear oscillatory behavior. It is easy to observe in Figure 4.51 that all steady-state values of the oscillating adaptive gains are greater than unity, so the plant would be in the so-called "asymptotically stable" region, if fixed gains rather than varying adaptive gains were used. As was already mentioned, variable gains do not necessarily guarantee asymptotic stability even if they are kept within the "stability bounds" that were established for fixed gains. Yet, in the particular case of the adaptive controller, this situation can be easily corrected.

Since the adaptive controllers presented in this book maintain stability at any rate of adaptation, one can reduce the burst phenomena by using faster adaptation [108], with higher adaptation coefficients, such as $\gamma_p = 100; \gamma_I = 100; \sigma = 0.03$ (Figure 4.52). It can be seen that while the gain oscillation is still remarkable, the tracking error is considerably reduced.

Thus, the adaptive gains (4.129) are generated by the tracking errors. If the error is zero, the adaptive gain supplied by (4.129) must also vanish. Yet, in order to guarantee stability, the steady-state gain must be nonzero even if the errors are zero. An adjustment of (4.129) is thus called for to overcome this contradiction. Indeed, the burst phenomena can be completely eliminated if one adjusts the adaptation rule using the prior knowledge on the lower bound for the admissible gain $K_{\min}$, whenever it is available. If $K_{\min}$ is not known in advance, the results of previous runs might be used to estimate this bound. From Figures 4.51 and 4.52 it should be clear that the bursting phenomena happen at or above the lower stability bound.

To avoid the burst phenomena, one could replace $K(t)$ in (4.129) and in (4 10) for the general case by

$$K_a(t) = K(t) + K_0 = K_e(t) + K_0 \tag{4.131}$$

where K_0 is any stabilizing gain that satisfies

$$K_0 \geq K_{\min}. \tag{4.132}$$

Figure 4.53 presents the behavior of the adaptive system with $\gamma_I = 100$, $\gamma_p = 100$, $\sigma = 0.03$, and $K_0 = 1$. The simulation is continued for 4000 sec to show that no bursting phenomena or oscillations appear. It can be seen that the errors vanish, and the adaptive gains reach the values $K_e = 0$, $K_a = K_o$. The reader who wants to be assured that these unwanted phenomena will not appear even after a much longer time must follow the

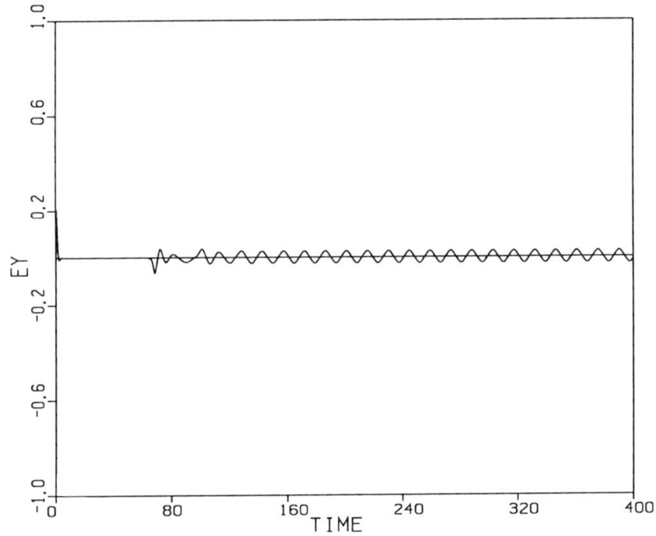

a. Output error

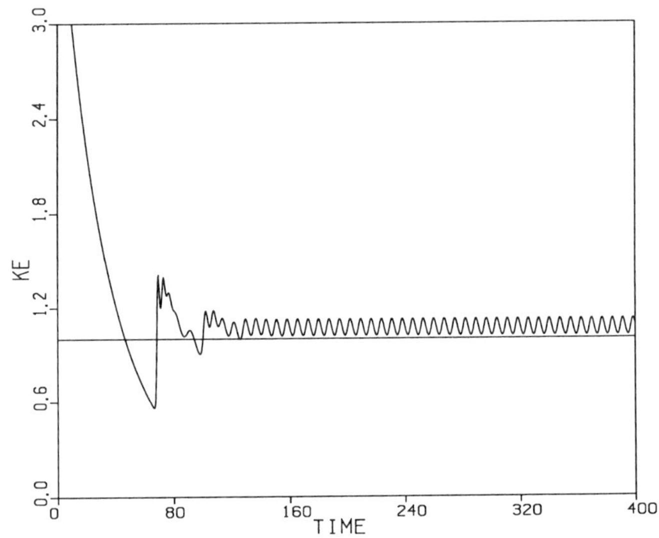

b. Adaptive error gain

FIGURE 4.52. Reduced Bursting: $\gamma_I = 100.; \gamma_p = 100, \sigma = 0.03$.

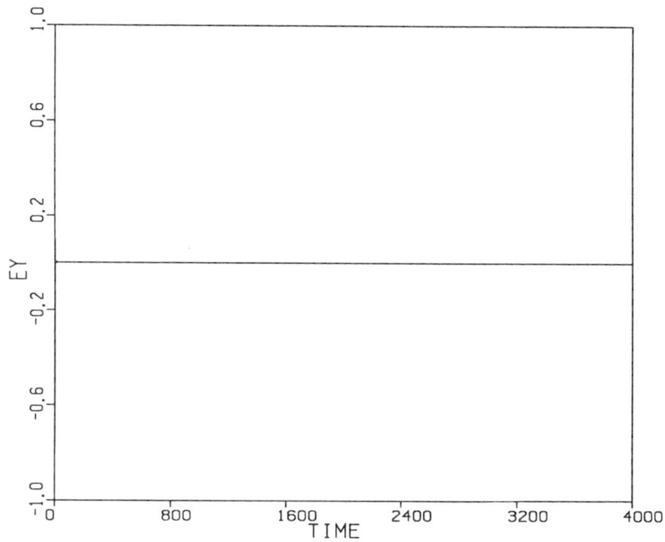

a. Output error

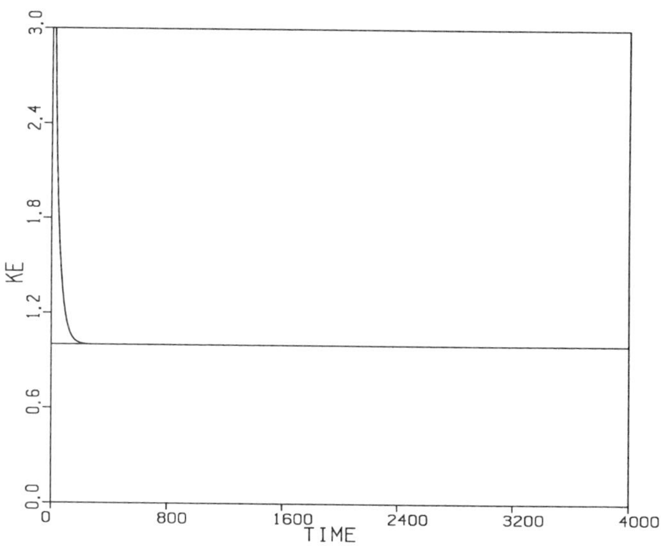

b. Adaptive error gain

FIGURE 4.53. Addition of Constant Gain. $\gamma_I = 100$, $\gamma_p = 100$, $\sigma = 0.03$, $K_0 = 1$.

detailed analysis presented in references [21], [22], [23] for both stationary and nonstationary systems. These references show that the bursting phenomena are eliminated when K_o is used, because if either $\| e_y(t) \|$, $\| K_e(t) \|$ or $\| K_e(t)e_y(t) \|$ becomes small, the adaptive system enters the domain of attraction of the stable equilibrium point $(e_y(t) = 0, K_e(t) = 0)$ or $(e_y(t) = 0, K_a(t) = K_o)$ and cannot escape any more. It is also shown that this is the only equilibrium point of the system, so the adaptive system has now an equilibrium point $(e_y(t) = 0, K_e(t) = 0)$ that is both asymptotically stable and unique.

4.6 Summary

This chapter has shown that the presence of disturbances, complex input signals, and/or noise may lead to an error signal that is persistently nonzero. Without any modification to the previously developed algorithms, this may result in an unbounded integral adaptive gain and possible system instability. To alleviate this effect, the integral gain update is in all cases replaced by the "leaky integrator" defined in (4.13).

Appendix 4A

Proof of Robust Stability, Theorem 1

Recall first that the undisturbed augmented ASPR plant satisfies the positive realness relations (2.125)–(2.126). The following quadratic Lyapunov function is then used to prove stability of the adaptive system described by (4.13) and (4.46):

$$V(t) = e_x^T(t)Pe_x(t) + tr\left[(K_I(t) - \tilde{K})T^{-1}(K_I(t) - \tilde{K})^T\right] \tag{4A.1}$$

where P is a positive definite matrix, and T is the adaptation coefficient defined in (4.13).

The derivative of the Lyapunov function (4A.1), becomes (as shown in Appendix 4B):

$$\dot{V}(t) = e_x^T(t)Qe_x(t) - 2\sigma tr\left[(K_I(t) - \tilde{K})T^{-1}(K_I(t) - \tilde{K})^T\right]$$
$$-2e_y^T(t)e_y(t)r^T(t)\bar{T}r(t)$$

$$-2\sigma\left[(K_I(t) - \tilde{K})T^{-1}\tilde{K}^T\right] - 2e_x^T(t)PF(t) - 2d_0^T(t)(K(t) - \tilde{K})r(t). \tag{4A.2}$$

One can substitute $K(t) = K_I(t) + K_p(t) = K_I(t) + e_y(t)r^T(t)\bar{T}$ and $r^T(t)\bar{T}r(t) = e_y^T(t)\bar{T}_{e_y}e_y(t) + x_m^T(t)\bar{T}_{x_m}x_m(t) + u_m^T(t)\bar{T}_{u_m}u_m(t)$ into (4A.2) to obtain

$$\begin{aligned}
\dot{V}(e_x, K_I) = \quad & -e_x^T(t)Qe_x(t) - 2\sigma tr\left[(K_I(t) - \tilde{K})T^{-1}(K_I(t) - \tilde{K})^T\right] \\
& -2e_y^T(t)e_y(t)e_y^T(t)\bar{T}_{e_y}e_y(t) \\
& -2e_y^T(t)e_y(t)\left[x_m^T(t)\bar{T}_{x_m}x_m(t) + u_m^T(t)\bar{T}_{u_m}u_m(t)\right] \\
& -2\sigma tr\left[(K_I(t) - \tilde{K})T^{-1}\tilde{K}^T\right] \qquad (4A.3) \\
& -2e_x^T(t)PF(t) - 2d_0^T(t)(K_I(t) - \tilde{K})r(t) \\
& -2d_0^T(t)e_y(t)e_y^T(t)\bar{T}_{e_y}e_y(t) \\
& -2d_0^T(t)e_y(t)\left[x_m^T(t)\bar{T}_{x_m}x_m(t) + u_m^T(t)\bar{T}_{u_m}u_m(t)\right].
\end{aligned}$$

Since (4A.3) contains nonnegative terms of higher order than two in the output error $e_y(t)$, and also cross–terms involving this term, one must take into consideration two kinds of trajectories that might lead to divergence.

a) First, consider only those trajectories where

$$\| e_y(t) \|$$

remains bounded in spite of the fact that

$$\| e_x(t) \|$$

and

$$\| K_I(t) - \tilde{K} \|$$

tend to increase without bound.

Since $d_i(t)$, $d_0(t)$, $u_m(t)$, $x_m(t)$, and thus $F(t)$ are all bounded by assumption, it is easy to see that there exist $\alpha_i > 0$, $i = 1, 2 \ldots, 10$, such that

$$\begin{aligned}
\dot{V}(t) < -\alpha_1 \| e_x(t) \|^2 -\alpha_2 \| K_I(t) - \tilde{K} \|^2 +\alpha_3 \| e_x(t) \| + \\
\alpha_4 \| K_I(t) - \tilde{K} \| +\alpha_5.
\end{aligned} \qquad (4A.4)$$

For any real scalars x and y there exists some positive finite coefficients, a_1, a_2, a_3, a_4, such that

$$x < a_1\sqrt{x^2 + y^2} \qquad (4A.5)$$

$$y < a_2\sqrt{x^2 + y^2} \tag{4A.6}$$

$$y^2 < a_3(x^2 + y^2) \tag{4A.7}$$

$$y^2 < a_4(x^2 + y^2) \tag{4A.8}$$

and since $V(e_x, K_I)$ is a positive definite quadratic function of $\| e_x(t) \|$ and $\| K_I(t) - \tilde{K} \|$, it is easy to see from (4A.8) that

$$\dot{V}(e_x, K_I) < -\alpha_6 V(t) + \alpha_7\sqrt{V(t)} + \alpha_8 \tag{4A.9}$$

or

$$\dot{V}(e_x, K_I) < -\alpha_6(\sqrt{V(t)} - \alpha_9)^2 + \alpha_{10}. \tag{4A.10}$$

It is easy to see from (4A.10) that if $V(e_x, K_I)$ takes any value larger than $V_1 = (\alpha_9 + \sqrt{\frac{\alpha_{10}}{\alpha_6}})^2$, then $\dot{V}(e_x, K_I)$ is negative, and since $V(e_x, K_I)$ is positive, this implies that $\dot{V}(e_x, K_I) < -\epsilon V(e_x, K_I)$ for any $V(e_x, K_I) > V_1$ and for some strictly positive scalar ϵ.

b) Second, consider those trajectories where $\| e_y(t) \|$ tends to increase without bound together with $\| e_x(t) \|$, and $\| K_I(t) - \tilde{K} \|$ and thus together with $V(t)$. In this case one gets from (4A.3)

$$\begin{aligned}
&\dot{V}(e_x, K_I) < -\alpha_1 \| e_x(t) \|^2 - \alpha_2 \| K_I(t) - \tilde{K} \|^2 - \\
&\alpha_3 \| e_y(t) \|^4 - \alpha_4 \| e_y(t) \|^2 (\| x_m(t) \|^2 + \alpha_5 \| u_m(t) \|^2) + \\
&+ \alpha_6 \| e_x(t) \| + \alpha_7 \| K_I(t) - \tilde{K} \| + \alpha_8 \| K_I(t) - \\
&\tilde{K} \| \| e_y(t) \| + \alpha_9 \| e_y(t) \|^3 + \\
&+ \alpha_{10} \| e_y(t) \| (\| x_m(t) \|^2 + \alpha_{11} \| u_m(t) \|^2)
\end{aligned} \tag{4A.11}$$

and then there exist $\beta_1 \ldots \beta_9 > 0$ such that

$$\begin{aligned}
\dot{V}(e_x, K_I) < -\beta_1(\sqrt{V(t)} - \beta_2)^2 - \beta_3(\| K_I(t) - \tilde{K} \| - \\
\beta_4 \| e_y(t) \|)^2 - \beta_5 \| e_y(t) \|^4 + \beta_6(\| e_y(t) \|^3 + \\
\beta_7 \| e_y(t) \|^2 + \beta_8 \| e_y(t) \|) + \beta_9.
\end{aligned} \tag{4A.12}$$

What happens on those trajectories where the increase of $V(e_x, K_I)$, due to the increase of $\| e_x(t) \|$ or $\| K_I(t) - \tilde{K} \|$, leads also to the increase of $\| e_y(t) \|$? For some value V_2 such that $V(e_x, K_I) > V_2$, the negative quartic term $-\beta_5 \| e_y \|^4$, along with the first two negative terms, becomes dominant over the not necessarily negative terms of lesser degree in (4A.12) and $\dot{V}(e_x, K_I)$ becomes again strictly negative. One can write again that $\dot{V}(e_x, K_I) < -\epsilon V(e_x, K_I)$ for any $V(e_x, K_I) > V_2$.

In conclusion, $\dot{V}(e_x, K_I) < -\epsilon V(e_x, K_I)$ for any $V(e_x, K_I) > \max(V_1, V_2)$, and therefore, $V(e_x, K_I)$ is uniformly bounded, and so are $\| e_x(t) \|$ and $\| K_I(t) - \tilde{K} \|$.

As a qualitative result, since the last four terms in (4A.3) that may be nonnegative are the direct effect of the disturbances and of the required sigma–term, one may expect that the magnitude of the final tracking errors tends to increase monotonically with the magnitude of $d_i(t)$, $d_0(t)$ and σ.

One may also notice that the third term in (4A.2), which is a direct result of the "proportional" gain (4.13), is negative definite quartic with respect to the output error and proportional to the adaptation coefficient $\bar{T}$. By appropriate selection of $\bar{T}$, the relative weight of this negative term can be increased, and the output error can thus be controlled and reduced, as a tradeoff between large adaptation coefficients and large errors. In order to prevent any improper expectations from this result, it is worth mentioning that only the disturbed output satisfies $y_a \approx y_m(t)$ and therefore the plant output $y_p(t)$ would usually track with bounded errors, as the simulations indeed show.

Appendix 4B

Development of Lyapunov Function Derivative

Let

$$V(e_x, K_I) = V_1(e_x, K_I) + V_2(e_x, K_I) \tag{4B.1}$$

where

$$V_1(e_x, K_I) = e_x^T(t) P e_x(t) \tag{4B.2}$$

$$V_2(e_x, K_I) = tr\left[(K_I(t) - \tilde{K})T^{-1}(K_I(t) - \tilde{K})^T\right] \tag{4B.3}$$

Then

$$\dot{V}_1(e_x, K_I) = \dot{e}_x^T(t) P e_x(t) + e_x^T(t) P \dot{e}_x(t). \tag{4B.4}$$

Substituting $\dot{e}_x(t)$ from (4.46) gives

$$\begin{aligned} \dot{V}_1(e_x, K_I) = e_x^T(t)(PA_c + A_c^T P)e_x(t) - 2e_x^T(t)PB(K(t)- \\ \tilde{K})r(t) - 2e_x^T(t)PF(t). \end{aligned} \tag{4B.5}$$

Applying the positive real properties gives:

$$\dot{V}_1(e_x, K_I) = -e_x^T(t)Qe_x(t) - 2e_x^T(t)C^T(K(t) - \tilde{K})r(t) - 2e_x^T(t)PF(t) \tag{4B.6}$$

$$\dot{V}_2(e_x, K_I) = 2tr\left[(K_I(t) - \tilde{K})T^{-1}\dot{K}_I^T(t)\right] \tag{4B.7}$$

or

$$\dot{V}_2(e_x, K_I) = 2tr\left[(K_I(t) - \tilde{K})T^{-1}(e_y(t)r^T(t)T - \sigma K_I(t))\right] \tag{4B.8}$$

and thus,

$$\dot{V}_2(e_x, K_I) = -2\sigma tr\left[(K_I(t) - \tilde{K})T^{-1}K_I^T(t)\right] + 2e_y^T(t)(K_I(t) - \tilde{K})r(t) \tag{4B.9}$$

Substituting in the second term of 4B.9

$$K_I(t) = K(t) - K_p(t) = K(t) - e_y(t)r^T(t)\bar{T}. \tag{4B.10}$$

gives

$$\dot{V}_2(e_x, K_I) = -2\sigma\ tr\left[(K_I(t) - \tilde{K})T^{-1}K_I^T(t)\right] + 2e_y^T(t)(K(t) - \tilde{K})r(t)$$
$$- 2e_y^T(t)e_y(t)r^T(t)\bar{T}r(t) \tag{4B.11}$$

Adding and subtracting $2\sigma tr[(K_r(t) - \tilde{K})T^{-1}E^T]$ and substitution of $e_y(t)$ from (4.42) into the second right hand term of (4B.11) gives:

$$\begin{aligned}\dot{V}_2(e_x, K_I) = &-2\sigma\ tr\left[(K_I(t) - \tilde{K})T^{-1}(K_I(t) - \tilde{K})^T(t)\right]\\ &-2\sigma\ tr\left[(K_I(t) - \tilde{K})T^{-1}\tilde{K}^T\right]\\ &-2e_y^T(t)e_y(t)r^T\bar{T}r(t) + 2e_x^T(t)C^T(K(t) - \tilde{K})r(t)\\ &-2d_0^T(t)(K_I(t) - \tilde{K})r(t)\end{aligned} \tag{4B.12}$$

Adding (4B.6) and (4B.12) gives (4A.2).

Appendix 4C

Proof of Theorem 2

Stability is analyzed by using the Lyapunov function from Appendix 4C of Chapter 3 which is given by

$$V(e_x, K_I) = V_1(e_x, K_I) + V_2(e_x, K_I) \tag{4C.1}$$

where

$$V_1(e_x(t)) = e_x^T(t)Pe_x(t) \tag{4C.2}$$

and

$$V_2(e_x, K_I) = tr[(K_I(t) - \tilde{K})T^{-1}(K_I(t) - \tilde{K}^T)] \tag{4C.3}$$

Then,

$$\dot{V}_1 = \dot{e}_x^T Pe_x + e_x^T P\dot{e}_x. \tag{4C.4}$$

Substitute $\dot{e}_x$ from (4.111) to obtain

$$\dot{V}_1 = e_x^T(PA_c + A_c^T P)e_x - z^T B^T Pe_x - e_x^T PBz - F_1^T Pe_x - e_x^T PF_1 \tag{4C.5}$$

where F_1 is defined by (4.112)

Next, substitute (4.113) into (4C.5) to obtain

$$\dot{V}_1 = -e_x^T Re_x - e_x^T LL^T e_x - 2e_x PBz - 2e_x^T PF_1. \tag{4C.6}$$

Then, add and subtract $-2e_x^T LWz + z^T W^T Wz$ in (4C.6) to obtain

$$\begin{aligned}\dot{V}_1 &= -e_x^T Re_x - (L^T e_x - Wz)^T(L^T e_x - Wz)\\ &+2e_x^T(-LW - PB)z + z^T W^T Wz - 2e_x^T PF_1.\end{aligned} \tag{4C.7}$$

Next, use (4C.3) to obtain

$$\dot{V}_2 = tr[\dot{K}_1 T^{-1}(K_I - \tilde{K})^T] + tr[(K_I - \tilde{K})T^{-1}(\dot{K}_I)^T]. \tag{4C.8}$$

Substitute $\dot{K}_I$ from (4.96) into (4C.8) to obtain

$$\dot{V}_2 = tr\left[(vr^T T - \sigma K_I)T^{-1}(K_I - \tilde{K})^T\right] + tr\left[(K_I - \tilde{K})T^{-1}(vr^T T - \sigma K_I)^T\right]. \tag{4C.9}$$

Substitute $K_I = K - K_p = K - vr^T T$ into (4C.9) and use the definition $z = (K - \tilde{K})r$ to obtain

$$\dot{V}_2 = tr[vz^T] + tr[zv^T] - 2v^T vr^T \bar{T} r - 2\sigma tr[(K_I - \tilde{K})T^{-1}(K_I - \tilde{K})^T] - 2\sigma tr[(K_I - \tilde{K})T^{-1}\tilde{K}^T]. \tag{4C.10}$$

Substitute $v = Qe_{yv} + GKr = QCe_v - Q\tilde{d}_0 + GKr$ into (4C.10) to obtain

$$\begin{aligned}
\dot{V}_2 &= tr\left[(QCe_v - Q\tilde{d}_0 + GKr)z^T\right] + tr\left[z(QCe_v - Q\tilde{d}_0 + GKr)^T\right] \\
&- 2v^T vr^T \bar{T} r - 2\sigma tr\left[(K_I - \tilde{K})T^{-1}(K_I - \tilde{K})^T\right] \\
&- 2\sigma tr\left[(K_I - \tilde{K})T^{-1}\tilde{K}^T\right] = \\
&\quad 2e_v^T C^T Q^T z + tr\left[G(Kr - \tilde{K}r\tilde{K}r)z^T\right] \\
&+ tr\left[z(Kr - \tilde{K}r + \tilde{K}r)^T G^T\right] \\
&- 2v^T vr^T \bar{T} r - 2\tilde{d}_0^T Q^T z \\
&- 2\sigma tr\left[(K_I - \tilde{K})T^{-1}(K_I - \tilde{K})^T\right] - 2\sigma tr\left[(K_I - \tilde{K})T^{-1}\tilde{K}^T\right] \\
&= 2e_v^T C^T Q^T z + z^T(G + G^T)z + 2z^T G(\tilde{K}_a Ce_v - \tilde{K}_a \tilde{d}_0 \\
&+ \tilde{K}_x x_m + \tilde{K}_u u_m) - 2v^T vr^T \bar{T} r - 2\tilde{d}_0^T Q^T z \\
&- 2\sigma tr\left[(K_I - \tilde{K})T^{-1}(K_I - \tilde{K})^T\right] - 2\sigma tr\left[(K_I - \tilde{K})T^{-1}\tilde{K}^T\right] \\
&= 2e_v^T C^T (Q^T + \tilde{K}_a^T G^T)z + z^T(G + G^T)z + 2z^T G(-\tilde{K}_a \tilde{d}_0 \\
&+ \tilde{K}_x x_m + \tilde{K}_u u_m) - 2v^T vr^T \bar{T} r - 2\tilde{d}_0^T Q^T z
\end{aligned} \tag{4C.11}$$

$$- \quad 2\sigma tr\left[(K_I - \tilde{K})T^{-1}(K_I - \tilde{K})^T\right]$$

$$- \quad 2\sigma tr\left[(K_I - \tilde{K})T^{-1}\tilde{K}^T\right].$$

Substitute (4.110) for $e_v(t)$ into (4C.12) to obtain

$$\begin{aligned}
\dot{V}_2 &= 2(e_x^T - x_0^{*T})C^T(Q^T + \tilde{K}_a^T G^T)z + z^T(G + G^T)z + 2z^T G(-\tilde{K}_a\tilde{d}_0 \\
&+ \tilde{K}_x x_m + \tilde{K}_u u_m) - 2v^T v r^T \bar{T} r \\
&- 2\tilde{d}_0^T Q^T z - 2\sigma tr\left[(K_I - \tilde{K})T^{-1}(K_I - \tilde{K})^T\right] \\
&- 2\sigma tr\left[(K_I - \tilde{K})T^{-1}\tilde{K}^T\right] \\
&= 2e_x^T C^T(Q^T + \tilde{K}_a G^T)z + z^T(G + G^T)z \\
&- 2v^T v r^T \bar{T} r + 2z^T[G(\tilde{K}_x x_m + \tilde{K}_u u_m - \tilde{K}_a\tilde{d}_0) \\
&- Q\tilde{d}_0 - (Q + G\tilde{K}_a)Cx_0^*] \\
&- 2\sigma tr\left[(K_I - \tilde{K})T^{-1}(K_I - \tilde{K})^T\right] - 2\sigma tr\left[(K_I - \tilde{K})T^{-1}\tilde{K}^T\right] \\
&= 2e_x^T C^T(Q^T + \tilde{K}_a G^T)z + z^T(G + G^T)z - 2v^T v r^T \bar{T} r - 2z^T F_2 \\
&- 2\sigma tr\left[(K_I - \tilde{K})T^{-1}(K_I - \tilde{K})^T\right] \\
&- 2\sigma tr\left[(K_I - \tilde{K})T^{-1}\tilde{K}^T\right] \qquad (4C.12)
\end{aligned}$$

where

$$F_2 = -G(\tilde{K}_x x_m + \tilde{K}_u u_m - \tilde{K}_a\tilde{d}_0) + Q\tilde{d}_0 + (Q + G\tilde{K}_a)Cx_0^*. \qquad (4C.13)$$

Combine (4C.7) and (4C.12) to obtain

$$\begin{aligned}\dot{V} &= -e_x^T R e_x - (L^T e_x - Wz)^T (L^T e_x - Wz) \\ &+ 2e_x^T \left[-LW - PB + C^T(Q^T + \tilde{K}_a^T G^T)\right] z \\ &+ z^T(W^T W + G + G^T)z - 2v^T v r^T \bar{T} r - 2e_x^T P F_1 \\ &- 2z^T F_2 - 2\sigma tr\left[(K_I - \tilde{K})T^{-1}(K_I - \tilde{K})^T\right] \\ &- 2\sigma tr\left[(K_I - \tilde{K})T^{-1}\tilde{K}^T\right].\end{aligned} \tag{4C.14}$$

Finally, substitute (4.114) and (4.115) into (4C.15) and use $e_x^T R e_x = e_x^T(R + R^T)e_x/2$ to obtain

$$\begin{aligned}\dot{V}(e_x(t), K_I(t)) &= -e_x^T(t)(R + R^T)e_x(t)/2 - (L^T e_x(t) \\ &-Wz(t))^T(L^T e_x(t) - Wz(t)) + z^T(t)(J + J^T + G + G^T)z(t) \\ &-2v^T(t)v(t)r^T(t)\bar{T}r(t) - 2e_x^T(t)PF_1(t) - 2z^T(t)F_2(t) \\ &-2\sigma tr\left[(K_I(t) - \tilde{K})T^{-1}(K_I(t) - \tilde{K})^T\right] \\ &-2\sigma tr\left[(K_I(t) - \tilde{K})T^{-1}\tilde{K}^T\right].\end{aligned} \tag{4C.15}$$

The rest of the theorem follows by using the discussion in Appendix 4A to show that $V(e(t), K_I(t))$ is uniformly bounded.

Chapter 5

Adaptive Control of Time–Varying and Nonlinear Systems

This chapter contains theoretical concepts that are more advanced than those used in the preceding chapters. These concepts and several references are used to show that the preceding adaptive controllers still maintain stability with respect to boundedness even for certain classes of time varying and nonlinear systems. An early illustration of this behavior was shown in the example of Section 1.2. To prevent this presentation from being overburdened with new and complex material, this chapter gives only a brief presentation of the main proofs and results. More elaborate proofs and some illustrative examples are found in the references cited at the end of the chapter.

5.1 Introduction

The previous chapters have presented simple adaptive control algorithms that guarantee robust stability and satisfactory performance in linear time-invariant (LTI) systems with uncertain parameters. It was also shown that the stabilizability property of these systems can be used to satisfy the "almost positive real" conditions that are required to guarantee stability of the simple adaptive controllers.

It is useful to remember that the concept of 'positive realness' and 'strict positive realness' originated in the special properties of the transfer functions of LTI systems. When the transfer function is positive real, the corresponding plant is called 'passive' and when the transfer function is strictly positive real, the corresponding plant is called 'strictly passive'. Thus, a plant is called 'almost strictly passive' if there exists some constant output gain such that the resulting closed-loop plant is strictly passive.

As in many practical applications the uncertain plant is not necessarily linear-time-invariant, this section attempts to extend the applicability of the simple adaptive controllers to include linear time-varying (LTV) systems and also classes of so-called "nonlinear systems linear in control (SLIC)."

The section presents only the main definitions and results. The mathematical treatment that is necessarily long and complex in LTV and nonlinear systems is left for the appendices or can be found in the references of this section.

While one cannot use transfer functions and positive realness any more, Section 5.2 shows that the 'passivity' and 'almost passivity' concepts can be defined and used in LTV systems. In Section 5.3 it is shown that the 'almost strictly passive' configurations have properties that guarantee robust stability with adaptive controllers. In Section 5.4 it is shown that stabilizability of the original plant and parallel feedforward are sufficient to guarantee the implementation of 'almost strictly passive' configurations. Thus, the stability of the adaptive control configurations is ensured. Section 5.5 then presents the main results for nonlinear systems that are linear in the control.

5.2 Passivity and Almost Passivity of Nonstationary Systems

A dynamic time-varying linear plant is given in the state-space representation

$$\dot{x}_p(t) = A_p(t)x_p(t) + B_p(t)u_p(t) \tag{5.1}$$

$$y_p(t) = C_p(t)x_p(t) \tag{5.2}$$

$$y_a(t) = y_p(t) + D_p(t)u_p(t) = C_p(t)x_p(t) + D_p(t)u_p(t). \tag{5.3}$$

Here, $x_p(t)\epsilon\mathcal{R}^n, y_p(t)\epsilon\mathcal{R}^m, y_a(t)\epsilon\mathcal{R}^m$, and $u_p(t)\epsilon\mathcal{R}^m$. The matrices $A_p(t)$, $B_p(t)$, $C_p(t)$, $D_p(t)$ are of compatible dimensions, and the gain matrices $B_p(t)$, $C_p(t)$, $D_p(t)$ are uniformly bounded. The system (5.1)–(5.3) is not (necessarily) stable. The representation (5.1)–(5.3) has been selected to include both the case of strictly causal systems (with $D_p(t) = 0$), and the case of causal but not strictly causal systems, $D_p(t) \neq 0$.

It is desired to define those conditions that will allow the plant (5.3) to be "almost passive." Assume first that $D_p(t) \neq 0$ and that an output stabilizing feedback controller (Figure 5.1) is used , i.e.

$$u_p(t) = -K_e(t)y_a(t) + u_{pc}(t) \tag{5.4}$$

where $K_e(t) > 0$.

The resulting closed-loop system (Figure 5.2) is

$$\dot{x}_p(t) = A_{pc}(t)x_p(t) + B_{pc}(t)u_{pc}(t) \tag{5.5}$$

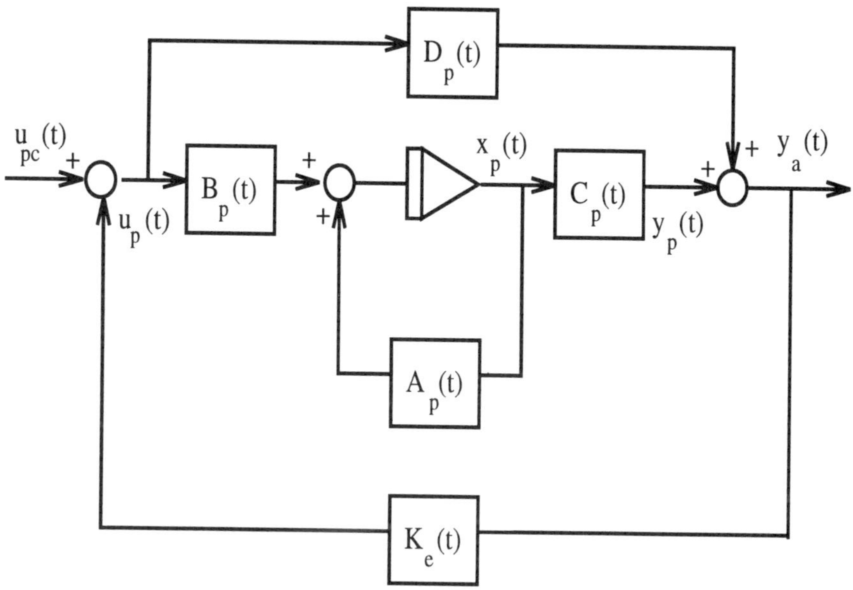

FIGURE 5.1. A Strictly Passive (SP) Configuration.

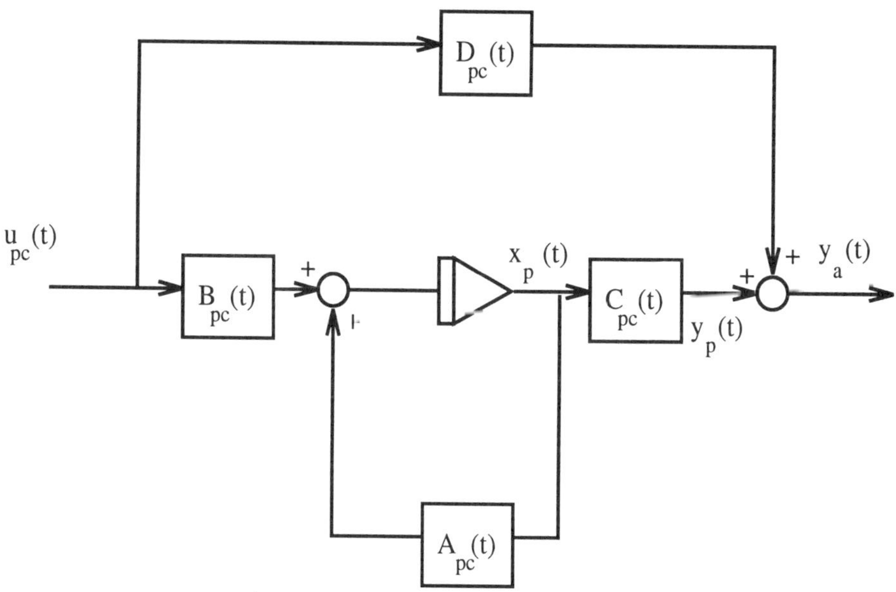

FIGURE 5.2. A First Equivalent Strictly Passive (SP) Configuration.

$$y_a(t) = C_{pc}(t)x_p(t) + D_{pc}(t)u_{pc}(t) \tag{5.6}$$

where

$$A_{pc}(t) = A_p(t) - B_p(t)K_{ec}(t)C_p(t) \tag{5.7}$$

$$\begin{aligned} K_{ec}(t) &= K_e(t)[I + D_p(t)K_e(t)]^{-1} \\ &= [I + K_e(t)D_p(t)]^{-1}K_e(t) \end{aligned} \tag{5.8}$$

$$B_{pc}(t) = B_p(t)[I + K_e(t)D_p(t)]^{-1} \tag{5.9}$$

$$C_{pc}(t) = [I + D_p(t)K_e(t)]^{-1}C_p(t) \tag{5.10}$$

and

$$D_{pc}(t) = [I + D_p(t)K_e(t)]^{-1}D_p(t) = D_p(t)[I + K_e(t)D_p(t)]^{-1}. \tag{5.11}$$

Definition: The system (5.5)–(5.6) is called "strictly passive (SP)" if it satisfies (as sufficient conditions) the following relations [86]:

$$\dot{P}(t) + P(t)A_{pc}(t) + A_{pc}^T(t)P(t) = -L^T(t)L(t) - Q(t) < 0 \tag{5.12}$$

$$P(t)B_{pc}(t) = C_{pc}^T(t) - L^T(t)W(t) \tag{5.13}$$

$$D_{pc}(t) + D_{pc}^T(t) = W^T(t)W(t) \tag{5.14}$$

for some uniformly bounded positive definite matrices, $P(t)$ and $Q(t)$. Condition (5.14) implies that $D_{pc}(t) > 0$. (The reader may want to use references [86], [110], [111], [112] for a better understanding of definitions and more general aspects of passivity in nonstationary and nonlinear systems.)

From (5.13) one gets

$$-L^T(t) = W(t)^{-1}[P(t)B_{pc}(t) - C_{pc}^T(t)]. \tag{5.15}$$

Substitute (5.15) into (5.12), and use (5.14) to show that relations (5.12)–(5.13) are equivalent to the Riccati equation

$$\begin{aligned} \dot{P}(t) &+ P(t)A_{pc}(t) + A_{pc}^T(t)P(t) \\ &+ [P(t)B_{pc}(t) - C_{pc}^T(t)][D_{pc}(t) + D_{pc}^T(t)]^{-1}[B_{pc}^T(t)P(t) - C_{pc}(t)] \\ &+ Q(t) = 0 \end{aligned} \tag{5.16}$$

where $L(t)\epsilon\mathcal{R}^{n\times p}, W(t)\epsilon\mathcal{R}^{p\times m}, p \geq m$. Thus the causal (but not strictly causal) system (5.5)–(5.6) with $D_p(t) > 0$ is called "strictly passive (SP)" if there exist some uniformly bounded positive definite matrices, $P(t)$ and $Q(t)$, that satisfy (5.12)–(5.14) or the equation (5.16).

In the strictly causal case, when $D_p(t) = 0$, the closed loop obtained from (5.1)–(5.2) with the feedback

$$u_p(t) = -K_e(t)y_p(t) + u_{pc}(t) \tag{5.17}$$

is

$$\dot{x}_p(t) = A_{pc}(t)x_p(t) + B_p(t)u_{pc}(t) \tag{5.18}$$

$$y_a(t) = C_p(t)x_p(t). \tag{5.19}$$

In this case $A_{pc}(t) = A_p(t) - B_p(t)K_e(t)C_p(t)$. The strictly causal plant (5.18)–(5.19) is called "strictly passive (SP)" if there exist some uniformly bounded positive definite matrices, $P(t)$ and $Q(t)$, that satisfy the SP relations

$$\dot{P}(t) + P(t)A_{pc}(t) + A_{pc}^T(t)P(t) = -Q(t) < 0 \tag{5.20}$$

$$P(t)B_p(t) = C_p^T(t). \tag{5.21}$$

Notice that only an output feedback gain, $K_e(t)$, separates the original plant (5.1)–(5.2) or (5.1)–(5.3) from strict passivity. Therefore such a plant is called "almost strictly passive" according to the following definition:

Definition: The system (5.1)–(5.3) (or (5.1)–(5.2)) is called "almost strictly passive (ASP)" if there exists a stabilizing positive definite feedback gain $K_e(t)$ (unknown and not needed for implementation), such that the resulting closed-loop system (5.5)–(5.6) (or (5.18)–(5.19)) is "strictly passive."

The next section will show now that, as in the case with linear time-invariant systems, nonstationary ASP systems can be robustly and efficiently controlled by nonlinear adaptive controllers.

5.3 Adaptive Control of ASP Plants

The following proof of stability for multivariable adaptive control systems extends the mathematical foundation to allow application of simple adaptive controllers to nonstationary systems. It is instructive to recall that "robust adaptive stabilization" means that all values involved in the adaptation process, namely, states, adaptive gains and errors, are bounded in the presence of any bounded input commands and any bounded input and output disturbances.

In this section it is assumed that the plant is almost strictly passive. In many realistic environments, the controlled plant has the representation

$$\dot{x}_p(t) = A_p(t)x_p(t) + B_p(t)u_p(t) + d_i(t) \tag{5.22}$$

$$y_a(t) = C_p(t)x_p(t) + D_p(t)u_p(t) + d_o(t) \tag{5.23}$$

where $d_i(t)$ and $d_o(t)$ are some bounded input and output disturbances, that can also represent the inaccuracies of the linear representation of the real plant. Except for the fact that it satisfies the ASP conditions, the plant can be very large and otherwise unknown. If $D_p(t) = 0$, then $y_a(t) = y_p(t) = C_p(t)x_p(t) + d_o(t)$. It is assumed first that $D_p(t) > 0$.

The output of the plant is required to follow the output of the (possibly) low-order model

$$\dot{x}_m(t) = A_m(t)x_m(t) + B_m(t)u_m(t) \tag{5.24}$$

$$y_m(t) = C_m(t)x_m(t) + D_m(t)u_m(t). \tag{5.25}$$

This model incorporates the desired input-output behavior of the plant. Here it is assumed to be nonstationary though in most practical cases the designer is expected to select a stable, well-behaved LTI model.

Let the output tracking error be defined as

$$e_y(t) = y_m(t) - y_a(t) \tag{5.26}$$

and use the following simple multivariable adaptive control algorithm (that was presented in [112], [58], [102], [70]):

$$u_p(t) = K_{e_y}(t)e_y(t) + K_{x_m}(t)x_m(t) + K_{u_m}(t)u_m(t) = K(t)r(t) \tag{5.27}$$

where

$$K(t) = [K_{e_y}(t) \quad K_{x_m}(t) \quad K_{u_m}(t)] \tag{5.28}$$

$$r^T(t) = [e_y^T(t) \quad x_m^T(t) \quad u_m^T(t)]. \tag{5.29}$$

The adaptive gains are a combination of "proportional" and "integral" gains,

$$\begin{aligned} K_p(t) &= e_y(t)r^T(t)\bar{T} = [e_y(t)e_y^T(t)\bar{T}_{e_y}, \quad e_y(t)x_m^T(t)\bar{T}_x, \\ & \qquad e_y(t)u_m^T(t)\bar{T}_u] && (5.30) \\ \dot{K}_I(t) &= e_y(t)r^T(t)T - \sigma K_I(t) && (5.31) \\ K(t) &= K_p(t) + K_I(t). && (5.32) \end{aligned}$$

Here T and $\bar{T}$ are (selected) positive definite scaling matrices of the form

$$T = \begin{bmatrix} T_{e_y} & 0 & 0 \\ 0 & T_x & 0 \\ 0 & 0 & T_u \end{bmatrix} \tag{5.33}$$

$$\bar{T} = \begin{bmatrix} \bar{T}_{e_y} & 0 & 0 \\ 0 & \bar{T}_x & 0 \\ 0 & 0 & \bar{T}_u \end{bmatrix}. \tag{5.34}$$

In general, the adaptive controller assumes that the controlled plant is of much larger order than the reference model that it tries to track. If the plant can move along some special trajectories (as discussed below) then perfect following can be achieved. In realistic cases when perfect following is not possible, the control system is required to keep the plant trajectories as "close" as possible to the desired trajectories, i.e. to guarantee tracking with bounded and preferably small errors.

The detailed analysis of stability and performance of the adaptive control is presented in Appendix 5A. First the existence of "ideal" bounded trajectories $x_p^*(t)$ and an "ideal" control $u_p^*(t)$ is investigated. They are to guarantee perfect following when no disturbances are present. In other words it is required that

$$y_a^*(t) = y_m(t) \tag{5.35}$$

where

$$y_a^*(t) = C_p(t)x_p^*(t) + D_p(t)u_p^*(t) \tag{5.36}$$

(5.35) gives the condition:

$$C_p(t)x_p^*(t) + D_p(t)u_p^*(t) = C_m(t)x_m(t) + D_m(t)u_m(t). \tag{5.37}$$

It is useful to represent $x_p^*(t)$ and $u_p^*(t)$ as linear combinations of $x_m(t)$ and $u_m(t)$

$$x_p^*(t) = X(t)x_m(t) + U(t)u_m(t) \tag{5.38}$$

$$u_p^*(t) = \tilde{K}_{x_m}x_m(t) + \tilde{K}_{u_m}u_m(t). \tag{5.39}$$

Note that $X(t)$ and $U(t)$ are time-varying coefficient matrices, while $\tilde{K}_{x_m}$ and $\tilde{K}_{u_m}$ are constant matrices. With this representation, Appendix 5A shows that condition (5.37) becomes:

$$\begin{aligned}&[C_p(t)X(t) + D_p(t)\tilde{K}_{x_m} - C_m(t)]x_m(t)\\ &+[C_p(t)U(t) + D_p(t)\tilde{K}_{u_m} - D_m(t)]u_m(t) = 0.\end{aligned} \tag{5.40}$$

Target ideal trajectories exist usually, since equation (5.40) has many more variables than equations, and thus has solutions in general for the unknown matrices $X(t)$, $U(t)$, $\tilde{K}_{x_m}$, and $\tilde{K}_{u_m}$. Thus the relation $y_a^*(t) = y_m(t)$ can be assumed to be satisfied. Yet, it should be understood that these are only target ideal trajectories and that the controlled plant may or may not reach them. For the plant to exactly move along these trajectories, (i.e. perfectly track), trajectories of the form (5.38)–(5.39) must also satisfy the plant differential equation (5.22) for $d_i(t) = 0$ and $d_0(t) = 0$. As shown in Appendix 5A the plant can reach perfect tracking if the matrices $X(t), U(t), \tilde{K}_{x_m}$, and $\tilde{K}_{u_m}$ satisfy the following two conditions simultaneously:

$$\begin{aligned}&[C_p(t)X(t) + D_p(t)\tilde{K}_{x_m} - C_m(t)]x_m(t)\\ &+[C_p(t)U(t) + D_p(t)\tilde{K}_{u_m} - D_m(t)]u_m(t) = 0\end{aligned} \tag{5.41}$$

$$\begin{aligned}&[A_p(t)X(t) + B_p(t)\tilde{K}_{x_m} - \dot{X}(t) - X(t)A_m(t)]x_m(t)\\ &+[A_p(t)U(t) + B_p(t)\tilde{K}_{u_m} - X(t)B_m(t)\\ &-\dot{U}(t)]u_m(t) - U(t)\dot{u}_m(t) = 0.\end{aligned} \tag{5.42}$$

For the general tracking case, it is assumed that (5.41) is satisfied exactly. However, solutions of (5.41) do not necessarily make (5.42) zero in general. Thus, whenever (5.42) does not vanish, it plays the role of an internal disturbance that may prevent the plant from tracking with zero error even when no external disturbances are present.

To monitor the difference between the actual plant state and the "ideal" state that allows perfect tracking, one defines the state error as

$$e_x(t) = x_p^*(t) - x_p(t) \tag{5.43}$$

The output error then becomes

$$e_y(t) = y_m(t) - y_a(t) = y_a^*(t) - y_a(t) \tag{5.44}$$

In the most general case when (5.42) is not exactly satisfied, and disturbances are present, the differential equation of the state error is (Appendix 5A)

$$\dot{e}_x(t) = A_{pc}(t)e_x(t) - B_{pc}(t)[K(t) - \tilde{K}]r(t) - F(t) \tag{5.45}$$

where $F(t)$ is the "residual" term

$$\begin{aligned} F(t) &= [A_p(t)X(t) + B_p(t)\tilde{K}_{x_m} - \dot{X}(t) - X(t)A_m(t)]x_m(t) \\ &+ [A_p(t)U(t) + B_p(t)\tilde{K}_{u_m} - X(t)B_m(t) - \dot{U}(t)]u_m(t) \\ &- U(t)\dot{u}_m(t) - B_p(t)\tilde{K}_{ec}d_0(t) - d_i(t) \end{aligned} \tag{5.46}$$

The residual term $F(t)$ is thus a combined result of all internal and external disturbances. When $F(t)$ is nonzero, bounded rather than vanishing error is obtained. $F(t)$ vanishes when no disturbances are present, i.e. $d_0(t) = 0$ and $d_i(t) = 0$, and, in addition, conditions (5.41)–(5.42) hold. The error equation (5.45) then becomes

$$\dot{e}_x(t) = A_{pc}(t)e_x(t) - B_{pc}(t)[\ K(t) - \tilde{K}]r(t) \tag{5.47}$$

It is shown in Appendix 5A that in this case the adaptive controller indeed can achieve perfect tracking.

Theorem 1: The model reference adaptive controller (5.27)–(5.32) guarantees robust adaptive stabilization of the ASP plant (5.22)–(5.23) in the presence of any bounded input commands and any bounded input and output disturbances. In those situations when perfect tracking is possible (i.e., $F(t)$ in (5.46) vanishes), the adaptive algorithm with $\sigma = 0$ reaches asymptotically perfect tracking.

Proof: The detailed proofs of stability are presented in Appendix 5A (for causal ASP systems) and in Appendix 5B (for strictly causal ASP systems).

This theorem gives the necessary theoretical background for the successful applications of the algorithm in nonstationary systems [58], [20], [61], [24].

Although perfect following is not guaranteed in the most general case, the adaptive gains increase whenever the errors tend to increase, thus reducing the tracking errors. In most practical cases, the plant will finally track with very small errors. In this context it is worth mentioning again that, while stability of other adaptive algorithms usually requires slow adaptation rates, the simple adaptive algorithm proposed here guarantees robust stability at any rate of adaptation. The designer is thus free to select those rates that fit the specific application with the immediate practical implication that the higher the adaptation coefficients T and $\bar{T}$, the faster the transient response and the smaller the steady-state error. In particular, reference [24] presents a good illustration of the performance of this adaptive with some examples from the control system literature, that had been selected because they were considered to be very difficult time-varying uncertain plants.

5.4 The "Almost Passivity" Lemmas

In the previous chapters and in references [113], [58], [114], the desired almost positive realness property has been shown to be related in continuous and discrete-time stationary linear systems to "almost stabilizability," namely, to output stabilizability via static or dynamic output feedback, as defined by Schumacher [115]. This section extends this important relationship to "almost passivity" of time-varying systems. As in stationary systems, the extension will first include only those systems that can be stabilized via output feedback.

While in stationary systems, passivity and stabilizability could be intuitively related [113], [58], [114] using the pole-zero description of the systems [116], the state-space representation will be used in the time varying case.
Lemma 1: Consider now the strictly causal plant

$$\dot{x}_p(t) = A_p(t)x_p(t) + B_p(t)u_p(t) \tag{5.48}$$

$$y_p(t) = C_p(t)x_p(t) \tag{5.49}$$

Assume that there exists a positive definite symmetric output feedback matrix $K_y(t)$, and, thus a control feedback $u_p(t) = -K_y(t)y_p(t)$, such that the closed loop system

$$\dot{x}_p(t) = A_y(t)x_p(t) \tag{5.50}$$

$$y_p(t) = C_p(t)x_p(t) \tag{5.51}$$

is globally and uniformly asymptotically stable, where

$$A_y(t) = A_p(t) - B_p(t)K_y(t)C_p(t). \tag{5.52}$$

Under the above stability assumptions the augmented open-loop system (5.1)–(5.3) with $D_p(t) = K_y^{-1}(t)$ is "almost strictly passive (ASP)."

Proof: The detailed proof of Lemma 1 is presented in Appendix 5C.

Lemma 1 proclaims that if a plant can be stabilized by some output feedback, the inverse of this feedback can be used in parallel with the plant to obtain an augmented ASP system. Yet, the relationship between stabilizability and passivity is more general. The applicability of "almost passivity" can be extended to general nonstationary systems that need some dynamic output feedback in order to reach stability.

Lemma 2 ("The almost passivity lemma"): Let G_p be a time-varying system of the form (5.48)–(5.49). Let $H : \{A_f(t), B_f(t), C_f(t), D_f(t)\}$ be a dynamic stabilizing linear feedback for G_p with the inverse $H^{-1} : \{A_f - B_f D_f^{-1} C_f,\ B_f D_f^{-1},\ D_f^{-1} C_f,\ D_f^{-1}\}$. Then, the augmented system $G_a : \{A(t),\ B(t),\ C(t),\ D(t)\}$ defined as $G_a = G_p + H^{-1}$ is "almost strictly passive (ASP)".

Proof: The detailed proof of Lemma 2 can be found in [21].

5.5 Adaptive Control of Nonlinear Systems

The unified treatment of SAC techniques (for both SISO and multivariable linear systems) was made possible by the state-space representation, and by the Lyapunov stability analysis of the adaptive algorithms. This tools will also allow extensions of the SAC techniques to nonlinear systems that are linear in control. This class includes, among others, robots and missiles.

To develop the SAC tools for these systems, consider nonlinear plants (not necessarily stable) that are modeled by the following state-space representation

$$\dot{x}_p(t) = A_p(x_p)x_p(t) + B_p(x_p)u_p(t) \tag{5.53}$$

$$y_p(t) = C_p(x_p)x_p(t) \tag{5.54}$$

$$y_a(t) = y_p(t) + D_p(x_p)u_p(t) = C_p(x_p)x_p(t) + D_p(x_p)u_p(t) \tag{5.55}$$

Here $x_p(t)\epsilon R^n$, $y_p(t)\epsilon R^m$, $y_a(t)\epsilon R^m$, $u_p(t)\epsilon R^m$, and $A_p(x_p)$, $B_p(x_p)$, $C_p(x_p)$, $D_p(x_p)$ are uniformly bounded matrices of appropriate dimensions. To keep the derivation general, one may use the (not necessarily strictly causal) form $\{A_p(x_p),\ B_p(x_p),\ C_p(x_p),\ D_p(x_p)\}$. Results are perfectly valid if the systems are strictly causal, (namely, $D_p(x_p) = 0$).

The development is very similar to the analysis of linear time-invariant systems, only instead of including time-varying entries, the system matrices $A_p(x_p)$, $B_p(x_p)$, $C_p(x_p)$, $D_p(x_p)$ are comprised of nonlinear functions of the state.

Definition 1: The system (5.53)–(5.55) is called "almost strictly passive (ASP)" if there exists a positive definite static feedback matrix K_e such that the resulting closed-loop system is strictly passive (SP).

By closing the loop with the controller

$$u_p(t) = -K_e y_a(t) + u_{pc}(t) \tag{5.56}$$

one gets the following closed-loop system

$$\dot{x}_p(t) = A_{pc}(x_p)x_p(t) + B_{pc}(x_p)u_{pc}(t) \tag{5.57}$$

$$y_a(t) = C_{pc}(x_p)x_p(t) + D_{pc}(x_p)u_{pc}(t) \tag{5.58}$$

where

$$A_{pc}(x_p) = A_p(x_p) - B_p(x_p)K_{ec}(x_p)C_p(x_p) \tag{5.59}$$

$$K_{ec}(x_p) = K_e[\, I + D_p(x_p)K_e \,]^{-1} = [\, I + K_e D_p(x_p) \,]^{-1} K_e \tag{5.60}$$

$$B_{pc}(x_p) = B_p(x_p)[\, I + K_e D_p(x_p) \,]^{-1} \tag{5.61}$$

$$C_{pc}(x_p) = [\, I + D_p(x_p)K_e \,]^{-1} C_p(x_p) \tag{5.62}$$

and

$$D_{pc}(x_p) = [\, I + D_p(x_p)K_e \,]^{-1} D_p(x_p) = D_p(x_p)[\, I + K_e D_p(x_p) \,]^{-1}. \tag{5.63}$$

It is assumed first that $D_p(x_p) > 0$ and $D_{pc}(x_p) > 0$, which means that $D_p(x_p)$ and $D_{pc}(x_p)$ are positive definite (and hence, also nonsingular) for all $x_p \epsilon R^n$.

The closed-loop system (5.57)–(5.58) is called "strictly passive" if there exists two uniformly bounded positive definite matrices, $P(x_p)$ and $Q(x_p)$, and two matrices $L(x_p)\epsilon R^{m\times n}$ and $W(x_p)\epsilon R^{m\times m}$, that satisfy the following relations:

$$\dot{P}(x_p) + P(x_p)A_{pc}(x_p) + A_{pc}^T(x_p)P(x_p) = -Q(x_p)$$

$$-L^T(x_p)L(x_p) \tag{5.64}$$

$$P(x_p)B_{pc}(x_p) = C_{pc}^T(x_p) - L^T(x_p)W(x_p) \tag{5.65}$$

$$D_{pc}(x_p) + D_{pc}^T(x_p) = W^T(x_p)W(x_p) \tag{5.66}$$

where $\dot{P}(x_p) = \frac{dP(x_p)}{dt} = \frac{\partial P(x_p)}{\partial x_p}\frac{dx_p}{dt}$.

It is easy to see that relations (5.64)–(5.66) are equivalent to the Riccati equation [86]

$$\dot{P}(x_p) + P(x_p)A_{pc}(x_p) + A_{pc}^T(x_p)P(x_p) +$$

$$+[P(x_p)B_{pc}(x_p) - C_{pc}^T(x_p)][D_{pc}(x_p) + D_{pc}^T(x_p)]^{-1}$$

$$[B_{pc}^T(x_p)P(x_p) - C_{pc}(x_p)] < -Q(x_p) < 0 \tag{5.67}$$

In strictly causal systems, when $D_p(x_p) = 0$, the strict passivity relations are:

$$\dot{P}(x_p) + P(x_p)[A_p(x_p) - B_p(x_p)K_eC_p(x_p)]$$

$$+[A_p(x_p) - B_p(x_p)K_eC_p(x_p)]^T P(x_p) = -Q(x_p) < 0 \tag{5.68}$$

$$P(x_p)B_p(x_p) = C_p^T(x_p) \tag{5.69}$$

One can use the stabilizability of the nonlinear system in order to implement "almost passive" systems:

Lemma 3 (The almost passivity lemma): Let G_p be any *nonlinear*, (possibly unstable) system with a state-space representation of the form (5.53)–(5.55). Let $H : \{A_f, B_f, C_f, D_f\}$ be a stabilizing *linear* controller for G_p with the inverse $H^{-1} : \{A_f - B_fD_f^{-1}C_f, B_fD_f^{-1}, D_f^{-1}C_f, D_f^{-1}\}$. Let G_a be the augmented system $G_a = G_p + H^{-1}$.

Then $G_a : \{A(x), B(x), C(x), C(x)\}$ is "almost strictly passive (ASP)." In other words, G_a satisfies the almost strict passivity relations of the form (5.64)–(5.67) [23].

In a manner similar to that by which stability in nonstationary plants was proved, one can prove the following theorem on stability:

Theorem 2: The model reference adaptive controller (5.27)–(5.32) guarantees robust adaptive stabilization of ASP plants of the form (5.53)–(5.55) in the presence of any bounded input commands and any bounded input or output disturbances [117], [22], [23].

Appendix 5A

Proof of Stability for the Algorithm (5.27)–(5.32)

This section shall first question the existence of such "ideal" bounded trajectories $x_p^*(t)$ and "ideal" controls $u_p^*(t)$ that allow perfect following when no disturbances are present. Perfect tracking requires that

$$y_a^*(t) = y_m(t) \tag{5A.1}$$

This gives the condition

$$C_p(t)x_p^*(t) + D_p(t)u_p^*(t) = C_m(t)x_m(t) + D_m(t)u_m(t) \tag{5A.2}$$

It is useful to represent $x_p^*(t)$ and $u_p^*(t)$ as linear combinations of $x_m(t)$ and $u_m(t)$

$$x_p^*(t) = X(t)x_m(t) + U(t)u_m(t) \tag{5A.3}$$

$$u_p^*(t) = \tilde{K}_{x_m}x_m(t) + \tilde{K}_{u_m}u_m(t) \tag{5A.4}$$

Substituting (5A.3) and (5A.4) into (5A.2) gives the equivalent condition

$$\begin{aligned} &C_p(t)X(t)x_m(t) + C_p(t)U(t)u_m(t) + D_p(t)\tilde{K}_{x_m}x_m(t) \\ &+D_p(t)\tilde{K}_{u_m}u_m(t) = C_m(t)x_m(t) + D_m(t)u_m(t) \end{aligned} \tag{5A.5}$$

or

$$\begin{aligned} &[C_p(t)X(t) + D_p(t)\tilde{K}_{x_m} - C_m(t)]x_m(t) + [C_p(t)U(t) \\ &+D_p(t)\tilde{K}_{u_m} - D_m(t)\,]u_m(t) = 0 \end{aligned} \tag{5A.6}$$

Note that target ideal trajectories exist almost everywhere, because equation (5A.6) has many more variables than equations, and, in general, it has solutions for the unknown matrices $X(t), U(t), \tilde{K}_{x_m}$, and $\tilde{K}_{u_m}$. To illustrate this, note that (5A.6) has solutions if

$$C_p(t)X(t) = C_m(t) - D_p(t)\tilde{K}_{x_m} \tag{5A.7}$$

$$C_p(t)U(t) = D_m(t) - D_p(t)\tilde{K}_{u_m} \tag{5A.8}$$

Equations (5A.7) and (5A.8) have solutions for $X(t)$ and $U(t)$, for any selection of $\tilde{K}_{x_m}$, and $\tilde{K}_{u_m}$, if $C_p(t)$ is maximal rank for all values of t.

However, one cannot expect that solutions to (5A.6) will automatically satisfy the plant differential equation (5.22). To test the possibility of perfect following, differentiate (5A.3) to get

$$\dot{x}_p^*(t) = \dot{X}(t)x_m(t) + X(t)\dot{x}_m(t) + \dot{U}(t)u_m(t) + U(t)\dot{u}_m(t) \tag{5A.9}$$

Substituting $\dot{x}_m(t)$ from (5.24) gives:

$$\dot{x}_p^*(t) = \dot{X}(t)x_m(t) + X(t)A_m(t)x_m(t) + X(t)B_m(t)u_m(t) +$$

$$\dot{U}(t)u_m(t) + U(t)\dot{u}_m(t) \tag{5A.10}$$

Adding and subtracting $A_p(t)x_p^*(t) + B_p(t)u_p^*(t)$ gives

$$\begin{aligned}\dot{x}_p^*(t) &= \dot{X}(t)x_m(t) + X(t)A_m(t)x_m(t) + X(t)B_m(t)u_m(t) + \dot{U}(t)u_m(t)\\ &+ U(t)\dot{u}_m(t) + A_p(t)x_p^*(t) + B_p(t)u_p^*(t)\\ &- A_p(t)x_p^*(t) - B_p(t)u_p^*(t)\end{aligned} \tag{5A.11}$$

Substituting $x_p^*(t)$ and $u_p^*(t)$ in the last terms of (5A.11) gives

$$\begin{aligned}\dot{x}_p^*(t) &= \dot{X}(t)x_m(t) + X(t)A_m(t)x_m(t) + X(t)B_m(t)u_m(t) + \dot{U}(t)u_m(t)\\ &+ U(t)\dot{u}_m(t) + A_p(t)x_p^*(t) + B_p(t)u_p^*(t)\\ &- A_p(t)[X(t)x_m(t) + U(t)u_m(t)]\\ &- B_p(t)[\tilde{K}_{x_m}x_m(t) + \tilde{K}_{u_m}u_m(t)]\end{aligned} \tag{5A.12}$$

or

$$\begin{aligned}\dot{x}_p^*(t) &= A_p(t)x_p^*(t) + B_p(t)u_p^*(t) \\ &- [A_p(t)X(t) + B_p(t)\tilde{K}_{x_m} - \dot{X}(t) - X(t)A_m(t)]x_m(t) \\ &- [A_p(t)U(t) + B_p(t)\tilde{K}_{u_m} - X(t)B_m(t) - \dot{U}(t)]u_m(t)] \\ &+ U(t)\dot{u}_m(t) \end{aligned} \tag{5A.13}$$

By comparing (5A.13) with (5.22), when $d_0(t) = 0$ and $d_i(t) = 0$, a condition for the target ideal trajectories to be solutions of the plant is obtained. In the absence of any external disturbances (namely, when $d_0(t) = 0, d_i(t) = 0$), the condition is:

$$\begin{aligned}&[A_p(t)X(t) + B_p(t)\tilde{K}_{x_m} - \dot{X}(t) - X(t)A_m(t)]x_m(t) \\ &+[A_p(t)U(t) + B_p(t)\tilde{K}_{u_m} - X(t)B_m(t) - \dot{U}(t)]u_m(t)] \\ &-U(t)\dot{u}_m(t) = 0\end{aligned} \tag{5A.14}$$

Thus, perfect tracking is possible if the matrices $X(t), U(t), \tilde{K}_{x_m}$, and $\tilde{K}_{u_m}$ satisfy the following two conditions simultaneously:

$$\begin{aligned}&[C_p(t)X(t) + D_p(t)\tilde{K}_{x_m} - C_m(t)]x_m(t) \\ &+[C_p(t)U(t) + D_p(t)\tilde{K}_{u_m} - D_m(t)]u_m(t) = 0\end{aligned} \tag{5A.15}$$

$$\begin{aligned}&[A_p(t)X(t) + B_p(t)\tilde{K}_{x_m} - \dot{X}(t) - X(t)A_m(t)]x_m(t) \\ &+[A_p(t)U(t) + B_p(t)\tilde{K}_{u_m} - X(t)B_m(t) - \dot{U}(t)]u_m(t) \\ &-U(t)\dot{u}_m(t) = 0\end{aligned} \tag{5A.16}$$

It is difficult to find the most general plant and tracking conditions that would guarantee satisfaction of relations (5A.15)–(5A.16).

In the constant input case, $u_m(t) = u_m$ =constant, $\dot{u}_m(t) = 0$, the output of the stable model driven by the constant input u_m eventually reaches the constant value x_{ss}, and the conditions for asymptotic tracking become

$$\begin{aligned}&[C_p(t)X(t) + D_p(t)\tilde{K}_{x_m} - C_m(t)]x_{ss} \\ &+[C_p(t)U(t) + D_p(t)\tilde{K}_{u_m} - D_m(t)]u_m = 0\end{aligned} \tag{5A.17}$$

$$
\begin{aligned}
&[A_p(t)X(t) + B_p(t)\tilde{K}_{x_m} - \dot{X}(t) - X(t)A_m(t)]x_{ss} \\
&+[A_p(t)U(t) + B_p(t)\tilde{K}_{u_m} - X(t)B_m(t) - \dot{U}(t)]u_m = 0
\end{aligned} \tag{5A.18}
$$

A solution for constant input exists if the coefficients of $x_m(t)$ and $u_m(t)$ in (5A.17)–(5A.18) vanish giving:

$$C_p(t)X(t) + D_p(t)\tilde{K}_{x_m} - C_m(t) = 0 \tag{5A.19}$$

$$C_p(t)U(t) + D_p(t)\tilde{K}_{u_m} - D_m(t) = 0 \tag{5A.20}$$

$$A_p(t)X(t) + B_p(t)\tilde{K}_{x_m} - \dot{X}(t) - X(t)A_m(t) = 0 \tag{5A.21}$$

$$A_p(t)U(t) + B_p(t)\tilde{K}_{u_m} - X(t)B_m(t) - \dot{U}(t) = 0 \tag{5A.22}$$

These are only sufficient conditions, because the relations (5A.17)–(5A.18) consists of fewer equations, and thus will have solutions in more situations than do (5A.19)–(5A.22).

Perfect tracking may not be possible in time-varying systems, either because of internal and external disturbances, or because of the complexity of the input. In the following development, whenever perfect following is not possible, the various matrices $X(t), U(t), \tilde{K}_{x_m}$, and $\tilde{K}_{u_m}$ are considered to satisfy only the output tracking equation (5A.15) exactly.

Notice that none of the matrices whose existence was tested above are needed for implementation of the adaptive algorithm (5.27)–(5.32). Only the conditions of existence are subsequently used for the stability proof.

Since it is desired that the plant states reach the "ideal" states that allow perfect tracking, one defines the state error as

$$e_x(t) = x_p^*(t) - x_p(t) \tag{5A.23}$$

and the output error then becomes

$$e_y(t) = y_m(t) - y_a(t) = y_a^*(t) - y_a(t) \tag{5A.24}$$

The following algebraic computations lead to the differential equation (5A.41) for the state error. One obtains from (5A.24)

$$e_y(t) = C_p(t)x_p^*(t) + D_p(t)u_p^*(t) - C_p(t)x_p(t) - D_p(t)u_p(t) - d_0(t) \tag{5A.25}$$

$$
\begin{aligned}
&e_y(t) = C_p(t)e_x(t) + D_p(t)[\tilde{K}_{x_m}x_m(t)\tilde{K}_{u_m}u_m(t)] \\
&+D_p(t)\tilde{K}_{e_y}e_y(t) - D_p(t)\tilde{K}_{e_y}e_y(t) - D_p(t)K(t)r(t) - d_0(t)
\end{aligned} \tag{5A.26}
$$

and

$$e_y(t) = C_p(t)e_x(t) - D_p(t)[K(t) - \tilde{K}]r(t) - D_p(t)\tilde{K}_{e_y}e_y(t) - d_0(t) \quad (5A.27)$$

Then, from (5A.27)

$$[I + D_p(t)\tilde{K}_{e_y}]e_y(t) = C_p(t)e_x(t) - D_p(t)[K(t) - \tilde{K}]r(t) - d_0(t) \quad (5A.28)$$

$$\begin{aligned} e_y(t) &= [I + D_p(t)\tilde{K}_{e_y}]^{-1}C_p(t)e_x(t) \\ &- [I + D_p(t)\tilde{K}_{e_y}]^{-1}D_p(t)[K(t) - \tilde{K}]r(t) \\ &- [I + D_p(t)\tilde{K}_{e_y}]^{-1}d_0(t) \end{aligned} \quad (5A.29)$$

and finally

$$e_y(t) = C_{pc}(t)e_x(t) - D_{pc}(t)[K(t) - \tilde{K}]r(t) - [I + D_p(t)\tilde{K}_{e_y}]^{-1}d_0(t) \quad (5A.30)$$

where

$$\tilde{K} = [\tilde{K}_{e_y} \quad \tilde{K}_{x_m} \quad \tilde{K}_{u_m}] \quad (5A.31)$$

and where $C_{pc}(t)$ and $D_{pc}(t)$ were defined in (5.10)–(5.11).

For the state error equation, differentiate $e_x(t)$ in (5A.23) and substitute (5A.13) and (5.22) to get

$$\begin{aligned} \dot{e}_x(t) &= \dot{x}_p^*(t) - \dot{x}_p(t) \\ &= A_p(t)x_p^*(t) + B_p(t)u_p^*(t) \\ &- [A_p(t)X(t) + B_p(t)\tilde{K}_{x_m} - \dot{X}(t) - X(t)A_m(t)]x_m(t) \\ &- [A_p(t)U(t) + B_p(t)\tilde{K}_{u_m} - X(t)B_m(t) - \dot{U}(t)]u_m(t)] + U(t)\dot{u}_m(t) \\ &- A_p(t)x_p(t) - B_p(t)u_p(t) - d_i(t) \end{aligned} \quad (5A.32)$$

Substituting $u_p^*(t)$ from (5A.4) and $u_p(t)$ from (5.27) into (5A.36) gives

$$\begin{aligned}
\dot{e}_x(t) &= A_p(t)x_p^*(t) + B_p(t)\tilde{K}_{x_m}x_m(t) + B_p(t)\tilde{K}_{u_m}u_m(t) \\
&- [A_p(t)X(t) + B_p(t)\tilde{K}_{x_m} - \dot{X}(t) - X(t)A_m(t)]x_m(t) \\
&- [A_p(t)U(t) + B_p(t)\tilde{K}_{u_m} - X(t)B_m(t) - \dot{U}(t)]u_m(t)] \\
&+ U(t)\dot{u}_m(t) - A_p(t)x_p(t) - B_p(t)K(t)r(t) - d_i(t) \qquad (5A.33)
\end{aligned}$$

or

$$\begin{aligned}
\dot{e}_x(t) &= A_p(t)[x_p^*(t) - x(t)] - B_p(t)K(t)r(t) + B_p(t)\tilde{K}_{x_m}x_m(t) \\
&+ B_p(t)\tilde{K}_{u_m}u_m(t - [A_p(t)X(t) + B_p(t)\tilde{K}_{x_m} - \dot{X}(t) \\
&- X(t)A_m(t)]x_m(t) - [A_p(t)U(t) + B_p(t)\tilde{K}_{u_m} \\
&- X(t)B_m(t) - \dot{U}(t)]u_m(t)] + U(t)\dot{u}_m(t) - d_i(t) \qquad (5A.34)
\end{aligned}$$

Adding and subtracting $B_p(t)\tilde{K}_{e_y}e_y(t)$ and substituting $e_y(t)$ from (5A.29) gives after some algebra:

$$\dot{e}_x(t) = A_{pc}(t) - B_{pc}(t)[K(t) - \tilde{K}]r(t) - F(t) \qquad (5A.35)$$

Here $F(t)$ is the residual term

$$\begin{aligned}
F(t) &= [A_p(t)X(t) + B_p(t)\tilde{K}_{x_m} - \dot{X}(t) - X(t)A_m(t)]x_m(t) \\
&+ [A_p(t)U(t) + B_p(t)\tilde{K}_{u_m} - X(t)B_m(t) - \dot{U}(t)]u_m(t) - U(t)\dot{u}_m(t) \\
&- B_p(t)\tilde{K}_{ec}d_0(t) - d_i(t) \qquad (5A.36)
\end{aligned}$$

Having finally developed a differential equation for the error, it will be shown below that perfect tracking can be reached whenever the residual term $F(t)$ vanishes. In this case, the error equation (5A.35) becomes

$$\dot{e}_x(t) = A_{pc}(t)e_x(t) - B_{pc}(t)[K(t) - \tilde{K}]r(t) \qquad (5A.37)$$

Proof of stability

The following quadratic Lyapunov function will be used for proof of stability for the adaptive system represented by the dynamic equations (5A.35) and (5.31):

$$V(t) = e_x^T(t)P(t)e_x(t) + tr\{ [K_I(t) - \tilde{K}]T^{-1}[K_I(t) - \tilde{K}]^T\}. \quad (5A.38)$$

In (5A.38), $P(t)$ is the positive definite matrix defined in (5.12)–(5.14) and T is the positive definite scaling matrix defined in (5.31). The derivative of $V(t)$ "along the trajectories" of (5A.37) and (5.31) is:

$$\begin{aligned}\dot{V}(t) &= e_x^T(t)\dot{P}(t)e_x(t) + \dot{e}_x^T(t)P(t)e_x(t) + e_x^T(t)P(t)\dot{e}_x(t) + \\ &+2tr\{[K_I(t) - \tilde{K}]T^{-1}\dot{K}_I^T(t)\}.\end{aligned} \quad (5A.39)$$

Substituting $\dot{e}_x(t)$ from (5A.32) and $\dot{K}_I(t)$ from (5.25) gives

$$\begin{aligned}\dot{V}(t) &= e_x^T(t)\dot{P}(t)e_x(t) + \dot{e}_x^T(t)A_{pc}^T(t)P(t)e_x(t) \\ &-r^T(t)[K(t) - \tilde{K}]^T B_{pc}^T(t)P(t)e_x(t) - F^T(t)P(t)e_x(t) + \\ &+e_x^T(t)P(t)A_{pc}(t)e_x(t) - e_x^T(t)P(t)B_{pc}(t)[K(t) - \tilde{K}]r(t) \quad (5A.40) \\ &+e_x^T(t)P(t)F(t) + 2tr\{[K_I(t) - \tilde{K}]T^{-1}[e_y(t)rT(t)T \\ &-\sigma K_I(t)]\}\end{aligned}$$

or

$$\begin{aligned}\dot{V}(t) &= e_x^T(t)[\dot{P}(t) + P(t)A_{pc}(t) + A_{pc}^T(t)P(t)]e_x(t) \\ &-e_x^T(t)P(t)B_{pc}(t)[K(t) - \tilde{K}]r(t) \\ &-r^T(t)[K(t) - \tilde{K}]^T B_{pc}^T(t)P(t)e_x(t) \\ &-2e_x^T(t)P(t)F(t) - e_y^T(t)[K_I(t) - \tilde{K}]r(t) \\ &+r^T(t)[K_I(t) - \tilde{K}]^T e_y(t) \\ &-2\sigma tr\{[K_I(t) - \tilde{K}]T^{-1}K_I^T(t)\end{aligned} \quad (5A.41)$$

Substituting

$$K_I(t) = K(t) - K_p(t) = K(t) - e_y(t)r^T(t)\bar{T} \tag{5A.42}$$

gives

$$\begin{aligned}\dot{V}(t) &= e_x^T(t)[\dot{P}(t) + P(t)A_{pc}(t) + A_{pc}^T(t)P(t)]e_x(t) \\ &-e_x^T(t)P(t)B_{pc}(t)[K(t) - \tilde{K}]r(t) \\ &-r^T(t)[K(t) - \tilde{K}]^T B_{pc}^T(t)P(t)e_x(t) \\ &-2e_x^T(t)P(t)F(t) + e_y^T(t)[K(t) - \tilde{K}]r(t) + r^T(t)[K(t) \\ &-\tilde{K}]^T e_y(t) - 2e_y^T(t)e_y(t)r^T(t)\bar{T}r(t) \\ &-2\sigma tr\{[K_I(t) - \tilde{K}]T^{-1}[K_I(t) - \tilde{K}]^T\} \\ &-2\sigma tr\{[K_I(t) - \tilde{K}]T^{-1}\tilde{K}^T\}.\end{aligned} \tag{5A.43}$$

Using the passivity relations (5.12) - (5.14) and the output error equation (5A.30) gives

$$\begin{aligned}
\dot{V}(t) &= -e_x^T(t)Q(t)e_x(t) - e_x^T(t)L^T(t)L(t)e_x(t) \\
&-e_x^T(t)C_{pc}^T(t)[K(t) - \tilde{K}]r(t) \\
&-r^T(t)[K(t) - \tilde{K}]^T C_{pc}(t)e_x(t) \\
&-e_x^T(t)L^T(t)W(t)[K(t) - \tilde{K}]r(t) \\
&+r^T(t)[K(t) - \tilde{K}]^T W^T(t)L(t)e_y(t) - 2e_x^T(t)P(t)F(t) \\
&+e_x^T(t)C_{pc}^T(t)[K(t) - \tilde{K}]r(t) \\
&+r^T(t)[K(t) - \tilde{K}]^T C_{pc}(t)e_x(t) \\
&-r^T(t)[K(t) - \tilde{K}]^T D_{pc}^T(t)[K(t) - \tilde{K}]r(t) \\
&-r^T(t)[K(t) - \tilde{K}]^T D_{pc}(t)[K(t) - \tilde{K}]r(t) \\
&-d_0^T(t)[I + D_p(t)\tilde{K}_{e_y}]^{-1}[K(t) \\
&-\tilde{K}]r(t) - r^T(t)[K(t) - \tilde{K}]^T[I + D_p(t)\tilde{K}_{e_y}]^{-1}d_0(t) \\
&-2e_y^T(t)e_y(t)r^T(t)\bar{T}r(t) \\
&-2\sigma tr\{[K_I(t) - \tilde{K}]T^{-1}[K_I(t) - \tilde{K}]T\} \\
&-2\sigma tr\{[K_I(t) - \tilde{K}]T^{-1}\tilde{K}^T\}
\end{aligned} \tag{5A.44}$$

Rearranging and using (5.14) gives

$$\begin{aligned}\dot{V}(t) &= -e_x^T(t)Q(t)e_x(t)\\ &-\{L(t)e_x(t) - W(t)[K(t) - \tilde{K}]r(t)\}^T\{L(t)e_x(t) - W(t)[K(t)\\ &-\tilde{K}]r(t)\} - 2e_y^T(t)e_y(t)r^T(t)\bar{T}r(t)\\ &-2\sigma tr\{[K_I(t) - \tilde{K}]T^{-1}[K_I(t) - \tilde{K}]^T\} \qquad (5A.45)\\ &-2\sigma tr\{[KI(t) - \tilde{K}]T^{-1}\tilde{K}^T\}\\ &-2r^T(t)[K(t) - \tilde{K}]^T[I + D_p(t)\tilde{K}_{e_y}]^{-1}d_0(t)\\ &+2e_x^T(t)P(t)F(t).\end{aligned}$$

It is now shown that in the particular cases when perfect following is possible and when neither input nor disturbances are present, the adaptive algorithm guarantees perfect tracking if $\sigma = 0$ in the adaptive gain (5.31). Substituting $F(t) = 0, d_0(t) = 0, \sigma = 0$ in (5A.45) gives

$$\begin{aligned}\dot{V}(t) &= -e_x^T(t)Q(t)e_x(t)\\ &- \{L(t)e_x(t)W(t)[K(t) - \tilde{K}]r(t)\}^T\{L(t)e_x(t) - W(t)[K(t)\\ &- \tilde{K}]r(t)\} - 2e_y^T(t)e_y(t)r^T(t)\bar{T}r(t). \qquad (5A.46)\end{aligned}$$

In this case all terms of $\dot{V}(t)$ are either negative definite or negative semidefinite. According to LaSalle's Invariance Principle [83] all states, gains and errors are bounded, and the system reaches asymptotically the region that guarantees $\dot{V}(t) \equiv 0$, which, from (5A.46), is equivalent to $e_x(t) = 0$ and $e_y(t) = 0$. In other words, perfect tracking is achieved.

For the general case, when perfect following is not possible, one can see that there exist some positive coefficients $\alpha_1, \alpha_2, \alpha_3, \ldots, \alpha_9$ such that one gets from (5A.45)

$$\begin{aligned}\dot{V}(t) &\geq \alpha_1||e_x(t)||^2 - \alpha_2||[K(t) - \tilde{K}]r(t)||^2 - \alpha_3||e_y(t)||^4\\ &-\alpha_4||e_y(t)||^2||x_m(t)||^2\\ &-\alpha_5||e_y(t)||^2||u_m(t)||^2 - \alpha_6\sigma||K_I(t) - \tilde{K}||^2 + \alpha_7\sigma||K_I(t) - \tilde{K}|| \qquad (5A.47)\\ &+\alpha_8||[K(t) - \tilde{K}]r(t)|| + \alpha_9||e_x(t)||\end{aligned}$$

If either $||e_x(t)||, ||(K(t) - \tilde{K})r(t)||$, or $||K_I(t) - \tilde{K})||$ increases beyond some bound, the negative definite quadratic terms in (5A.47) become dominant, and thus $\dot{V}(t)$ becomes negative. The positive definite quadratic form of the Lyapunov function $V(t)$ then guarantees that all the dynamic values (namely $e_x(t), K_I(t)$, and $e_y(t)$) are bounded. Note that, if the σ-terms were missing in (5.25), then the adaptive controller could not have avoided those situations when the nonnegative definite term that include $||K_I(t) - \tilde{K}||$ might increase in spite of $||[K(t) - \tilde{K})]r(t)||$ being small. In this case, $\dot{V}(t)$ could be positive for large $||K_I(t) - \tilde{K}||$, leading to possible divergence. The existence of the negative quadratic σ-term in (5A.47) shows that for large values of $||K_I(t) - \tilde{K}||$ the negative terms are dominant and all values are therefore bounded.

On the other hand, because of the nonnegative σ-term in (5A.47), one cannot guarantee that $\dot{V}(t)$ is negative for low values of the variables. The price for robustness in nonideal situations, is thus paid by (possibly) nonzero tracking errors even in ideal situations. Observe the effect of the negative definite term $-2e_y^T(t)e_y(t)r^T(t)\bar{T}r(t)$. The larger the adaptation matrix $\bar{T}$, the more dominant is this term relative to the possibly nonnegative terms in (5A.45) and the smaller are the final tracking errors.

Appendix 5B

Strictly Causal Almost Passive Systems

The adaptive controller (5.27)–(5.32) can also be applied to strictly causal ASP systems. Such systems satisfy the relations

$$\dot{P}(t) + P(t)[A_p(t) - B_p(t)K_e(t)C_p(t)]$$

$$+[A_p(t) - B_p(t)K_e(t)C_p(t)]^T P(t) = -Q(t) < 0 \qquad (5B.1)$$

$$P(t)B_p(t) = C_p^T(t) \qquad (5B.2)$$

Assume that $B_p(t)$ is maximal rank. In this case, relation (5B.2) requires that $C_p(t)$ be also maximal rank and that

$$C_p(t)B_p(t) = B_p^T(t)P(t)B_p(t) = B_p^T(t)C_p^T(t) > 0 \qquad (5B.3)$$

which is the nonstationary equivalent of the Linear Time-Invariant conditions (which require that strictly causal ASP systems have $n-m$ minimum-phase finite zeros and n arbitrary poles).

For the proof of stability in this case, use the results of Appendix 5A, and substitute $D_p(t) = 0, L(t) = 0, W(t) = 0$. Then

$$\begin{aligned}\dot{V}(t) &= -e_x^T(t)Q(t)e_x(t) - 2e_y^T(t)e_y(t)r^T(t)\bar{T}r(t) \\ &-2\sigma tr\{[K_I(t) - \tilde{K}]T^{-1}[K_I(t) - \tilde{K}]^T\} - 2\sigma tr\{[K_I(t) \qquad (5B.4)\\ &-\tilde{K}]T^{-1}\tilde{K}^T\} - 2r^T(t)[K(t) - \tilde{K}]^T d_0(t) + 2e_x^T(t)P(t)F(t)\end{aligned}$$

and the stability results following (5A.45) or (5A.46) will also be perfectly valid here.

Appendix 5C

Proof of Lemma 1

Since $A_y(t)$ in (5.52) is uniformly asymptotically stable by assumption, there exist some uniformly positive definite matrices, $P(t)$ and $Q_1(t)$, that satisfy the following differential Lyapunov equation [21].

$$\dot{P}(t) + P(t)A_y(t) + A_y^T(t)P(t) = -Q_1(t) < 0 \qquad (5C.1)$$

Because $Q_1(t)$ is strictly positive definite and because $P(t), B_{pc}(t), C_{pc}(t)$ are uniformly bounded, there exists some positive definite matrix $K_{yc}(t)$, sufficiently small, such that

$$Q(t) = Q_1(t) - \frac{1}{2}[P(t)B_{pc}(t) + C_{pc}^T(t)]K_{yc}(t)[B_{pc}^T(t)P(t) + C_{pc}(t)] > 0 \qquad (5C.2)$$

Substituting $Q_1(t)$ from (5C.2) in (5C.1) gives

$$\begin{aligned} &\dot{P}(t) + P(t)A_y(t) + A_y^T(t)P(t) + \\ &\frac{1}{2}[P(t)B_{pc}(t) + C_{pc}^T(t)]K_{yc}(t)[B_{pc}^T(t)P(t) \\ &+C_{pc}(t)] + Q(t) = 0 \end{aligned} \qquad (5C.3)$$

First define for convenience

$$K_y(t) = K_{yc}(t) + K_{ec}(t) \qquad (5C.4)$$

where $K_{ec}(t)$ is the gain matrix defined in (5.7), namely

$$K_{ec}(t) = [I + K_e(t)K_y^{-1}(t)]^{-1}K_e(t) = [I + K_e(t)D_p(t)]^{-1}K_e(t) \qquad (5C.5)$$

Define similarly

$$K_{yc}(t) - [I + K_e(t)K_y^{-1}(t)]^{-1}K_y(t) = [I + K_e(t)D_p(t)]^{\ 1}K_y(t) \qquad (5C.6)$$

and then it is easy to see that indeed

$$\begin{aligned} K_{yc}(t) + K_{ec}(t) &= [I + K_e(t)K_y^{-1}(t)]^{-1}[K_y(t) + K_e(t)] \\ &= K_y(t)[K_y(t) + K_e(t)]^{-1}[K_y(t) + K_e(t)] = K_y(t) \end{aligned} \tag{5C.7}$$

By using (5C.5) and (5C.6) one gets from (5C.3)

$$\begin{aligned} \dot{P}(t) \quad & +P(t)[A_p(t) - B_p(t)K_y(t)C_p(t)] \\ & +[A_p(t) - B_p(t)K_y(t)C_p(t)]^T P(t) \\ & +\tfrac{1}{2}[P(t)B_{pc}(t) + C_{pc}^T(t)]K_{yc}(t)[B_{pc}^T(t)P(t) + C_{pc}(t)] \\ & +Q(t) = 0 \end{aligned}$$

$$\begin{aligned} \dot{P}(t) \quad & +P(t)[A_p(t) - B_p(t)(K_{ec}(t) + K_{yc}(t))C_p(t)] \\ & +[A_p(t) - B_p(t)(K_{ec}(t) + K_{yc}(t))C_p(t)]^T P(t) \\ & +\tfrac{1}{2}[P(t)B_{pc}(t) + C_{pc}^T(t)]K_{yc}(t)[B_{pc}^T(t)P(t) + C_{pc}(t)] \\ & +Q(t) = 0 \end{aligned} \tag{5C.8}$$

$$\begin{aligned} \dot{P}(t) \quad & +P(t)[A_p(t) - B_p(t)K_{ec}(t)C_p(t)] + [A_p(t) \\ & -B_p(t)K_{ec}(t)C_p(t)]^T P(t) \\ & -P(t)B_p(t)K_{yc}(t)C_{pc}(t) - C_{pc}^T(t)K_{yc}(t)B_{pc}^T(t)P(t) \\ & +\tfrac{1}{2}[P(t)B_{pc}(t) + C_{pc}^T(t)]K_{yc}(t)[B_{pc}^T(t)P(t) + C_{pc}(t)] \\ & +Q(t) = 0 \end{aligned}$$

or

$$\begin{aligned} \dot{P}(t) \quad & +P(t)A_{pc}(t) + A_{pc}^T(t)P(t) \\ & +\tfrac{1}{2}[P(t)B_{pc}(t) - C_{pc}^T(t)]K_{yc}(t)[B_{pc}^T(t)P(t) - C_{pc}(t)] \\ & +Q(t) = 0 \end{aligned} \tag{5C.9}$$

$$\begin{aligned} K_{yc}(t) &= [I + K_e(t)D_p(t)]^{-1}K_y(t) = \\ & [I + K_e(t)D_p(t)]^{-1}\left[\frac{D_p(t)}{2} + \frac{D_p^T(t)}{2}\right]^{-1} \\ &= [I + K_e(t)D_p(t)]^{-1}\left[\frac{D_p(t)}{2)} + \frac{D_p^T(t)}{2}\right]^{-1}[I+ \\ & D_p(t)K_e(t)][I + D_p(t)K_e(t)]^{-1} \end{aligned} \tag{5C.10}$$

Substituting (5C.10) into (5C.9) finally gives

$$\begin{aligned} \dot{P}(t) & +P(t)A_{pc}(t) + A_{pc}^T(t)P(t) \\ & +[P(t)B_{pc}(t) - C_{pc}^T(t)][D_{pc}(t) + D_{pc}^T(t)]^{-1}[B_{pc}^T(t)P(t) \\ & -C_{pc}(t)] + Q(t) = 0 \end{aligned} \tag{5C.11}$$

Relation (5C.11) is identical with the passivity relation (5.15), and therefore it shows that the closed-loop system $\{A_{pc}(t), B_{pc}(t), C_{pc}(t), D_{pc}(t)\}$ is strictly passive (Figure 5.2). Since then the equivalent closed-loop system of Figure 5.1 is also strictly passive, the augmented open-loop system (5.1)–(5.2) (without the feedback gain $K_e(t)$) is ASP; this concludes the proof.

Chapter 6

Design of Model Reference Adaptive Controllers

6.1 Algorithm Overview

Prior to the case study presentations in Chapter 6, it is useful at this point to review the algorithms that have been presented in the previous three chapters. These algorithms vary in complexity according to the need for satisfying positivity and/or robustness constraints that guarantee asymptotic tracking or stability at the very least.

The basic direct model reference adaptive controller (DMRAC) is defined in Chapter 2 and summarized in Section 2.6. This algorithm, which is the simplest for implementation, requires that the original system itself be almost strictly positive real (ASPR) as defined in Section 2.3.3.

However, since most systems are not themselves ASPR, Chapter 3 presents various forms of compensation, designed using basic stabilizability properties of the plant, which result in an augmented system which is ASPR and thus suitable for DMRAC. The most straightforward augmentation results from the addition of a feedforward component to the plant output. The feedforward component is the result of passing the control input signal through a transfer function. This transfer function is selected to be the inverse of a stabilizing configuration such that the augmented transfer function (i.e. the sum of the original system transfer function and the feedforward compensator) is ASPR. This procedure is discussed in Section 3.2 and summarized in Section 3.2.2. Because asymptotic tracking can now only be guaranteed for the augmented output, satisfactory tracking will only be possible if the feedforward component is negligible (at least in steady state) when compared with the true output.

If, however, the feedforward component is such that the augmented output differs significantly from the true output, then additional compensation is needed. To this effect Section 3.3 introduces the concept of using additional feedforward around the reference model. A summary of the algorithm is given in Section 3.3.3. As shown in Section 3.3, this configuration will result in asymptotic output tracking provided that the original system with feedforward compensation is ASPR.

A more general approach based upon the development of a metasystem is presented in Section 3.4 and summarized in Section 3.4.3. Three configurations are developed by inserting supplementary compensation into either a feedback path, a parallel path or in a cascade path. Additional adaptive gains are then included with the supplemental dynamics. Asymptotic output tracking can result from use of the parallel or cascade supplemental paths.

Because the above algorithms have been developed for linear systems with perfect measurements and in the absence of disturbances, it is important to also note the modifications that have been developed for more realistic situations. A specific type of nonlinearity of known form is treated in Section 3.5, and modifications to account for input and output disturbances are presented in Chapter 4.

Basically in order to account for disturbances, the integral gain adjustment is modified to include an additional term which is composed of the integral gain itself multiplied by a (small in magnitude) negative factor. This prevents the integral gain from growing without bound in the presence of persistent (small but not zero) output error.

Finally, Chapter 5 contains a theoretical development that shows that the basic adaptive algorithms still maintain stability with respect to boundedness for certain classes of nonlinear and time varying systems.

Application of these algorithms requires a certain degree of design prior to implementation. Relevant issues include compensator design to satisfy the constraints, weight selection, reference model selection, digital implementation, and treatment of time varying commands. These issues are subsequently discussed and further illustrated in the case studies of Chapter 7.

6.2 Constraint Satisfaction

As noted in Chapter 3, in many cases in order to satisfy the positivity constraints, a feedforward compensator is incorporated into the system in a manner such that the augmented combination is almost strictly positive real (ASPR) for all possible process parameter values. Relative to the notation of Chapter 3, this means that $G_a(jw) = G_p(jw) + R_p(jw)$ is ASPR.

As discussed previously in 3.2, this condition can be satisfied if:

1. $R_p(s)$ is itself strictly positive real and
2. The closed loop system shown in Figure 6.1, which is comprised of the original process G_p in the feedforward path and the inverse of R_p in the feedback loop, is stable.

Thus, an alternate procedure for evaluating the ASPR condition is a stability analysis of the related closed loop system. Since this stability property must hold for all expected variations in the process parameters, a robust stability analysis is needed. If, as in the problem definition of 1.1, upper and lower bounds are available for each of the process parameters, then it

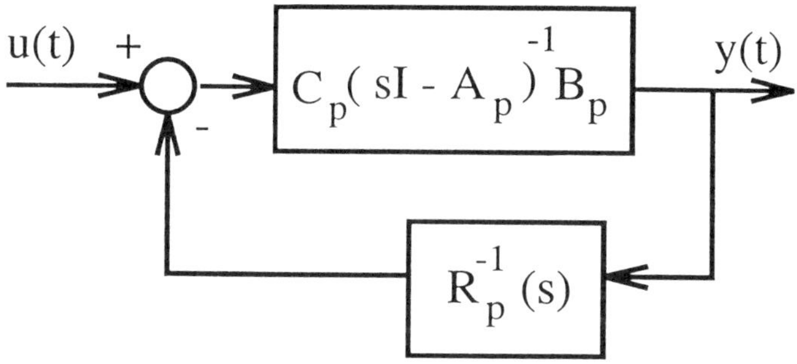

FIGURE 6.1. System Stabilized by Inverse Feedforward Network

is possible to perform such an analysis with the Kharitonov polynomials [118].

To illustrate such an approach, consider first the single input–single output system:

$$y_p(s)/u_p(s) = ka(s)/b(s) = G_p(s) \tag{6.1}$$

For simplicity, assume that the feedforward compensator is defined by the positive real transfer function:

$$R_p(s) = D/(\tau s + 1) \tag{6.2}$$

The characteristic equation for the corresponding closed loop system shown in Figure 6.1 and comprised of $G_p(s)$ in the feedforward path and $R_p^{-1}(s)$ in the feedback path is:

$$C(s) = Db(s) + ka(s)(\tau\ s + 1) = \sum_{i=0}^{N} C_i s^i \tag{6.3}$$

where the coefficients C_i are functions of the process parameters.

Two situations will now be considered:

1. Given a set of feedforward compensator parameters, D and τ, determine if the above characteristic equation has stable roots for all possible parameter variations.

2. Find compensator parameters, D and τ, such that the above characteristic equation is stable for all possible process parameters.

Assuming that the first situation is of interest, the Kharitonov polynomial based analysis requires that an upper and lower bound be computed for each coefficient of the characteristic polynomial. Given values for the feedforward compensator parameters, these bounds may be determined by maximizing and then minimizing each of the coefficients of the characteristic polynomial with respect to the process parameters, subject to the given upper and lower bounds of these process parameters.

For example:

$$\underline{C}_k = \min\ [C_k(a_p(i,j), b_p(i,j))] \tag{6.4}$$

$$\bar{C}_k = \max\ [C_k(a_p(i,j), b_p(i,j))] \tag{6.5}$$

$$\underline{a}_p(i,j) \leq a_p(i,j) \leq \bar{a}_p(i,j) \tag{6.6}$$

$$\underline{b}_p(i,j) \leq b_p(i,j) \leq \bar{b}_p(i,j) \tag{6.7}$$

Once the upper $\overline{C}_k$ and lower $\underline{C}_k$ values are available, the following four so–called Kharitonov polynomials are constructed:

$$\begin{aligned}
p_{00}(s) &= \underline{C}_0 + \underline{C}_1 s + \overline{C}_2 s^2 + \overline{C}_3 s^3 + \underline{C}_4 s^4 + \underline{C}_5 s^5 + \ldots \\
p_{11}(s) &= \overline{C}_0 + \overline{C}_1 s + \underline{C}_2 s^2 + \underline{C}_3 s^3 + \overline{C}_4 s^4 + \overline{C}_5 s^5 + \ldots \\
p_{01}(s) &= \underline{C}_0 + \overline{C}_1 s + \overline{C}_2\, s^2 + \underline{C}_3 s^3 + \underline{C}_4 s^4 + \overline{C}_5 s^5 + \ldots \\
p_{10}(s) &= \overline{C}_0 + \underline{C}_1 s + \underline{C}_2\, s^2 + \overline{C}_3 s^3 + \overline{C}_4 s^4 + \underline{C}_5 s^5 + \ldots
\end{aligned} \tag{6.8}$$

If each of these polynomials has all its zeros in the left half plane, the original characteristic equation will have its roots in the left half plane for all possible process parameter combinations [118].

However, if one or more of the Kharitonov polynomials has an unstable root,and/or the feedforward parameters are to be selected, then the evaluation of the Kharitonov polynomial roots must be incorporated into a design procedure.

This might be accomplished by defining an appropriate optimization problem that incorporates the stability constraints. Possibilities include:

1. Minimization, with respect to the compensator paramcters (i.e., D and τ), of the maximum real part of all roots of the four Kharitonov polynomials. Note that in this case, at each iteration of the numerical

optimization procedure, the maximum and minimum values of each of the characteristic polynomial coefficients have to be computed to define the four Kharitonov polynomials. Furthermore, as soon as all roots of all the polynomials have negative real parts, the current compensator, although not optimal, will be acceptable.

2. Minimization of an index involving the compensator parameters (e.g. minimization of D alone if τ is prespecified) subject to the constraint that all roots of the four Kharitonov polynomials be in the left half plane.

Note that in each of the above optimization methods, at each iteration of the numerical optimization procedure, the maximum and minimum values of each of the characteristic polynomial coefficients have to be computed to define the four Kharitonov polynomials. Furthermore, as soon as all roots of all the polynomials have negative real parts, the current compensator, although not optimal, will be acceptable.

In the general multiple input–multiple output situation, a more complex compensator defined by a strictly positive real (SPR) transfer function matrix must be considered. Furthermore, direct analytic expressions of the characteristic polynomial coefficients in terms of the process parameters may be quite arduous, except through the use of a symbolic manipulator. In this case, it may be more expedient to use a nongradient procedure for finding the extreme values for each of the characteristic polynomial coefficients.

Because such a multiple input–multiple output compensator must itself be SPR, this constraint must be included in any optimization approach. Assuming a symmetric compensator, the parameters (i.e., D_{ij}, τ_{ij}) must be such that each of the principle minors is positive. Since this constraint can be quite demanding, the following recommendations are suggested:

1. Consider diagonal compensators with (i, i)th element:

$$d_{ii}/(\tau_{ii}s + 1) \tag{6.9}$$

where $d_{ii} > 0$.

If such a compensator is not adequate, additional off-diagonal terms can be incorporated in a manner such that the positive realness is preserved. However, starting the optimization with the best possible diagonal positive real compensator should certainly improve the conditioning of the computation. A recent analysis of the application of Kharitonov polynomials to first order compensators [119] is relevant if upper and lower bounds are given for the plant numerator and denominator coefficients.

2. Consider symmetric positive definite constant compensators of the form

$$R_p(j\omega) = \mathbf{D} = \mathbf{D}^T > 0 \tag{6.10}$$

Once such a compensator is found, consideration should be given to an implementable version formed by dividing each component by

$$(\tau_{ij}s + 1) \tag{6.11}$$

If this version is not acceptable, then the optimization procedures should be used to recompute the d_{ij} terms and/or the time constants, τ_{ij}.

6.3 Weight Selection

Selection of the weighting matrices T and $\bar{T}$ as defined in (2.60) and (2.61) can often be an arduous trial and error procedure. However, this selection can be expedited if it is possible to determine an initial set of weights with the proper magnitude and ordering. One possible approach is to base the initial weight selection on the CGT solution for the nominal process parameters $(A_p^\circ, B_p^\circ, C_p^\circ)$. Then $\bar{T}$ would be selected such that K_p, for a representative value of x_m^0 of the model state vector at a representative error e_y°, will be equal to the CGT gain. Then the T elements should be chosen as the corresponding $\bar{T}$ elements divided by the reference model settling time (which may represent the time required for K_I to integrate up to the CGT gain).

To illustrate these procedures, assume that the CGT problem defined in Section 2.2. can be solved using the nominal plant matrices $(A_p^\circ, B_p^\circ, C_p^\circ)$ to give the feedforward gains S_{21}° and S_{22}° as defined in (2.17). The feedback gain K_e° might be selected using procedures such as in [120] or [121]. Using the results of [120], it may be possible to select K_e° such that $(A_p^\circ - B_p^\circ K_e^\circ C_p^\circ)$ is stable and $(N_u - 1)$ of the eigenvalues of $(A_p^\circ, B_p^\circ, C_p^\circ)$ are equal to the most dominant $(N_u - 1)$ eigenvalues of A_m. Using the procedure of [121], it may be possible to generate quadratic performance index weighting matrices such that the eigenvalues of $(A_p^\circ - B_p^\circ K_e^\circ C_p^\circ)$ are at satisfactory locations.

Now to select $\bar{T}$, consider the relation:

$$K_p = e_y r^T \bar{T} \tag{6.12}$$

or

$$(K_{pe}, K_{px}, K_{pu}) = e_y(e_y^T\ x_m^T\ u_m^T) \begin{pmatrix} \bar{T}_e & 0 & 0 \\ 0 & \bar{T}_x & 0 \\ 0 & 0 & \bar{T}_u \end{pmatrix} \tag{6.13}$$

Then for representative values (say maximums or midpoints of e_y, x_m, u_m), choose $\bar{T}$ so that $(K_{pe}, K_{px}, K_{pu}) = (K_e^\circ, S_{21}^\circ, S_{22}^\circ)$
Then

$$e_y^\circ(e_y^\circ)^T\ \bar{T}_e = K_e^\circ \tag{6.14}$$

$$e_y^\circ(x_m^\circ)^T\ \bar{T}_x = S_{21}^\circ \tag{6.15}$$

$$e_y^\circ(u_m^\circ)^T\ \bar{T}_u = S_{22}^\circ \tag{6.16}$$

For (6.14) and (6.16) it is important to note that $\bar{T}_e, K_e^\circ, \bar{T}_u$, and S_{22}° are all square matrices with dimension $(N_u \times N_u)$. Then it is recommended that $\bar{T}_e$ and $\bar{T}_u$ be designed with elements initially set equal to:

$$\bar{T}_e(i,i) = \|K_e^\circ(i,i)\|/(e_y^\circ(i))^2 \tag{6.17}$$

$$\bar{T}_u(i,i) = \|S_{22}^\circ(i,i)\|/\|e_y^\circ(i) \times u_m^\circ(i))\| \tag{6.18}$$

Then (6.15) for diagonal $\bar{T}_x$ would consist of $N_u \times N_{xm}$ equations in N_{xm} unknowns. If $N_u = N_{xm}$, then select $\bar{T}_x(i,i)$ as above, i.e.,

$$\bar{T}_x(i,i) = \|S_{21}^\circ(i,i)\|/\|(e_y^\circ(i) \times x_m^\circ(i))\| \tag{6.19}$$

Otherwise, for say the jth component of e$_y^\circ$, one might consider the relationship:

$$e_y^\circ(j)\ u_m^\circ(i)\ \bar{T}_u(i,i) = S_{22}^\circ(j,i)\ \ j = 1, \ldots, n_y \tag{6.20}$$

Then, for each value of j, it is possible to find the diagonal entries of $\bar{T}_n$ to satisfy the above relations. One might next select $\bar{T}_u$ by:

1. Using the average of the $\bar{T}_u(i,i)$ values as computed for each component of e_y°.

   or

2. Using the maximum value of $\bar{T}_u(i,i)$ over all the components of e_y°.

Following the selection of $\bar{T}$, T would then be computed as:

$$T = \bar{T}/\tau_s \tag{6.21}$$

where τ_s is the settling time of the reference model.

6.4 Reference Model Selection

Selection of the reference model dynamics is a very important part of the adaptive system design procedure that ultimately defines the desired system behavior. Clearly, it can be quite difficult to tune the weights of a system that is commanded to operate at time constants much faster or slower than those of the uncontrolled process. If indeed, such characteristics are required, then it is recommended that either:

- Tuning proceed in stages with somewhat more demanding reference model specifications in each successive step.

  or

- The adaptation be applied to a closed loop system consisting of the process and a compensator designed for the nominal system.

Suggestions for selecting the reference model dynamics include:

- Construct the reference model to correspond to the nominal system cascaded with a compensator designed to satisfy system specifications. In this case, the reference model order would be at least equal to the order of the plant. The initial conditions of the reference model would then have to be set such that the initial plant and reference model output vectors have the same values. Since this might be done in an infinity of different settings, it may be advantageous, to attempt matching the corresponding plant and model states. If this is not possible, some sort of trial and error adjustment is required. Of course, if the plant starts from rest, all initial plant and model states would be zero.

- For each component of the output vector, use a first or second order linear transfer function forced by a step (possibly of zero magnitude) to define the desired behavior of that component. In this case, the initial position state of the model transfer function would be set to the initial value of the output itself. If a second order model is used, then in the absence of any other information, the velocity state component should be set to zero.

6.5 Digital Implementation

Implementation of the simple adaptive controllers defined in Chapters 2 and 3 will, in general, require some degree of digitization since:

- the outputs y_p are available only at specified sampling times
- the control u_p is computed and applied using a digital computer.

Thus, consideration must be given to various possible digital implementation procedures specific to the following portions of the algorithm:

- Reference model response: The reference model output and state vectors can be computed digitally at any assigned sampling frequency.
- Feedforward dynamics: The response of the feedforward dynamics may have to be computed at a rate faster than the gain or control update frequency. This is especially true if the feedforward time constants are small fractions of the reference model time constants.
- Gain update: The gains may be computed at the same frequency as the control update frequency or, if desired, at a slower rate. In practice, an Euler type algorithm should be adequate for computing the integral gain K_I.
- Control update: The control vector can be updated at each sample time that the output vector is sensed. If a more frequent update is needed, some type of interpolation will be necessary.

Ideally, if the outputs are sensed sufficiently fast, and computing power is adequate, all functions should be computed at the same frequency. However, if the reference model time constants are relatively small and demand a control frequency faster than the sensor frequency, then the sensed outputs may have to be interpolated to give values at intermediate times.

6.6 Time Varying Commands

When the command u_m is not a step or constant signal, it is still possible to achieve satisfactory model following by incorporating some modifications and/or assumptions.

6.6.1 Command Generated as Output of Linear System

If as illustrated in Section 2.5, the command can be generated as the solution of a linear system, then the corresponding equations would be augmented to the reference model. Thus, if u_m can be generated from:

$$\dot{v}_m = A_v v_m + B_v u_o \tag{6.22}$$

$$u_m = C_v v_m \tag{6.23}$$

where u_o is a step command, then the augmental reference model would become:

$$\begin{pmatrix} \dot{x}_m \\ \dot{v}_m \end{pmatrix} = \begin{pmatrix} A_m & B_m C_v \\ 0 & A_v \end{pmatrix} \begin{pmatrix} x_m \\ v_m \end{pmatrix} + \begin{pmatrix} 0 \\ B_v \end{pmatrix} u_o. \tag{6.24}$$

6.6.2 Command Variations Slow Compared with Reference Model

If the dominant frequencies in the command are much lower than the cutoff frequency of the reference model, it may be possible to treat the command as a step input to the reference model dynamics. In this manner, even though u_m is varying, the same adaptive algorithms derived in Chapters 2 and 3 could be used, but with the instantaneous values of u_m replacing the fixed value.

Chapter 7

Case Studies

This chapter presents a series of significant case studies that illustrate the procedures for applying the direct model reference adaptive algorithms that have been developed and discussed in the previous chapters. These examples discuss items such as reference model selection, constraint satisfaction, tuning, and performance.

Section 7.1 demonstrates the applicability to robotic dynamics. Whereas initial tuning was performed using a representative simulation of the PUMA 560 manipulator, final adjustment and implementation were carried out with the actual robotic hardware.

Section 7.2 discusses the application to large flexible structures using as illustration the JPL/PL Large Spacecraft Control Laboratory Ground Experiment Facility at JPL.

Application to automated drug delivery is then considered in Section 7.3, which discusses the embedding of the direct model reference algorithm into a hierarchical system designed to yield course to fine control of blood pressure dependent upon the available information.

Finally, Section 7.4 presents the application of direct model reference control to aircraft stability augmentation.

7.1 Direct Model Reference Adaptive Control of a PUMA Manipulator

This material is based upon research performed by David C. Swift, at the Rensselaer Polytechnic Institute, Troy, NY for his Masters Thesis entitled "Command Generator Tracker Based Direct Model Reference Adaptive Control of a PUMA 560 Manipulator," August 1992

7.1.1 Introduction

Direct adaptive control offers the potential for uniform control of robotic manipulators in the presence of uncertain flexibilities, changing dynamics due to unknown and varying payloads, and nonlinear joint interactions without explicit parameter identification.

One such direct adaptive algorithm that is especially attractive for robotic control is the direct model reference adaptive controller (DMRAC) discussed in Sections 2.4 and 3.3

This procedure has been previously used to control a single link flexible robotic joint and a nonlinear model of a two link Puma [73], [122]. In view of the excellent tracking results demonstrated in these papers, it was concluded that this adaptive algorithm should be used to control an actual Puma arm. This effort has consisted of two main thrusts: namely, control of the representative simulation model developed in [123], and the transition of the tuned algorithm to the actual robotic arm.

Results in all cases indicated that the performance of the DMRAC algorithm is robust with respect to representative load variations.

7.1.2 Puma Model Development

In order to test the performance of the Direct Model Reference Adaptive Controller (DMRAC), an accurate non-linear coupled model of the PUMA manipulator was needed. A full explicit dynamic model of the PUMA 560 manipulator, derived by Armstrong, Khatib, and Burdick [123], was selected. The formulation of the PUMA model was computationally efficient using 25% fewer calculations than a 6 degree of freedom recursive Newton-Euler method. The algebraic formulation of the model also allowed for the easy addition of a load by modifying the link 6 mass, center of mass, and inertia parameters.

Figure 7.1 shows the six rotational joint axis, $\{z_1, \ldots, z_6\}$, for the PUMA 560. Only the rotational, z_1, axis is shown in the figure. Positive rotations follow the right hand rule - counter-clockwise looking down the z axis. The six joints of the PUMA 560 are as follows:

- *Joint 1.* A vertical rotation about the base, z_1.
- *Joint 2.* A horizontal rotation about the shoulder, z_2.
- *Joint 3.* A horizontal rotation about the elbow, z_3.
- *Joint 4.* A twist of the wrist, z_4.
- *Joint 5.* An inclination of the wrist, z_5.
- *Joint 6.* A twist of the mounting flange, z_6.

The position of the manipulator in Figure 7.1 illustrates the zero position. Note that when Joint 5 is at zero, axis z_4 and z_6 coincide.

The dynamic equations of motion used to model the PUMA are:

$$A(\theta)\ddot{\theta} + B(\theta)[\dot{\theta}\dot{\theta}] + C(\theta)[\dot{\theta}^2] + g(\theta) = \Gamma \tag{7.1}$$

where

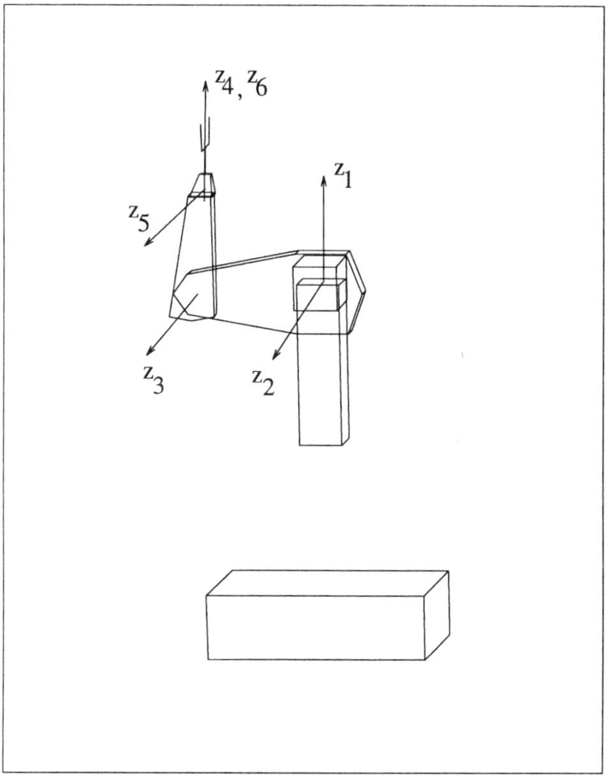

FIGURE 7.1. PUMA 560 Coordinate Frame Assignments

$A(\theta)$ is the 6 x 6 positive definite kinetic energy matrix,

$B(\theta)$ is the 6 x 15 matrix of coriolis torques,

$C(\theta)$ is the 6 x 6 matrix of centrifugal torques,

$g(\theta)$ is the 6 vector of gravity torques,

$\ddot{\theta}$ is the 6 vector of joint accelerations,

$[\dot{\theta}\dot{\theta}]$ is the 15 vector of velocity products, where

$$[\dot{\theta}\dot{\theta}] = [\dot{\theta}_1\dot{\theta}_2, \dot{\theta}_1\dot{\theta}_3, ..., \dot{\theta}_1\dot{\theta}_6, \dot{\theta}_2\dot{\theta}_3, ..., \dot{\theta}_4\dot{\theta}_6, \dot{\theta}_5\dot{\theta}_6]^T$$

$[\dot{\theta}^2]$ is the 6 vector of squared velocities, where

$$[\dot{\theta}^2] = [\dot{\theta}_1^2, \dot{\theta}_2^2, ..., \dot{\theta}_6^2]^T$$

and Γ is the 6 vector of joint torques.

The above model can be cast into state space form by solving Equation (7.1) for $\ddot{\theta}$ and choosing the following 12 x 1 state vector,

$$x^T = [\theta^T, v^T] \tag{7.2}$$

where

$$\begin{aligned} \theta &= [\theta_1, ..., \theta_6]^T \\ v &= [\dot{\theta}_1, ..., \dot{\theta}_6] \end{aligned}$$

Thus, the state space model is as follows,

$$\begin{aligned} \dot{\theta} &= v \\ \dot{v} &= A^{-1}(\theta)[\Gamma - B(\theta)[\dot{\theta}\dot{\theta}] - C(\theta)[\dot{\theta}^2] - g(\theta)] \end{aligned}$$

The controlled output vector for the plant is

$$y_{plant} = \theta + \alpha v \tag{7.3}$$

where α is a diagonal 6 x 6 matrix of velocity weighting factors.

This velocity term is present to help remove high frequency oscillations caused by the controller. Furthermore, inclusion of such a term can force certain systems to satisfy the ASPR conditions. The maximum allowable torques (in n-m) were [97.6, 186.4, 89.4, 24.2, 20.1, 21.3]

7.1.3 Implementation Issues

Reference Model

The first decision to be made in implementing the DMRAC algorithm is the choice of reference model order. If one chooses the order too low, then excessively large gains may occur even in a well-tuned controller. This may produce greater than desired accelerations in the robot arm joints resulting in joint torque saturations leading to poor model following. If one chooses the order too high, then excessive response delays may be incurred. For the PUMA 560, an independent second order reference model was selected for each of the six joints. This is not unreasonable since in a PUMA 560, as with many manipulators, the mass matrix is approximately diagonal for all θ making the system nearly decoupled.

Thus, for each joint, the reference model transfer function was:

$$y_{m_i}(s)/u_{m_i}(s) = \frac{w_{n_i}^2}{s^2 + 2\zeta_i w_{n_i} s + w_{n_i}^2} \tag{7.4}$$

where

$$w_{n_i} = 5 \tag{7.5}$$

and

$$\zeta_i = 1 \qquad i = \{1, 2, 3, 4, 5, 6\}. \tag{7.6}$$

Critical damping was selected so as to reduce the possibility of joint angle overshoots. This conforms to a standard safety feature of robot arm controllers which tends to avoid obstacle collisions. Of course, once the choice of critical damping is made, the choice of natural frequency governs the speed of model response to inputs. A choice of $w_n = 5$ yields a 90% rise time of about 0.8 sec.

Command Generation

For Testing purposes a minimum jerk trajectory was generated through the following positions at the noted times.

Joint Position (deg)						
1	2	3	4	5	6	Time (sec)
0	-45	180	0	45	90	0
90	-90	90	45	0	45	6
0	0	180	0	90	90	13
0	-45	180	0	45	90	18

The resulting angular paths for each joint were then used as the reference model commands $u_{m_i}(t)$.

Bias Introduction

For the PUMA 560 manipulator, the origin of the coordinate system should be such that the adaptation gains have a non-zero excitation throughout the range of interest. For example, assume that in order to maintain an output of $y_p = [0, ..., 0]^T$, a non-zero input, u_p, is required. However a zero command to the reference model, $u_m = [0, ..., 0]^T$, will result in a zero model output and a zero state vector. Thus in this case $e_y = y_m - y_p$ will also be zero, and the vector, $r(t)$, defined by (3.10) will be zero resulting in a zero control. Since the plant requires a non-zero control to maintain a zero output, the DMRAC algorithm requires a non-zero error signal in order to apply a non-zero control which will result in a steady-state error at the zero output position.

If the reference model coordinates are shifted by a constant bias term, then a zero command to the reference model, $u_m = [0, ..., 0]^T$ will produce non-zero outputs for the model state and output vectors which, in turn, will produce a non-zero command to the plant. This bias term is subtracted

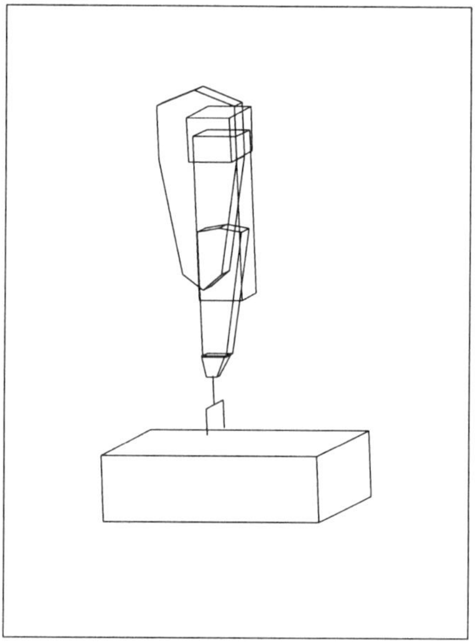

FIGURE 7.2. Stable Equilibrium for the PUMA 560

from the model command, u_m, and the plant output, y_p, as follows,

$$u_m(t) = \hat{u}_m(t) - q_{bias} \quad (7.7)$$

$$y_p(t) = \hat{y}_p(t) - q_{bias} \quad (7.8)$$

where $\hat{u}_m(t)$ is the original model command in the original coordinate system, $u_m(t)$ is the new biased model command to be applied to the model dynamics, $\hat{y}_p(t)$ is the actual plant output, $y_p(t)$ is the new biased plant output to be used to form the error signal, and q_{bias} is a constant bias term. For robotic manipulators, q_{bias} should be selected such that a new plant output of $y_p = [0, ..., 0]$ corresponds to an equilibrium position. By examining the zero position of the robot, Figure 7.1, it is clear that $y_p = \{0, 0, 0, 0, 0, 0\}$ is not an equilibrium. However a bias of,

$$q_{bias} = \{0, 90, 90, 0, 0, 0\}\text{degrees} \quad (7.9)$$

will shift the zero position to that shown in Figure 7.2.

Feedforward Design

The feed-forward filter dynamics for joint i as defined in (3.5)–(3.6) is

$$R_{p_i}(s) = \frac{K_{d_i}}{1 + \tau_i s} \quad (7.10)$$

where K_{d_i} is the DC gain, and τ_i is the time constant.

TABLE 7.1. Tunable Parameters

Parameters	Description	Values
T	24 x 24 integral weighting matrix	576
$\bar{T}$	24 x 24 proportional weighting matrix	576
w_{n_i}	Undamped natural frequency for Joint i model	6
ζ_i	Damping ratio for Joint i model	6
α	6 x 6 diagonal plant derivative weighting matrix	6
K_{d_i}	DC gain of Joint i supplementary feed-forward block	6
τ_i	Time constant of Joint i supplementary feed-forward block	6
	Total	1182

7.1.4 Simulation Results

In this section, we briefly discuss the tuning process and present plots of a simultaneous, six joint response of the PUMA 560 under DMRAC control.

Tuning

Once the reference model has been chosen, one must choose values for the various DMRAC parameters. Specifically these are

$\bar{T} \equiv$ proportional gain weighting matrix, (2.201)

$T \equiv$ integral gain weighting matrix, (2.202)

$K_{d_i} \equiv$ plant/model feedforward gain, (7.10)

$\tau_i \equiv$ plant/model feedforward time constant, (7.10)

$\underline{\alpha} \equiv$ 6 vector of plant rate feedforward gains, (7.3)

For the fully centralized DMRAC algorithm with the plant derivative output term and the supplementary feed-forward in the reference model and plant, there are 1182 parameters to be selected as shown in Table 7.1. At first, this number seems very intimidating, but as will be shown, the number of tuning parameters can be greatly reduced by some simplifications and by adjusting the parameters in groups rather than individually.

The most drastic reduction in the number of tuning parameters can be achieved by forcing the integral and proportional adaptation weighting matrices, T and $\bar{T}$ to be diagonal. This reduces the number of tuning parameters from 1182 to 78.

The reference model dynamics have 12 tuning parameters, six w_n's and six ζ_i's.

It is customary in robotic applications to tune controllers for critical damping so that there is no overshoot. Overshoot may cause a robot end effector to penetrate the surface of the work environment. The undamped natural frequency terms, w_{n_i} are chosen such that the reference model will have a specified step response. Typically, the reference model dynamics are chosen such that they are "reasonable" for the plant to follow since the DMRAC algorithm will try to force the plant to follow the model output. For the case of a PUMA 560 Manipulator, all of the w_{n_i} were initially set to 5.0. The model's dynamic parameters can be changed as needed if the robot is having problems tracking the model.

Initially, the plant output derivative weights, α, were set to zero. These weights are used to remove high frequency components from the plant control signal, u_m, and should only be used when needed as they will affect the transient response.

The feed-forward filter has 12 tuning parameters, six gains K_{d_i} and six time constants τ_i. A good first choice for the τ_i is approximately one-tenth the model time constant.

Initially the value of τ_i were all set to 0.1 s, and the six DC filter gains K_{d_i} were set to 1.0. Increasing the filter gain was seen to typically improve the tracking performance.

The diagonal components of $\bar{T}$ and T were initially, all set to unity. A reasonable method of tuning a DMRAC controller is to start the plant at an equilibrium position and apply small step inputs. After a reasonable performance is achieved with the step inputs, the DMRAC should be fine tuned using typical plant trajectories.

If the closed loop system is very sensitive to initial conditions, start with small steps as described above, let the system reach steady-state, and then save all of the DMRAC controller state information (integral adaptation matrix, K_I; reference model state vector, x_m; and the filter state vector) to be used as initial conditions for the next run. This will significantly cut down the adaptation time required for the gains to reach their steady-state values.

In order to compare the tuning results, some criterion must be established. For this example, the goal was to keep the peak model following errors small and to keep the error trajectory as close to zero as possible. Small errors were tolerable during motion. It was also desired to achieve zero error in steady-state.

The step response with the initial tuning values was sluggish for Joints 1, 4, 5, and 6 with overshoot and oscillations. Joints 2 and 3 settled into

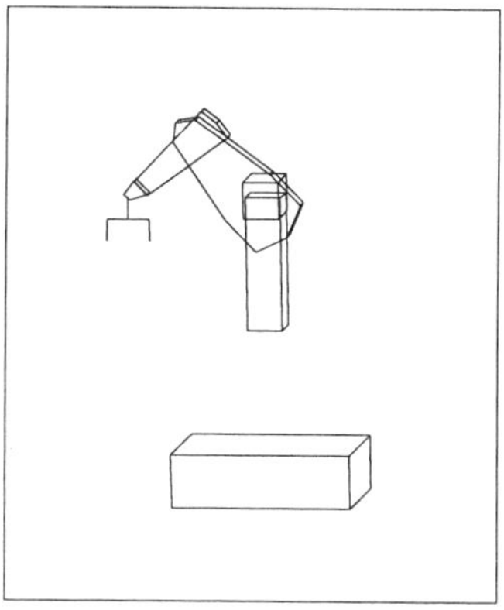

FIGURE 7.3. Shutdown Position, $\{0, -45, 180, 0, 45, 90\}$ degrees

their steady-state values quickly but with very large steady-state errors. The process used to complete the tuning was as follows:

1. Refine the tuning for a 10 degree step from the equilibrium position.
2. Using the refined parameter values, move the robot to the shutdown position of Figure 7.3 and save the DMRAC internal state values at that position for use as initial conditions.
3. Refine the tuning for a 10 degree step from the shutdown position using the initial conditions from Step 2.
4. Refine the tuning from typical min-jerk trajectories from the shutdown position.

The final tuning parameter values after Step 4 are shown in Table 7.2. The weighting matrix values for Joints 1, 2, and 3 differ from the weighting matrix values for the last three joints by a factor of about 100 which reflects the mass/inertia difference between the upper arm and the wrist. The weighting matrix values which are multiplied by the "x_{m_2}"products are about a factor of seven lower than the values multiplying the "x_{m1}"products since the second state variable of each decoupled reference model had a higher peak value in a transient. The Joint 1, 2, and 3 reference models have an undamped natural frequency of 4.0 rad/sec where the wrist model used 7.0 rad/sec which again reflected the inertia difference between the upper arm and the wrist. The feedforward filter values were set to $K_d = 6.0$ and $\tau = 0.1$ for all joints. The alpha values were increased from the initial values of zero to damp out some high frequency oscillations.

TABLE 7.2. Final Parameter Values

$\bar{T}$ (diag component)	"e_z"	20	40	22	0.2	0.2	0.2
	"x_m"	140	20	140	35	100	22
		1.4	0.2	1.4	0.2	1.4	0.2
	"u_m"	140	160	110	1.4	1.4	1.4
T (diag component)	"e_z"	20	60	25	0.2	0.2	0.2
	"x_m"	140	20	150	35	140	25
		1.4	0.2	1.4	0.2	1.4	0.2
	"u_m"	140	160	130	1.4	1.4	1.4
Joint		1	2	3	4	5	6
Model	w_n	4	4	4	7	7	7
	ζ	1	1	1	1	1	1
Feed Forward	K_d	6	6	6	6	6	6
	τ	0.1	0.1	0.1	0.1	0.1	0.1
alpha	α	0.035	0.02	0.02	0.01	0.01	0.01

RESPONSE

Simulation results of the PUMA 560 dynamics responding to the tuned DMRAC controller are displayed in Figure 7.4. Both the initial and final shutdown position were (0, -45, 180, 0, 45, 90) degrees as shown in Figure 7.3.

Note that the model following is excellent for all 6 joints. Furthermore it was observed that all joint torques were smooth and below their saturation limits. In addition, for this specific case, the use of the feedforward component did not significantly affect the response, although in other cases (eg. step response) use of the feedforward resulted in significant improvements.

7.1.5 EXPERIMENTAL RESULTS

Because the simulator results of the previous section indicated that DMRAC should be useful for robot control, a set of experiments was performed on an actual PUMA 560 manipulator. The tuning process was similar to that described in the previous section. All parameters were initialized at those values from the simulation studies. Only minor variations were required. Final values are in Table 7.3.

Examples presented illustrate performance of the DMRAC for tracking various trajectories in the presence of static and dynamic load changes. In all cases the robot starts at the shutdown position and follows a trajectory which finishes at the shutdown position.

TABLE 7.3. Parameter Values for 3 Joint Trajectory Tracking Runs

$\bar{T}$ (diag component)	"e_z"	20	40	40
	"x_m"	140	20	200
		30	200	30
	"u_m"	140	200	200
T (diag component)	"e_z"	30	30	40
	"x_m"	200	30	400
		60	400	60
	"u_m"	200	400	400
joint		1	2	3
Model	w_n	10	10	10
	ζ	1	1	1
Feed Forward	K_d	6	6	6
	τ	0.05	0.05	0.05
alpha	α	0.02	0.02	0.02

TABLE 7.4. First Three Joint Tracking Test Trajectory

Knot Point	Joint Positions (deg) 1	2	3	Time (sec)
0	0	-45	180	–
1	-90	-90	90	6
2	0	0	180	8
3	0	-45	180	6

Three Joint Trajectory Tracking Study

The trajectory listed in Table 7.4 is very similar to the one used in the previous simulation. The arm first moves to a straight up position, curls up, and then moves back to the safe position. The wrist joints remain locked in their shutdown positions of $\{0.0,45.0.90.0\}$ degrees.

The response to the first trajectory is shown in Figure 7.5. The response is quite good. The effects of stiction can be seen on Joint 2 at $t = 15$ seconds in Figure 7.5. Figures 7.6–7.8 show the model following error and the link torques for Joints 1, 2, and 3 respectively. Figure 7.6 shows that the Joint 1 torque signal was quite noisy. This noise did not have a physically detectable effect on the actual arm motion. Typically one can feel or hear

TABLE 7.5. Static Load Change Trajectory

Knot Point	Joint Positions (deg) 1	2	3	Time (sec)
0	0	-45	180	–
1	0	-45	180	3
2	45	0	0	10
3	0	-45	180	10

TABLE 7.6. Dynamic Load Change Trajectory

Knot Point	Joint Positions (deg) 1	2	3	Time (sec)
0	0	-45	180	–
1	0	-45	180	3
2	45	-90	90	10
3	0	-45	180	10

a noisy torque signal on the actual arm.

The stiction effect mentioned above for Joint 2 can also be seen in Figure 7.7 at $t = 15$ *sec* near the 'X' at the peak error location. When stiction grabs a joint, the error ramps up as does the torque.

Static Load Changes

This section describes the ability of the DMRAC algorithm to adjust to static load variations. The trajectory of Table 7.4 will be run with different loads in the gripper. The algorithm will first be allowed to adjust to the load, and then the trajectory will be started.

The wrist joints remained locked in their shutdown positions of {0.0, 45.0, 90.0} degrees. Five different loads were run for the trajectory - $0kg$, $1kg$, $2kg$, $3kg$, and $4kg$.

For Joint 1, the error signals did not vary by more than 0.1 degrees between the five different load cases. Figures 7.9 and 7.10 show the response for Joints 2 and 3 respectively. The numbers on the plots are to help identify which curve represents which payload. For Joint 3, the peak errors varied from 2.4390 degrees for the no load case to 3.9972 degrees for the $4kg$ load case. The load changes make up only about 50% of the error. The other

50% is due to the adaptation to the changing arm dynamics. For Joint 2, the peak errors are around 0.8-1.0 degrees. As with Joint 3, the portion of the error due to the load change for Joint 2 is small compared to the no load case.

Dynamic Load Changes

To illustrate the effects of dynamic load change, the trajectory of Table 7.6 was considered. While running the same trajectory, various loads were added to the gripper while the robot was in motion. The same loads used in the previous section were employed. The wrist joints remained locked in their shutdown positions of $\{0.0, 45.0, 90.0\}$ degrees. Note: The $1kg$ and $4kg$ loads were added at about $t = 6.76$ seconds and the $2kg$ and $3kg$ loads were added at about $t = 7.34 sec$.

Figure 7.11 shows the model following error for Joint 2 for all loads. The numbers on the graphs indicate which peaks in the error plots match up with the various loads. This figure shows that the DMRAC algorithm has a good load disturbance rejection. The transient period only lasts about 2 seconds.

Figure 7.12 shows the error for Joint 3 for the various loads. Joint 3 suffers more with a load disturbance having a peak error of almost 5 degrees when the $4kg$ load is added. Again, the transient period is roughly 2 seconds. After the transient, good tracking performance was achieved with the additional loads.

As with the static load case, the model following errors for Joint 1 did not vary by more than 0.1 degrees.

7.1.6 Conclusions and Recommendations

In summary, the DMRAC algorithm was found to be an effective robotic control algorithm in both simulation and on the actual robotic manipulator. Its performance was robust with respect to static and dynamic load variations and also disturbances.

A logical extension of this work would be to control all six joints of the actual PUMA 560 Manipulator. The existing version of the Testbed Motion Control System did not easily support a distributed controller. The ability to control six joints in a fully centralized fashion will require increased computing power. One way to get this increased power might be to incorporate transputers into the control systems. Another method for achieving a centralized six joint controller would be to calculate the gain adaptation updates at a lower frequency than the control servo rate.

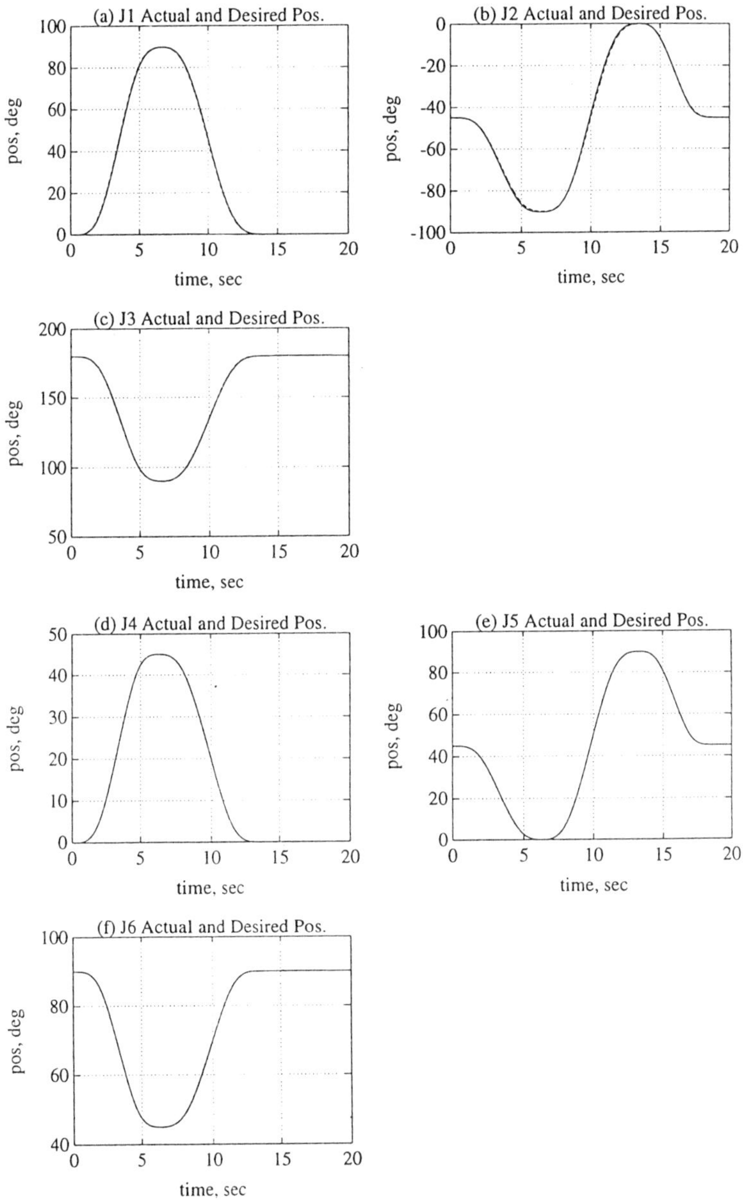

FIGURE 7.4. Response Using Final Tuning Parameter Values. (a) Joint 1. (b) Joint 2. (c)Joint 3. (d) Joint 4. (e) Joint 5. (f) Joint 6.

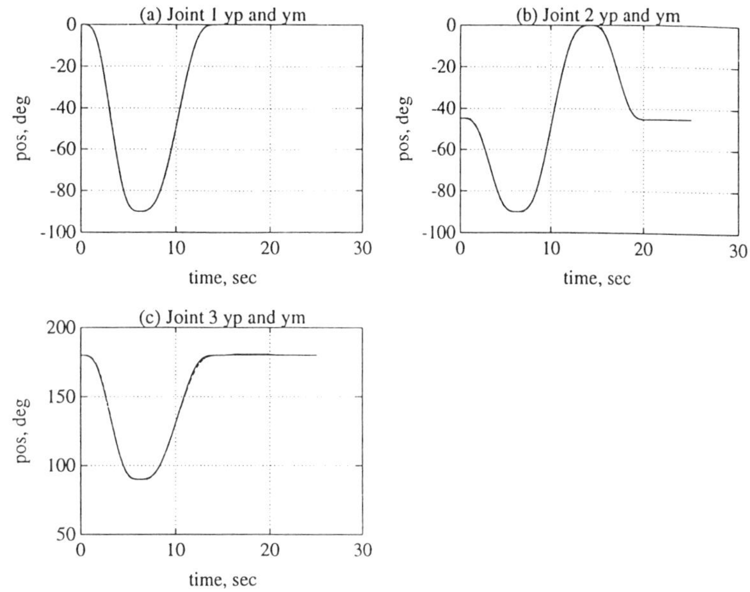

FIGURE 7.5. Plant and Model Output for First Trajectory. (a) Joint 1. (b) Joint 2. (c) Joint 3.

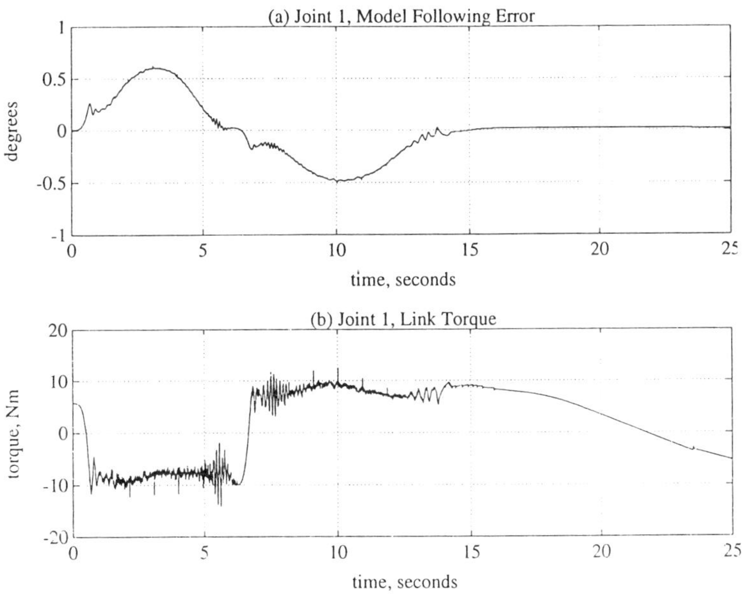

FIGURE 7.6. Joint 1 Data for First Trajectory. (a) Model Following Error. (b) Joint Torque.

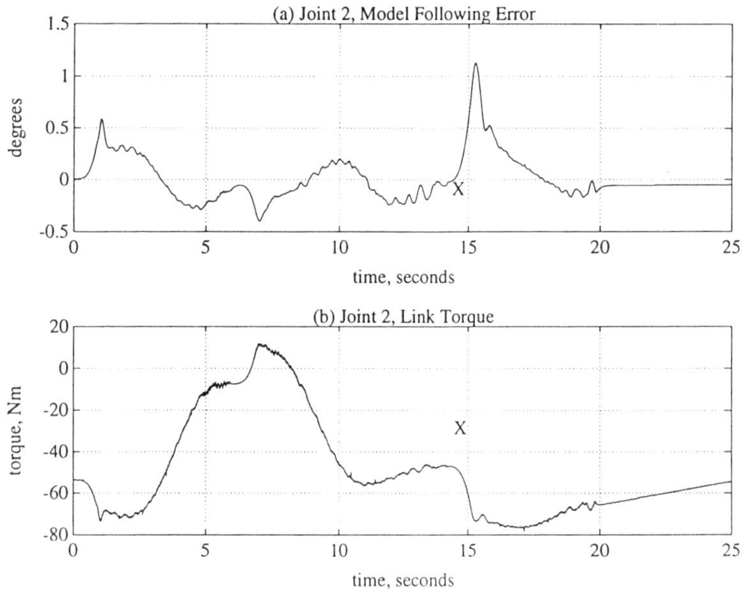

FIGURE 7.7. Joint 2 Data for First Trajectory. (a) Model Following Error. (b) Joint Torque.

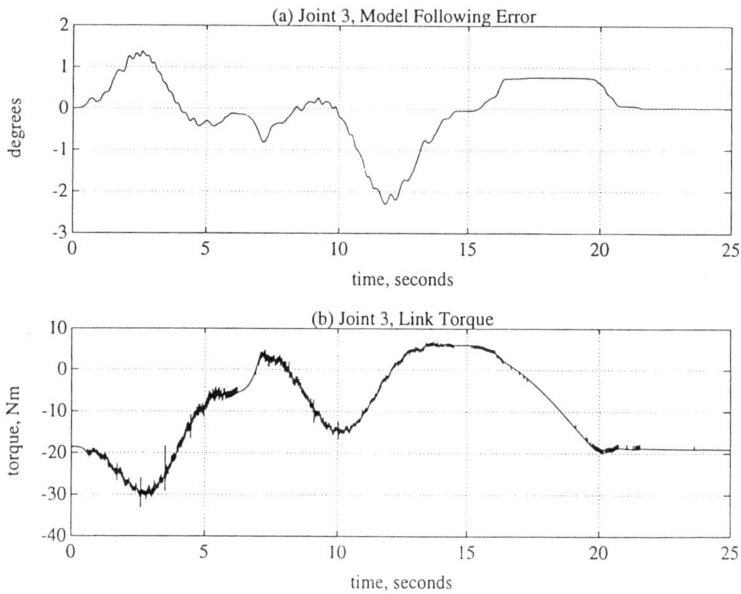

FIGURE 7.8. Joint 3 Data for First Trajectory. (a) Model Following Error. (b) Joint Torque.

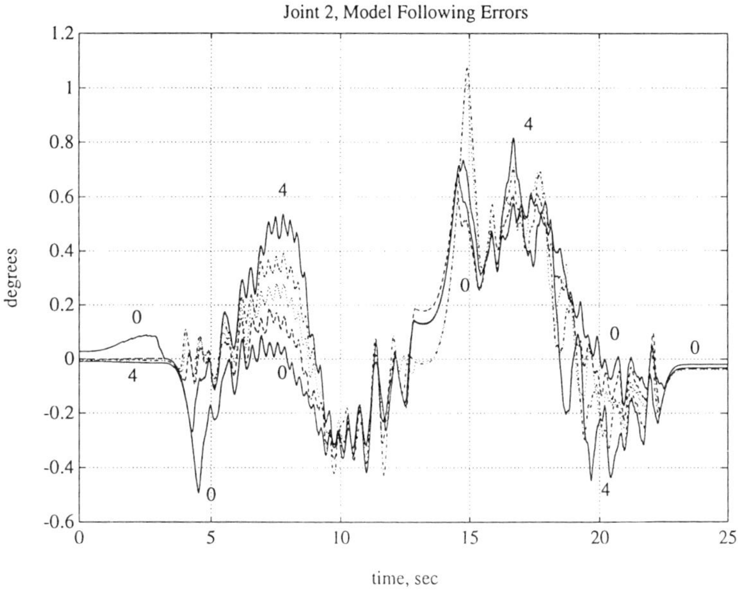

FIGURE 7.9. Joint 2 Static Load Model Following Error

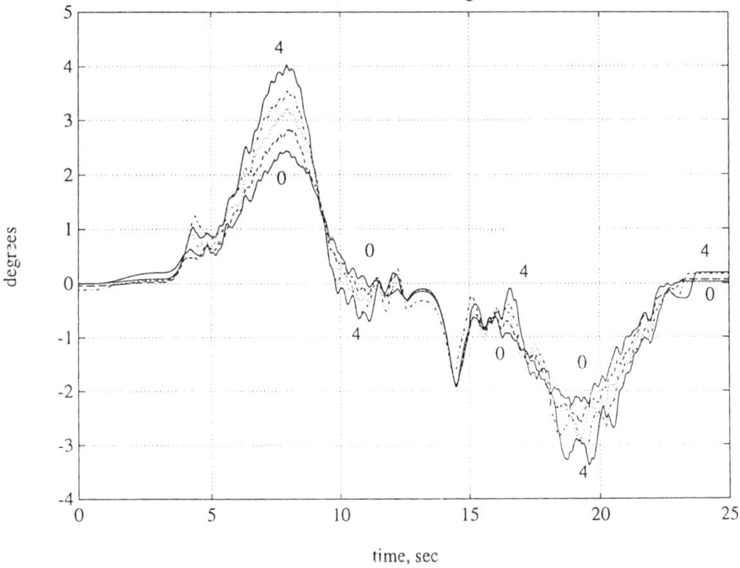

FIGURE 7.10. Joint 3 Static Load Model Following Error

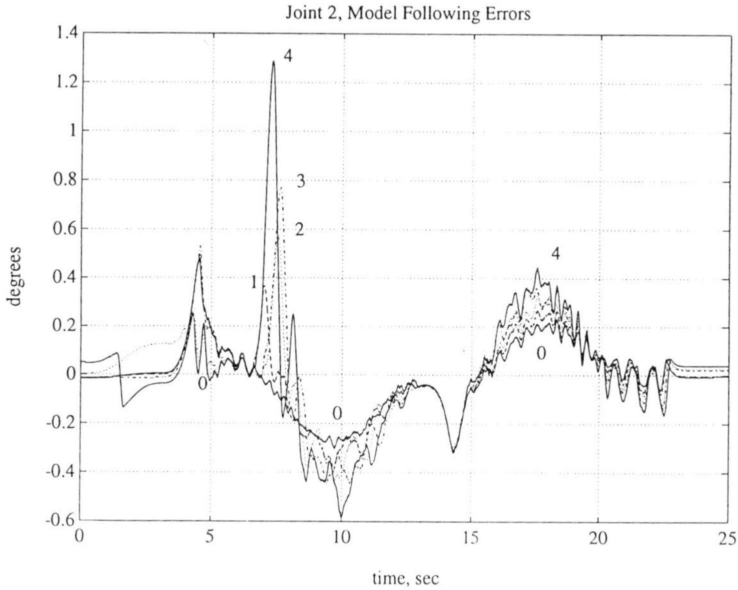

FIGURE 7.11. Joint 2 Dynamic Load Model Following Errors

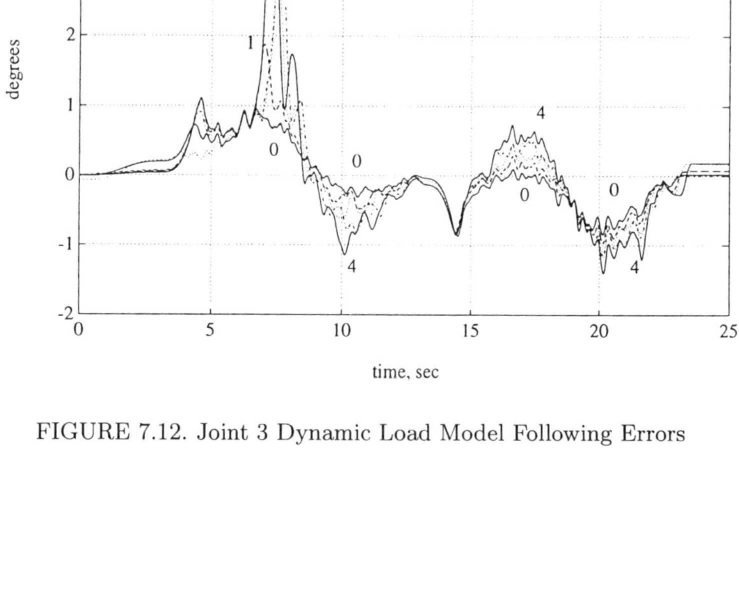

FIGURE 7.12. Joint 3 Dynamic Load Model Following Errors

7.2 Model Reference Adaptive Control of Large Structures

This section has been prepared by Dr. David S. Bayard, JPL, Pasadena, CA

7.2.1 Introduction

In this section, the direct model reference adaptive control methodology is applied to controlling large flexible structures (LFS). The test article for this study is the JPL/PL Large Spacecraft Control Laboratory (LSCL) Ground Experiment Facility at JPL. This test article is an antenna-like structure designed specifically to study the behavior of large flexible structures in a space environment. Earlier adaptive control experiments performed on this facility include a 2-input/2-output experiment [124], and several 6-input/6-output experiments involving vibration suppression [125], robustness [126], and tracking [127]. The material for the section is based primarily on the most recent experimental results from Boussalis, Bayard, Ih, Wang, and Ahmed [127], demonstrating multivariable DMRAC using 6 inputs and 6 outputs for tracking in 2 dimensions.

Various results have evolved to help focus the direct model reference adaptive approach towards the LFS application. This includes the use of leakage and plant feedforward modifications for robustness [128], [129], [130], input weighting to evenly distribute torque in multivariable systems with spatially distributed instrumentation [131], specific methods to ensure satisfaction of the ASPR conditions for adaptive convergence [132], [133], [134], branch filtering to suppress measurement noise [130], as well as systematic methods to choose the gain weights to optimize the adaptive transient [135], [136]. These results can be found in more detail in the references.

7.2.2 Large Flexible Structures (LFS) Dynamics

The dynamics of an elastic structure are most generally described by a set of partial differential equations. However, in practice it often suffices to utilize a finite-dimensional approximation based on the finite-element model [137],

$$M\ddot{x} + Kx = Bu_p \tag{7.11}$$

where $u_p \in R^m$ is a vector of nodal forces (or torques) on the structure, $x \in R^n$ is the state vector of nodal displacements, $M \in R^{n\times n}$ denotes the mass matrix, and $K \in R^{n\times n}$ denotes the stiffness matrix. For the present application to adaptive control, a measurement vector $y_p \in R^m$ is assumed to be available such that,

$$y_p = C(\alpha x + \dot{x}) = \alpha y_{pos} + y_{rate} \tag{7.12}$$

The form of (7.12) indicates that rate and position are measured simultaneously at each sensor location. The quantity α is a factor which scales the position measurement relative to the rate measurement, and can be freely chosen by the designer. In the special case where each actuator location coincides with a sensor location, the actuators and sensors are said to be "colocated", and the relation $B = C^T$ holds in the above model. It is noted that the plant is assumed to be square and m denotes both the number of inputs and outputs.

In order to simplify the dynamics, it is convenient to find a matrix Φ such that

$$\Phi^T M \Phi = I \tag{7.13}$$

$$\Phi^T K \Phi = \Omega^2 \tag{7.14}$$

where $I \in R^{n \times n}$ is an identity matrix and Ω is the diagonal matrix of the form,

$$\Omega = diag\{\omega_1, ..., \omega_n\} \tag{7.15}$$

The quantities $\omega_i, \quad i = 1, ..., n$ are denoted as *modal frequencies*. For notational convenience, it will be assumed that these frequencies are ordered by size in the definition of Ω, i.e.,

$$\omega_1 \leq \omega_2 \leq, ..., \leq \omega_n \tag{7.16}$$

A matrix Φ with the properties in (7.13) (7.14) is said to *simultaneously diagonalize* M and K, and will always exist if $M = M^T > 0$ and $K = K^T \geq 0$ [138]. The dynamics in (7.11) (7.12) are simplified by making the change of variable $x = \Phi\eta$ and using (7.13) (7.14) to give the modal form,

$$\ddot{\eta} + \Omega^2 \eta = \Phi^T B u_p \tag{7.17}$$

$$y_p = C(\alpha \Phi \eta + \Phi \dot{\eta}) \tag{7.18}$$

The modal form above represents a system of undamped resonances. However, due to various energy dissipation mechanisms which occur in real structures, it is common to add a dissipation term proportional to the rate $\dot{\eta}$. One choice is to add the damping term $2D\Omega\dot{\eta}$ which gives,

$$\ddot{\eta} + 2D\Omega\dot{\eta} + \Omega^2 \eta = \Phi^T B u_p \tag{7.19}$$

$$D = diag\{\zeta_1, ..., \zeta_n\} \tag{7.20}$$

where $\zeta_i \geq 0$ can be specified separately for each mode. A discussion of general damping models is given in [138].

A state-space form of the modal model can be realized by letting $x_p = [\eta^T, \dot{\eta}^T]^T \in R^{n_p}$ (i.e., note here $n_p = 2n$), and rearranging (7.17) (7.18) to give,

$$\dot{x}_p = A_p x_p + B_p u_p \tag{7.21}$$

$$y_p = C_p x_p \tag{7.22}$$

where,

$$A_p = \begin{bmatrix} 0 & I \\ -\Omega^2 & -2D\Omega \end{bmatrix} \tag{7.23}$$

$$B_p = \begin{bmatrix} 0 \\ \Phi^T B \end{bmatrix} \tag{7.24}$$

$$C_p = [\alpha C\Phi | C\Phi] \tag{7.25}$$

The state space model $(A_p, B_p, C_p, 0)$ provides a valid plant for application of simple adaptive control strategies. This will be the topic for the remainder of the section.

It is emphasized that the model $(A_p, B_p, C_p, 0)$ defined above is only finite-dimensional, and while it may serve as a satisfactory approximation for engineering purposes, it does not completely capture the full infinite-dimensional dynamics of an elastic structure. The nature of this approximation has been examined in more detail in [139]. It is worth mentioning that rigorous extensions of the DMRAC algorithm to infinite dimensions have been developed using Hilbert space formulations [140], [141].

7.2.3 The ASPR Condition for Flexible Structures

In order to use the adaptive control formulation of Chapter 2, Sections 2.3 and 2.4, it is necessary to ensure that the state-space form of the LFS dynamics $(A_p, B_p, C_p, 0)$ given in (7.23) (7.24) (7.25) satisfies the sufficient conditions for adaptive stability. As discussed in Chapter 2, the essential condition is that the plant is almost strictly positive real (ASPR) as defined below.

DEFINITION 1 (ASPR) A system $\{A_p, B_p, C_p, 0\}$ is ASPR if there exists a gain matrix K_e and matrices $P, Q \in R^{n \times n}$ such that $P = P^T > 0$, $Q = Q^T > 0$, and the following conditions hold,

$$P(A_p - B_p K_e C_p) + (A_p - B_p K_e C_p)^T P = -Q \tag{7.26}$$

$$PB_p = C_p^T \tag{7.27}$$

Explicit solutions to (7.26) (7.27) have been obtained in the literature for the LFS dynamics (7.21)–(7.25). These results are briefly reviewed here.

Mufti, [133] makes the choices $K_e = 0$ and

$$P = \frac{1}{2}\begin{bmatrix} 2\alpha D\Omega + \Omega^2 & \alpha I \\ \alpha I & I \end{bmatrix} \tag{7.28}$$

in (7.26) (7.27). This leads to the following result.

Lemma 1 *(Mufti, [133]) Let $C = B^T$ (i.e., colocated actuators/sensors) and $\omega_i > 0$ $i = 1, ..., n$, (i.e., no rigid body modes). Then the LFS dynamics (7.21) (7.25) are ASPR if*

$$0 < \alpha \leq 2\min_i\{\zeta_i\omega_i\} \tag{7.29}$$

The condition (7.29) requires that α be chosen sufficiently small relative to the system frequencies and dampings. Specifically, for a constant system damping ζ (i.e., $\zeta_i = \zeta$ for all i), the size of α is determined by the lowest frequency mode. Clearly, if rigid body (i.e., zero frequency) modes are present, the lowest mode is given by $\omega_1 = 0$ and a nonzero value of α is not possible from condition (7.29). In contrast, the more general conditions considered next indicate that a nonzero value of α *is* possible in the presence of rigid body modes.

Bayard [134] makes the choices,

$$P = \begin{bmatrix} \Omega^2 + 2\alpha K_e + 2\alpha D\Omega + \epsilon\alpha I & \alpha I \\ \alpha I & I \end{bmatrix} \tag{7.30}$$

$$K_e = \kappa \cdot I \tag{7.31}$$

$$\kappa = (1+\rho^2)2\lambda_{min}\{D_2\Omega_2(I + (\frac{1+\rho^2}{\rho^2})R)^{-1}\}/\lambda_{min}\{\Phi_1^T BB^T\Phi_1\} \tag{7.32}$$

$$D_2 = diag\{\zeta_{m+1}, ..., \zeta_n\} \tag{7.33}$$

$$\Omega_2 = diag\{\omega_{m+1}, ..., \omega_n\} \tag{7.34}$$

$$\Phi = [\underbrace{\phi_1, ..., \phi_m}_{\Phi_1}, \underbrace{\phi_{m+1}, ..., \phi_n}_{\Phi_2}] \tag{7.35}$$

$$R = \frac{\Phi_2^T BB^T\Phi_2}{\lambda_{min}\{\Phi_1^T BB^T\Phi_1\}} \tag{7.36}$$

where ϵ is chosen sufficiently small, and the choice of $\rho^2 > 0$ is arbitrarily. This leads to the following result.

Lemma 2 *(Bayard, [134])*
Let $C = B^T \in R^{m\times n}$ (i.e., colocated actuators/sensors), and let the first m modes be controllable and observable (i.e., $\lambda_{min}\{\Phi_1^T BB^T \Phi_1\} > 0$). Then the LFS dynamics (7.21)- (7.25) are ASPR if,

$$\alpha < 2\lambda_{min}\{D_2\Omega_2(I + (\frac{1+\rho^2}{\rho^2})R)^{-1}\} \tag{7.37}$$

where the choice of $\rho^2 > 0$ is arbitrary.

The quantity m in Lemma 2 is equal to the number of inputs (and outputs since the system is square by assumption). If m is greater than the number of rigid body modes, a nonzero value of α can always be obtained from (7.37) since $\omega_{m+1} > 0$. This result is important, since it implies that zero asymptotic tracking can be obtained in the adaptive algorithm. Previous bounds in the literature did not allow nonzero values of α for systems with both rigid and flexible modes.

Note that while the choice of $\rho^2 > 0$ in Lemma 2 is arbitrary, choosing it large to maximize the allowable value of α in (7.37) requires large gains from (7.32) to satisfy the ASPR conditions. Hence, the choice of $\rho^2 > 0$ allows a trade-off between the size of α and the size of the gain which is required to make the system ASPR. If infinitely large gains are permitted (i.e., $\kappa \to \infty$), one can let $\rho^2 = \infty$ in which case the bound on α is maximized to give,

$$\alpha < 2\lambda_{min}\{D_2\Omega_2(I + R)^{-1}\} \tag{7.38}$$

In order to obtain reasonably large values of α it is required that R be sufficiently small. This is accomplished if the sensors/actuators are placed so that the rigid body modes are strongly observable and controllable. Such configurations are always possible since the rigid body mode shapes are known a-priori from the general shape of the structure.

The treatment here on the positivity of LFS has been necessarily brief. Related results can be found in Mufti [133], Bayard [134], Wen [140], Benhabib, Iwens, and Jackson [142], and Bar-Kana, Kaufman and Balas [132].

7.2.4 Adaptive Control Algorithm

As in Chapter 2, a stable reference model which specifies the desired performance of the plant is described by the following state space representation,

$$\dot{x}_m(t) = A_m x_m(t) + B_m u_m(t) \tag{7.39}$$

$$y_m(t) = C_m x_m(t) \tag{7.40}$$

where $x_m \in R^{n_m}$, $u_m \in R^m$, $y_m \in R^m$; A_m, B_m, and C_m are of appropriate dimensions; and u_m is assumed to be a step function. It is further assumed that the plant (7.21) (7.22) and model (7.39) (7.40) satisfy a command generator tracker (CGT) condition stated as follows,

DEFINITION 2 (CGT condition) The plant $\{A_p, B_p, C_p, D_p\}$ has no transmission zeros at the origin or in common with any eigenvalues of A_m. It is emphasized that the CGT condition does not require the model order n_m to be the same as the plant order n_p. This is important for large flexible structure applications since the plant is generally of very high order.

Define the output error between the plant and the model as,

$$e_y(t) = y_m(t) - y_p(t) \tag{7.41}$$

It is desired to adaptively control the plant so that it tracks the model asymptotically at the output,

$$\lim_{t\to\infty} e_y(t) = 0$$

The adaptive control may, as in Chapters 2 and 3, be written as:

$$u_p = K_r(t)r \tag{7.42}$$

where,

$$r^T = [e_y^T, x_m^T, u_m^T]^T \tag{7.43}$$

$$K_r(t) = [K_e(t), K_x(t), K_u(t)] \tag{7.44}$$

The adaptive gain $K_r(t)$ is chosen as the sum of a proportional and integral component,

$$K_r(t) = K_p(t) + K_I(t) \tag{7.45}$$

The gains K_I and K_p are adapted as follows,

$$\dot{K}_I = -\sigma K_I + e_y r^T T \tag{7.46}$$

$$K_p = e_y r^T \bar{T} \tag{7.47}$$

where $T, \overline{T}$ are weighting matrices of appropriate dimensions, and a "leakage" term $\sigma > 0$ is used as in Chapter 4, to prevent a pure integration effect in (7.46).

Sufficient conditions for global stability of the adaptive algorithm with "leakage" to a *residual set* are summarized below [52], [129], [140], and follow by arguments similar to those presented in Chapter 4,

1. The CGT condition
2. $T > 0$; $\bar{T} \geq 0$

3. The open-loop plant $P(s) = C_p(sI - A_p)^{-1}B_p$ is ASPR

As shown in Chapters 3 and 4, augmenting the plant with a small feedforward term can help to ensure satisfaction of the ASPR condition when $P(s)$ deviates from its ideal form due to actuator/sensor dynamics and/or data acquisition/computational delays [129], [128], [126]. A small plant feedforward of the form

$$D_p = d \cdot I$$

will then be used in the algorithm, for practical purposes.

An important property of the SAC algorithm is that the order of the reference model does not have to be chosen the same as the order of the plant. In fact, the dimension of the plant state is not used anywhere in the algorithm. This property is unusual in the adaptive control literature, and is important for the LFS application since the order of the plant is typically unknown, and can be expected to be very large in practice due to the distributed nature of the dynamics.

7.2.5 Experimental Set-Up

Experimental Structure

Several experiments are now described which were conducted on the JPL/PL Large Spacecraft Control Laboratory (LSCL) Ground Experiment Facility. This facility is depicted schematically in Figure 7.13, and was built for conducting technology experiments for on-orbit spacecraft control applications. The experiment facility is an antenna-like structure exhibiting characteristics of large flexible space structures, including densely packed modes, low frequencies, and three-dimensional structural coupling.

The main component of the apparatus consists of a central hub to which 12 ribs are attached. The diameter of the dish structure is 18.5 feet, the large size being necessary to achieve the low modal frequencies desired. The ribs are coupled together by two rings of pretensioned wires. Functionally, the wires provide coupling of motion in the circumferential direction which cannot be provided by the hub. The ribs, being quite flexible and unable to support their own weight without excessive droop, are each supported at two locations along their free length by levitators. Each levitator assembly consists of a pulley, a counterweight, and a wire attached to the counterweight which passes over the pulley and attaches to the rib. The hub is mounted to the backup structure through a gimbal platform, so that it is free to rotate about two perpendicular axes in the horizontal plane. A flexible boom is attached to the hub and hangs below it, and a feed mass, simulating the feed horn of an antenna, is attached at the free end of the boom. A 3-feet long boom is used for the experiments.

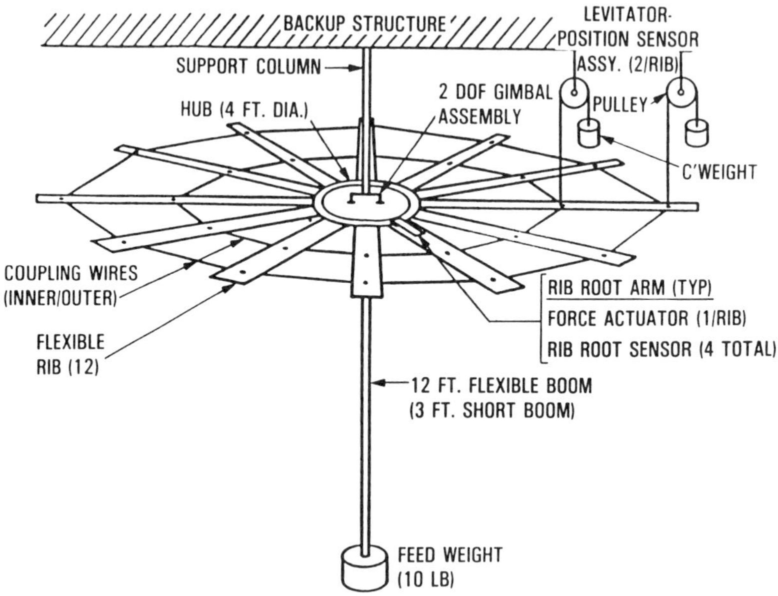

FIGURE 7.13. JPL/PL Large Spacecraft Control Laboratory (LSCL) Ground Experiment Facility at JPL

Instrumentation

A birds-eye view of the structure and associated instrumentation is depicted in Figure 7.14. As indicated, a large number of actuators and sensors have been distributed throughout the structure. For simplicity, the discussion here will only focus on the instrumentation used in the present experiment. Details on the full instrumentation can be found in [143].

For the experiments to be demonstrated in this section, the following colocated actuator/sensor pairs are used:

Hub
HA1/HS10
HA10/HS1
Ribs
RA1/RS1
RA4/RS4
RA7/RS7
RA10/RS10

The two hub torquers HAi, i=1,10 are linear force actuators that provide torques to the hub by pushing against an inner ring (inertially fixed). The torque provided is equal to the force times the lever arm about the axis of rotation. The hub sensors HSi, i=1,10 measure angular positions by rotary

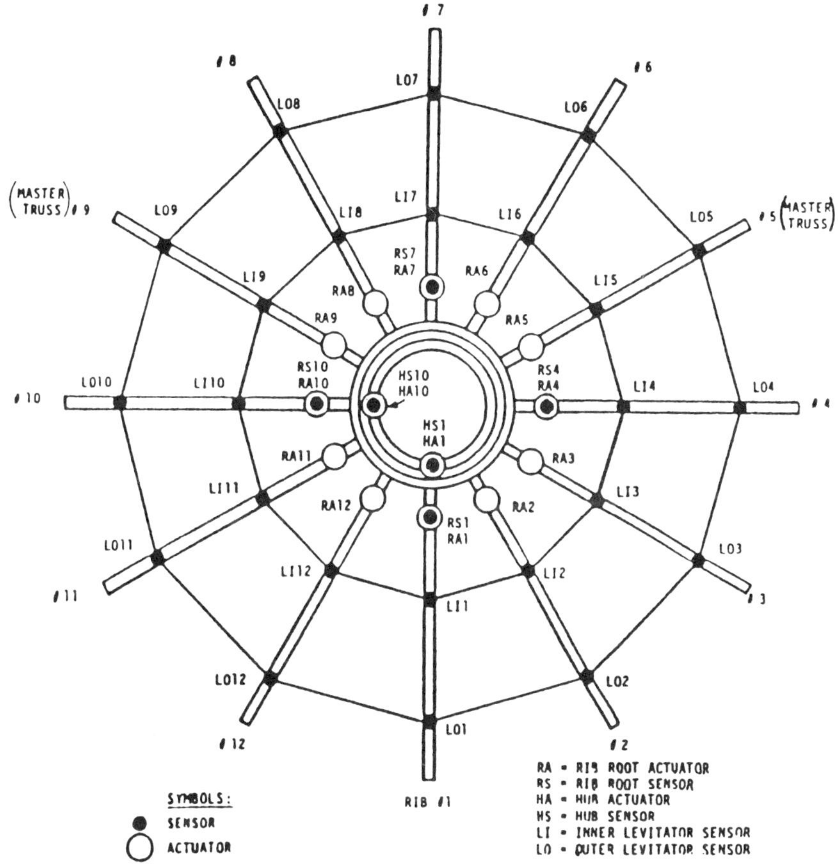

FIGURE 7.14. Bird's Eye View of LSCL Structure and Instrumentation

variable differential transformers (RVDTs) mounted directly on the gimbal axis. Note that each hub sensor measures the structural response to the actuator mounted orthogonal to itself. Hence, although the actuator/sensor pairs HA1/HS1 and HA10/HS10 are physically colocated, it is HA1/HS10 and HA10/HS1 that are colocated in the sense of "dual" variables which should be paired in the transfer function definition.

The rib root actuators RAi, i=1,4,7,10 are solenoid-based designs that provide the desired forces by reacting against the mounts that are rigidly attached to the hub. These actuators are colocated with the four rib root sensors RSi, i=1,4,7,10 that measure the linear rib displacements using linear variable differential transformers (LVDTs).

The computer facility for these experiments is the MicroVax II workstation with the RISC coprocessor. The insertion of the RISC coprocessor board increases the memory of the MicroVax II from 5 to 13 MB, and computational speed from 1 to 15 MIPS.

Dynamic Model

A finite element model is developed for the purpose of obtaining a modal model. The boom and each rib are divided into 10 beam-type elements, and the hub is modeled as a very stiff plate. This results is a 308x308 generalized eigenvalue problem to be solved. However, the symmetry of the structure makes it possible to reduce it to a 44x44 problem by separating variables and writing the dependence of the system mode shapes on a circular wave number k.

Mode shapes of the structure can be grouped according to their circular wave number k, which ranges from $k = 0$ to $k = 6$. Solutions with $k = 0, 2, 3, 4, 5$, and 6 are symmetric about the hub, in the sense that all reaction forces on the hub caused by the ribs exactly cancel out. In such modes which are called "dish modes", neither the hub nor the boom participate in modal motion. On the other hand, modes in which $k = 1$ are asymmetric with respect to reaction forces on the hub. These modes are called "boom-dish", and involve motion of the boom, hub, and dish structure together. The lower-frequency modes of the system are listed in Table 7.7. Note that the modal properties are for the purpose of analysis only, and they are not required for adaptive control design.

Rate Estimation

The output sensing for the adaptive controller requires both position and rate feedback (i.e., see (7.12)). Since the current facility does not include any rate sensors, the rate must be estimated using the noisy position measurements. In earlier studies, a 24-mode Kalman filter based on the finite element model of the plant was designed and successfully used in vibration suppression ([124], [125], [126]). While this filter gave very good performance in regulator experiments, it required plant knowledge and was not sufficiently accurate over the range of motions commensurate with tracking type problems.

In the present investigation, a second order rate estimator is used of the form,

$$\frac{\widehat{x}_2(z)}{y(z)} = \frac{0.0398975z^{-1}(1 - z^{-1})}{1 - 1.7899953z^{-1} + 0.809981z^{-2}} \tag{7.48}$$

This filter is derived as a Kalman filter for a fictitious double integrator system whose acceleration is chosen as noise. The covariances in the Kalman filter are chosen to give the desired bandwidth of the estimator. The filter is then discretized assuming a 45 ms sampling interval which is to be used in the application. This filter results in dramatically improved rate estimates

TABLE 7.7. Normal Modes of Vibration

Dish Modes		
Mode No.	Frequency Hz	k
1	0.210	0
2	0.253	2
3	0.290	3
4	0.322	4
5	0.344	5
6	0.351	6
7	1.517	0
8	1.533	2
9	1.550	3
10	1.566	4
11	1.578	5
12	1.583	6
13	4.656	0
14	4.658	2
15	4.660	3
16	4.661	4
17	4.662	5
18	4.663	6
19	9.474	0
20	9.474	2
21	9.474	3
22	9.474	4
23	9.474	5
24	9.474	6

Boom-Dish Modes			
Mode No.	Frequency Hz		k
	Axis 4-10 Subsystem	Axis 1-7 Subsystem	
1	0.091	0.091	1
2	0.616	0.628	1
3	1.685	1.687	1
4	2.577	2.682	1

compared to divided difference approximations (which are significantly distorted due to A/D quantization), and to model dependent Kalman filter rate estimates (which require plant knowledge). Furthermore, this estimator has the advantage of low dimensionality and plant model independence.

Experiment Design and Command Profile Generation

The objective of the experiment is to demonstrate the ability of the adaptive controller to track a specified reference trajectory. The selected profile is shown in Figure 7.15. Although an unlikely profile for actual slew maneuvering of space vehicles, it was chosen because it combines the continuous trajectory with rate discontinuities, thus, enabling testing of the adaptation in abrupt command changes. The profile in Figure 7.15 is generated

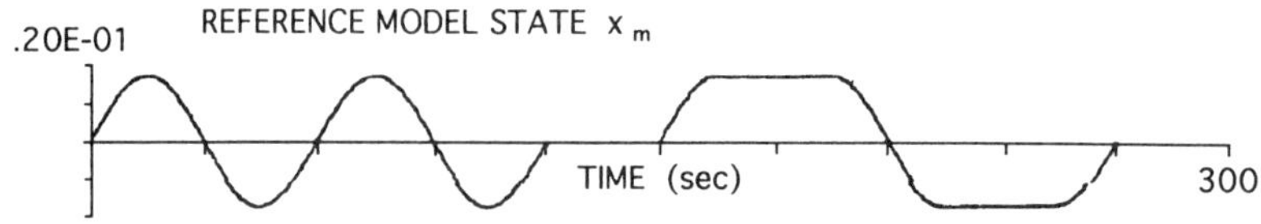

FIGURE 7.15. Reference Trajectory x_m for HA10/HS1 Actuator/Sensor Pair

by the first order reference model

$$\dot{x}_m = -12.5664 x_m + 12.5664 u_m \tag{7.49}$$

$$y_m = \alpha x_m + \dot{x}_m \tag{7.50}$$

where u_m is an "unsmoothed" version of x_m made up of segments of sinusoidal and constant amplitude signals. Due to the fast time constant in 7.49, x_m and u_m are practically identical. The command profile is applied to both hub axes with 90 degree phase shift, in order to effect the circular 2-dimensional trajectory. The performance is measured in terms of deviation of the actual hub position from the reference.

7.2.6 Experiment Results and Discussion

The adaptive control law described in Section 2 was coded in FORTRAN and integrated with the facility's experiment software. Since the plant is 6-input/6-output, it will not be possible to show all signals. For simplicity, plots will be restricted to hub instrumentation along one axis. Specifically, only data from the single actuator/sensor pair HA10/HS1 will be portrayed.

To start the experiments, the following initial choice of parameters is made,

$$\begin{aligned} T &= 10^5\,[I\ I\ I]^T \\ \overline{T} &= 60\,[I\ I\ I]^T \\ \sigma &= 0.5\ \text{(0.08 Hz Integral Gain Bandwidth)} \\ \alpha &= 0.5 \\ d &= 0.0045 \end{aligned}$$

Here, a leakage value of $\sigma = 0.5$ is used to prevent a pure integral action in the adaptation law, and hence provide some robustness to bounded disturbances. In addition, a small feedforward $d \cdot I$ is placed on the plant where $d = .0045$. As mentioned earlier, this feedforward has been theoretically and experimentally shown to help ensure satisfaction of the positivity conditions in non-ideal situations.

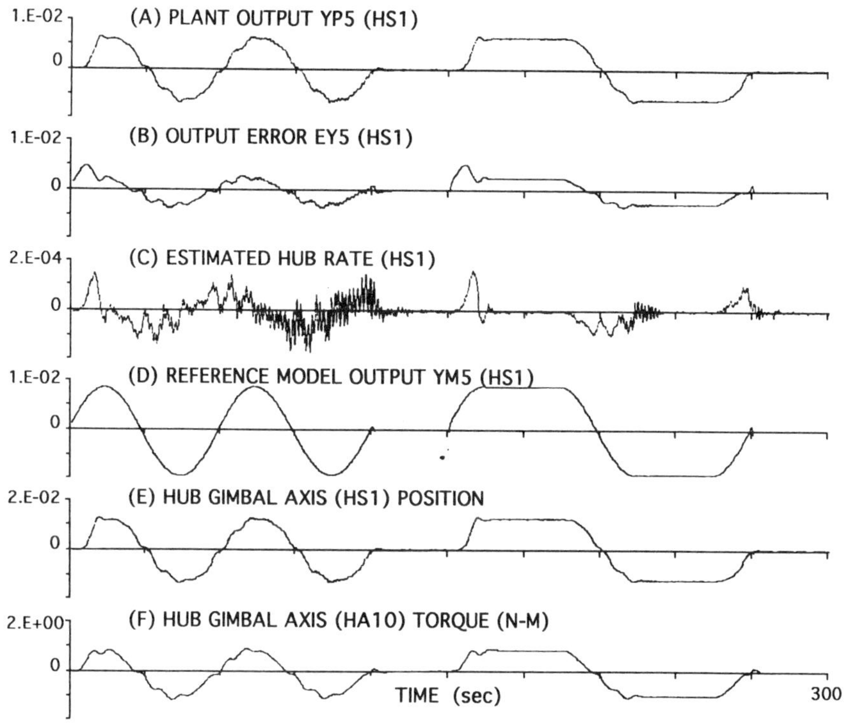

FIGURE 7.16. Closed-Loop Responses of Hub Gimbal Axis for First Run: a) Plant Output; b) Output Error; c) Estimated Rate (mm/sec); d) Reference Model Position (mm); e) Hub Gimbal Axis (HS1), Position (mm); f) Actuator (HA10) Commanded Torque (N-m).

The results of the first run are shown in Figure 7.16. The plant output in trace Figure 7.16(a), shows a fairly good tracking of the desired model reference profile of Figure 7.16(d). The presence of high frequency components in the rate estimate of Fig. 7.16(c) turn out to be due to saturation of the rib-root actuators (not shown here). However, there remains some residual output error (see trace Fig. 7.16(b) which can be improved upon.

It can be shown (cf. [127]) that there are three quantities contributing to the tracking error: σ, d, and u_{max}. Since d is already fairly small it will not be lowered further. The value of σ is lowered from the initial 0.5 to 0.001 in the second run to reduce the tracking error. This change, leads to the improved tracking response of Fig. 7.17 (e.g., compare tracking error in Fig. 7.17(c) with previous error in Fig. 7.16(b)). The results of the second run are summarized in Fig. 7.17 by directly overlaying the plant output and commanded reference trajectory.

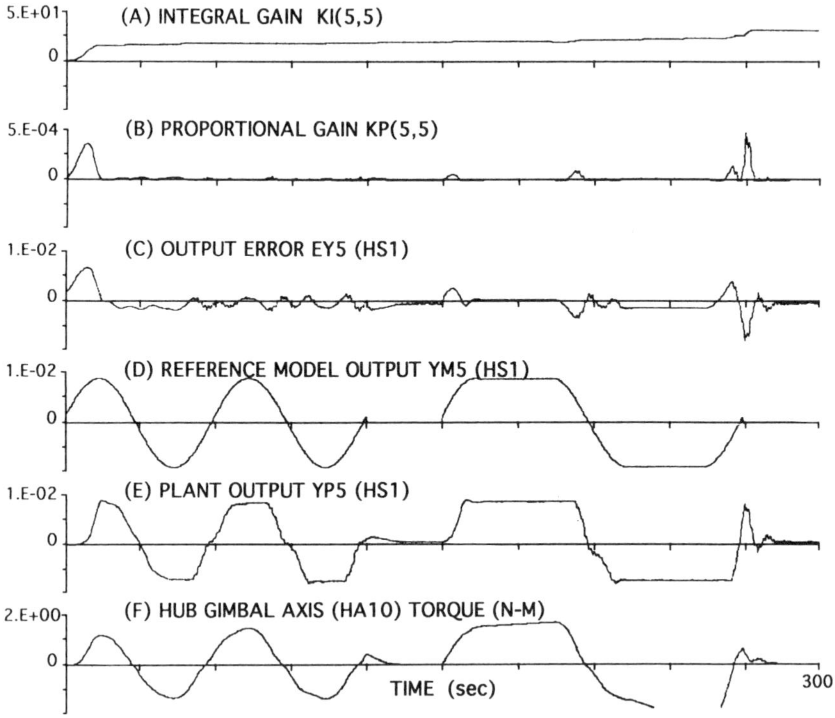

FIGURE 7.17. Closed-Loop Responses of Hub Gimbal Axis for Second Run: a) Component of Integral Gain; b) Component of Proportional Gain; c) Output Error; d) Reference Model Position (mm); e) Plant Output (HS1); f) Actuator (HA10) Commanded Torque (N-m).

It is seen in Fig. 7.18 that there is still some remaining tracking error, particularly in the extreme portions of the reference trajectory. The reason is that at the extreme off-axis positions the actuators have saturated. In order to reduce the error, torque requirements were relaxed in the third run by lowering the amplitude of the reference signal u_m by 70%. This reduces off-axis excursion and avoids actuator saturation. The tracking profiles for the third experimental run are summarized in Fig. 7.19. The improvement relative to the previous run can be seen directly by comparing Fig. 7.19 to Fig. 7.18.

Summary and Conclusions

Some useful lessons were learned in the course of the experimental investigation. These are mentioned below for completeness.

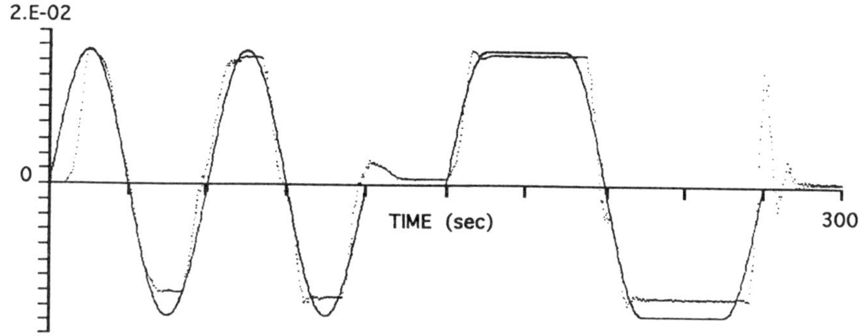

FIGURE 7.18. Summary of Tracking Performance for Second Run Showing Commanded Hub Position Superimposed on Actual Hub Position (HS1), (mm).

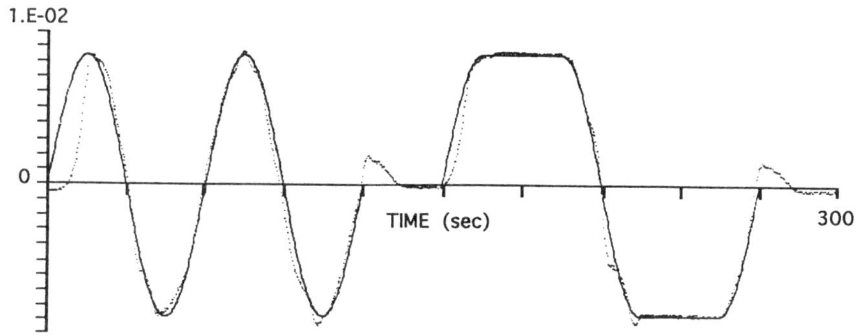

FIGURE 7.19. Summary of Tracking Performance for Third Run Showing Commanded Hub Position Superimposed on Actual Hub Position (HS1), (mm).

1. To achieve good pointing and tracking performance, it is important to maintain a moderate level of integral gain accumulation. In particular the leakage factor σ should be chosen to ensure integral action over a frequency range commensurate with the harmonic content of the reference trajectory.

2. The gain weighting matrices must be chosen sufficiently high to yield acceptable value of tracking error. In addition, results based on adaptive transient analysis (cf., [135], [136]) indicate that the integral gain weight matrix T should be chosen significantly larger than the proportional gain weight matrix $\overline{T}$ for faster adaptive response. This has provided a useful "rule of thumb" for choosing gains in the present study.

3. The tracking command profile used in this study is more challenging than those used in typical space applications. The high commanded rates and rate discontinuities for the two-axis tracking, resulted in stressful demand on the actuators and plant. Better performance is expected for smoother and slower tracking profiles.

4. Good rate measurements are essential for high performance tracking and vibration suppression. Since the quality of any rate estimator is signal dependent, it would be impractical to expect that one design will suit all applications. The Kalman filter and divided difference approximation methods used in previous studies to estimate rate were successfully replaced with a second order filter in the present study. This approach overcame quantization effects inherent in the differencing scheme, and allowed accurate rate estimation over a broader range of motion compared to that attainable using a Kalman filter.

5. The adaptive pointing and tracking control experiments conducted in the one-g environment with the LSCL facility resulted in a hard demand on actuation torque to overcome the gravitational bias torque created by the boom and tip mass when the structure is tilted at an angle. It is expected that the performance will improve in the zero-g environment.

6. Nonlinearities, hysteresis, static friction and other nonidealities have been neglected in the analysis and design of the adaptive system. While these nonidealities were not the major contributors to line-of-sight error in the present study, they can become the dominant issues in the design of systems with very stringent pointing requirements.

7.3 Adaptive Drug Delivery Control

This section has been prepared by Gregory W. Neat,
JPL, Pasadena, CA

7.3.1 INTRODUCTION

This case study describes the application of the direct model reference adaptive control procedure to the regulation of a drug delivery system. In its simplest form, the control objective is to adjust the infusion rate of one or more drugs based on measured physiological parameters. Automating this process with a control scheme provides an attractive alternative for the attending medical personnel who traditionally perform this task. Many factors contribute to the fact that the regulatory challenge faced by such a controller is a difficult one. These include: unmodeled dynamics of the plant, patient-to-patient variation in drug sensitivity, time-varying plant parameters, nonlinearities and cross-interaction of drug responses, narrow bounds of stable existence, and the inherent time delays of the patient's response to the drug.

To meet these challenging control objectives, the model reference adaptive control procedure is combined with a multiple model adaptive control procedure and a rule-based control algorithm in a complementary, hierarchical manner. A supervisory system selects the "best" controller configuration given the available information about the plant. This hybrid control architecture provides a control solution for a broad range of plant conditions including:

- a control solution at start-up when nothing is known about the patient's pharmacokinetics,
- a back-up control solution in case the sufficiency conditions of the model-based methods are violated,
- a stable, precise control solution which adapts to the variations in the patient's pharmacokinetics over time as well as inter-patient variations, and
- a control solution for regulating process time delays inherent to the patient's response to the drug infusion.

The following describes the application of this control strategy to a specific single-input-single-output drug delivery regulation problem. Experimental results demonstrate successful operation of this control approach.

7.3.2 Problem Statement

The performance of the hierarchical controller was assessed experimentally by using the approach to design a regulator for a drug delivery system. The specific control problem was to regulate mean arterial blood pressure by determining the infusion rate of sodium nitroprusside. Clinically, physicians use sodium nitroprusside to lower mean arterial pressure in the case of hypertensive physiological condition. Equation (7.51) represents a simplified model relating the change in mean arterial pressure in response to the infusion of sodium nitroprusside [144].

$$G(s) = \frac{Y(s)}{U(s)} = \frac{k_p exp(-T_d)}{\tau_p s + 1} \tag{7.51}$$

$$P(s) = Y(s) + P_o \tag{7.52}$$

$Y(s)$ is the change in blood pressure from its initial value P_o,
$U(s)$ is the drug infusion rate, and
$P(s)$ is the actual blood pressure.

The parameters in equation (7.51) were found to be [144]: k_p = -.25 to -9.0 (nominal = -1), τ_p = 30 to 60 (nominal = 40), T_d = 20 to 60 (nominal = 40). The variation in patient gain (k_p) and the time delay (T_d) represent the challenging dynamics to the regulator (both the physician and analytical methods).

7.3.3 Controller Design

This section delineates the application of the hierarchical controller by describing the design of its elemental controllers. The section concludes by describing how the components work together in the regulation process.

Rule-Based Control - Fuzzy Controller

The ideas of fuzzy control have grown from the work of Zadeh [145], [146], who introduced the concept of fuzzy sets and linguistic variables. This method of describing terms such as small, high, and low in mathematical terms has allowed for the "rules of thumb" that experts use in controlling poorly modeled processes to be encoded in a systematic manner. Thus, the application of a fuzzy controller to regulating a process is ideal when little is known about the plant and the goal is to model the actions of the expert. For the process presented here, the expert is the physician.

The fuzzy controller approach is successful in moving the state of the plant toward the set point, but once close, a steady-state oscillation frequently occurs about the set point arising from the nonlinear nature of the algorithm. In the hierarchical control structure, this controller makes gross adjustments to drive the plant's outputs to the region of the set point and is a backup controller in the case of an instability in the model-based approaches.

The fuzzy controller rules were defined by asking physicians what changes in infusion rate of sodium nitroprusside they would use (in terms of defined fuzzy numbers) to treat the spectrum of conditions defined in terms of $e(k)$ and $\Delta e(k)$, where e(k) = pressure - set point. Since the rules are a function of $\Delta e(k)$ and the intent of this method is to model accurately the physician's actions, the sample rate of the controller was set at 1.5 minutes (the approximate rate given by those interviewed).

Model-Based Methods

To achieve asymptotic tracking of the plant described by equation (7.51) without imposing stringent sufficiency conditions, this application utilizes a modified model reference adaptive control algorithm. An adaptive inner loop compensator (multiple model adaptive controller) augments and coarsely identifies the plant to satisfy sufficiency conditions required by the model reference adaptive controller (outer loop). The following discusses this two-staged model-based compensator.

Inner Loop - Multiple Model Adaptive Controller

The next level in this hierarchical control structure uses a dramatically different control scheme from that of the coarse fuzzy control approach. In contrast to modeling the operator, this method classifies the plant as one model out of a bank of predefined models and implements a controller designed about that nominal model. This adaptive approach is known as multiple model adaptive control. A bank of models with constant parameters is predefined so as to span the range of parameter variation expected by the plant. Calculated residuals represent the difference between each model in the bank and the plant response to the same applied control value. Each model in the model bank has a corresponding controller in the controller bank designed so that each controller-model pair meets specified design criteria. The actual control value applied to the plant and model bank is composed of a weighted combination of the controller bank outputs. The algorithm for determining the weights is given in [147]. The resulting control value, u_c, is given in equation (7.53), in which u_j is the individual controller value, W_j the corresponding weighting factor, and n the number of model-controller pairs in the model bank.

$$u_i(k) = \sum_{j=1}^{n} W_j(k) u_j(k) \tag{7.53}$$

In the ideal case, the plant would match a particular model in the model bank, and the resulting closed-loop system would meet precisely the predetermined design specifications. In actuality, there is a range of responses for a plant within one parameter space. The coarseness of the model bank dictates the ultimate performance of the controller.

Previous simulation studies [147] revealed that a successful multiple model controller design resulted by quantizing the plant parameter space according to patient gain. Accordingly, this was the approach used in this design. The process dynamics (equation (7.51)) were frozen at $\tau_p = 45.0$ (nominal value) and $T_d = 60.0$ (maximum value). Proportional integral controllers ($G_c(s)$) of the form given in equation (7.54) composed the controller bank.

$$G_c(s) = k_c(1 + \frac{1}{\tau_c s}) \tag{7.54}$$

A controller of this form in the closed loop with the process described in equation (7.51) provides the necessary integral action required by the model reference controller (see below). The study presented here utilizes four models in the model bank. The controller gains (k_c), were selected so that each controller regulating the process within its defined sub-space, satisfied specific rise time and settling time specifications. The settling time for this design was 10 minutes and the allowable range of phase margin was

TABLE 7.8. MMAC Controller and Model Parameters

number	k_c	τ_c	k_m	τ_m	T_{dm}	plant gain
1	.696	45.0	-.43	50.0	60.0	-.25 to -.61
2	.284	45.0	-.795	50.0	60.0	-.61 to -1.49
3	.116	45.0	-2.2	50.0	60.0	-1.49 to -3.63
4	.047	45.0	-5.06	50.0	60.0	-3.63 to -9.0

from 55-75 degrees within one plant sub-space. Models in the model bank ($G_m(s)$) took the form of equation (7.55).

$$G_m(s) = \frac{k_m \; exp(-T_{dm}s)}{1 + \tau_m(s)} \tag{7.55}$$

The model gains (k_m) were determined by guaranteeing that the residuals (W_j) represented in equation (7.53) met at the boundaries between two parameter spaces. Table 7.8 lists the resulting controller and model parameters. The sample rate of the controller was set at ten seconds.

Outer Loop - Model Reference Adaptive Controller

The model reference adaptive controller provides the most precise control for the hybrid scheme. The hierarchical control structure uses a modified version of the algorithm presented in Section 2.4; namely:

$$\dot{K}_e(t) = [e_y(t) + D_S(u^*(t) - u(t) + \tilde{K}_e \; e_y(t))]e_y(t)T_{e_y} \tag{7.56}$$

$$\dot{K}_x(t) = [e_y(t) + D_S(u^*(t) - u(t) + \tilde{K}_e \; e_y(t)]x_m(t)T_x \tag{7.57}$$

$$\dot{K}_u(t) = [e_y(t) + D_S(u^*(t) - u(t) + \tilde{K}_e \; e_y(t)]u_m(t)T_u \tag{7.58}$$

where $u^*(t)$ is the ideal plant input which provides perfect output tracking, $\tilde{K}_e$ is an output feedback gain which stabilizes the plant and T_{e_y}, T_x and T_u are positive definite weighting matrices regulating the rate of adaptation. Control input ($u(t)$) can then be determined from equation (7.59).

$$u(t) = K_e(t)(y_m(t) - y(t)) + K_x(t)x_m(t) + K_u(t)u_m(t) \tag{7.59}$$

For this version, asymptotic stability results when the following conditions are satisfied:

- **A1** The matrix

$$\Omega = \begin{bmatrix} A_p & B_p \\ C_p & 0 \end{bmatrix}^{-1} \tag{7.60}$$

is nonsingular.

- **A2** No transmission zero of the plant is equal to any eigenvalue of A_m.

- **A3** $D_S \geq J$ where

$$J + C_p(sI - A_p + B_p \widetilde{K}_e C_p)^{-1} B_p \tag{7.61}$$

must be strictly positive real for some choice of J and $\widetilde{K}_e$.

Unfortunately for this version of the algorithm, the ideal trajectory (in other words the answer) $u^*(t)$, must be known in order to insure asymptotic stability. Otherwise if approximations of $u^*(t)$ are used, a steady state error results. In addition, this algorithm requires the knowledge of J.

The algorithm modification requires plant augmentation with an inner loop compensator to solve the problem of requiring the knowledge of J and $u^*(t)$. This augmented plant then becomes the new plant from the vantage point of the model reference adaptive controller. The constant inner loop compensator must contain integral action and stabilize the plant. Such compensation provides knowledge of an upper bound on the value of J (equation (7.61)) for the new augmented plant. Knowledge of D_S then follows by selecting $D_S \geq J$. Requiring that the inner loop compensator stabilize the plant allows the selection of $\widetilde{K}_e = 0$ in equation (7.61). Requiring that the inner loop compensator have integral action solves the problem of knowing $u^*(t)$. This restriction forces the gain of the augmented plant to be unity when $t \to \infty$. Since it is desired at $t = \infty$ that $y(\infty) = y_m(\infty)$, u^* is set to $y_m(\infty)$ (which is *a priori* known).

With this two stage control structure, a division of labor exists between the constant gain inner loop and the adaptive outer loop. The functions of the inner loop include defining an upper bound for D_S and providing a value for $u^*(t)$. The adaptive outer loop compensation provides the necessary control adjustment to force the compensated plant to follow the desired reference model. In other words, its major function is to achieve the desired performance for the closed loop.

In view of the above discussion, equations (7.56-7.58) can be written as:

$$\dot{K}_e(t) = [e_y(t) + D_S(y_m(\infty) - u(t))]e_y(t)T_{e_y} \tag{7.62}$$

$$\dot{K}_x(t) = [e_y(t) + D_S(y_m(\infty) - u(t))]x_m(t)T_x \tag{7.63}$$

$$\dot{K}_u(t) = [e_y(t) + D_S(y_m(\infty) - u(t))]u_m(t)T_u \tag{7.64}$$

These equations represent the adaptive control used with the inner loop structure. For this version of the algorithm, asymptotic stability results when **A3** is replaced by the following conditions:

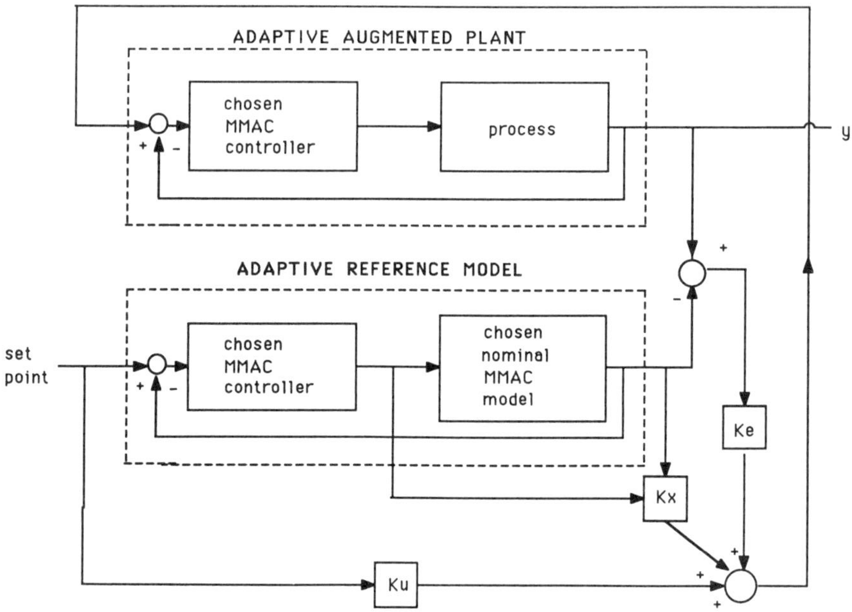

FIGURE 7.20. Hierarchical Controller Configured for the Model Reference Adaptive Controller.

- the existence and knowledge of an inner loop compensator which stabilizes the original plant and contains integral action, and
- $D_S \geq J$ as defined in equation (7.61).

The hierarchical control structure allows for the implementation of this modification. Once the multiple model adaptive controller identifies the process as one of the models in its bank, and the process state is close to the set point, the model reference controller can be called upon to perform precise, stable regulation. Fig. 7.20 presents a block diagram of the hierarchical controller at this stage of operation. The chosen multiple model compensator in combination with the process forms an inner loop which becomes the "augmented plant" for the model reference controller. Similarly, the reference model is formed by the closed loop combination of the selected nominal plant model (from the multiple model bank) and its corresponding controller. The model reference controller corrects for any variation that exists between the compensated process response and the compensated nominal model from the multiple model bank. Its precise control action adjusts the former set point command of the multiple model compensator.

In addition to satisfying the sufficiency conditions for the model reference controller, this control structure accommodates an adaptive reference model selection process. The dynamics of the reference model are a function of the multiple model classification. This design prevents a process with extremely fast dynamics from following a slow model, and similarly, prevents a process with extremely slow dynamics from following a fast model. Once the model reference controller begins its fine tuning adjustment of the augmented plant, the inner loop no longer changes in accordance with changes in the process. However, the multiple model controller continues to classify the process by calculating the weights (W_j) given in equation (7.53).

This hybrid control structure enables these two model-based control schemes to improve upon the abilities of each method used individually. The multiple model procedure used in the inner loop configuration satisfies the sufficiency conditions required by the model reference control algorithm. Correspondingly, the adjustment of the former set point by the model reference controller allows a decrease in the multiple model quantization, thus assuring a realizable control structure in order to meet performance specifications.

The closed loop combination of each controller with its corresponding nominal model was designed to meet the same performance specifications (rise time and range of phase margins for one parameter space). Therefore, independent of the selected model space, each augmented plant (formed by the closed loop combination of the process and the selected multiple model controller) was controlled to follow the same simplified reference model given by equation (7.65).

$$\frac{y_m(s)}{u_m(s)} = \frac{1}{\tau_{ref}s + 1} \tag{7.65}$$

where $u_m(s)$ represents the desired change in pressure from the initial value (P_o) and $\tau_{ref} = 200.0$ *sec.* This reference model represents the desired trajectory for the change in blood pressure for the augmented process, which includes the original process and the multiple model compensator. Thus, the adaptive control signal ($u(t)$ in equation (7.59)) is in a sense a set point command to the inner loop. Equation (7.59) represents the model reference control law with equations (7.62-7.64) describing the adaptive update laws. The $y_m(\infty)$ needed for calculation of the control law gains in equations (7.62-7.64) is $y_m(\infty) = u_m =$ commanded pressure change, since the gain of the reference model is unity. Given the allowable plant variation within one plant sub-space, the range for J from equation (7.61) for the compensated inner loop was .21-.64. For the studies shown, $D_S = 0.4$. Note that this value satisfies the sufficiency condition (equation (7.61)) for one half of the gain variation expected within a plant sub-space while it violates this condition in the other half range. This choice was made in order to test experimentally the robustness of such a value. Based on simulation studies [5], the weighting factors were selected as $T_{e_y} = .0001$, $T_x = 0.0001$

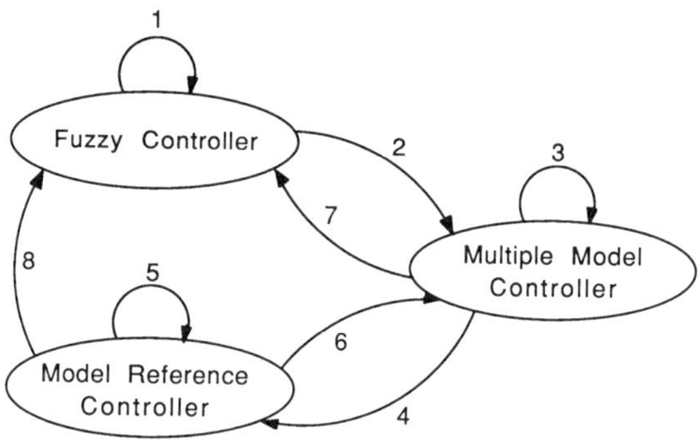

FIGURE 7.21. State Diagram Representing the Function of the Supervisory System.

and $T_u = .0001$, yielding the desired performance from the controller. For implementation, the controller sample time was ten seconds.

7.3.4 Operation of the Complete Hierarchical Controller

A supervisory system orchestrates the controller structure in accordance with the dynamic process. Fig. 7.21 presents a state diagram in which the supervisory function governs the switching between states. Two primary criteria determine changes in controller structure. The first of these conditions involves defining regions in the phase plane described by e and $\dot{e}$. Partitioning the phase plane into different regions provides a method of relating the vastly different control procedures in terms of a common performance measure. The second criteria involves the classification of the plant into a model sub-space by the multiple model identifier.

The hierarchical controller begins in the fuzzy controller state at start-up. The fuzzy controller determines control values during a transient when the state of the process is far from the set point (switch 1). If the state of process (defined by e and $\dot{e}$) approaches the origin in the phase plane and the multiple model controller classifies the plant as one of the models in the multiple model bank, the hierarchical controller will switch to the multiple model control strategy (switch 2). As long as the multiple model controller maintains the regulated parameter within a defined region of the set point, it will continue to determine control values (switch 3). However, if the state of the plant should vary substantially from the set point, the hierarchical controller calls upon the model reference controller (switch 4) to provide an aggressive nonlinear control action to drive the state back to the set point (switch 5). The hierarchical controller initializes the reference model to the value of the plant at the time of this switch. The model reference controller remains in command until the process state returns to the region of the set point (switch 6). In this case the model reference controller gains reinitialize to their initial values ($K_e = 0$, $K_x = 0$ and $K_u = 1$). If the state of the process described in terms of e and $\dot{e}$ deviates excessively from the set point while either of the model-based methods determine the control value, the hierarchical controller reverts to the fuzzy controller for control values (switches 7,8).

The complete hybrid control structure is shown in Fig. 7.22. The system builds on itself in a "learning" type manner. At the outset of controller operation, when the system knows nothing about the process, the fuzzy controller provides the initial crude control by imitating the actions of the operator (switch in position 1). This initial control algorithm allows the multiple model adaptive controller to determine which model in its model bank most closely matches the processes' dynamic response. Control switches to the multiple model controller when the supervisory function determines that conditions warrant the change. At this point, the compensator designed *a priori* for the chosen model dictates the control value (switch in position 2). The multiple model controller and plant at this stage become the inner loop (outlined) for the model reference adaptive controller in order to satisfy sufficiency conditions for its application. The final switch, as determined by the supervisory function, allows the model reference controller to adjust in a fine-tuning manner the set point command to the inner loop in order that the process follows the desired trajectory given by the reference model (switch in position 3).

The effect on the process of switching between different control schemes is an important operational consideration. Most likely, the control signal determined by the rule-based method will differ from that of the model-based methods. Therefore, consideration must be given in making the transition from one control scheme to another a smooth transition. Both the fuzzy controller and the multiple model controller calculate the change in control at each sample time. For the fuzzy controller, this requires obtaining rules from the operator in terms of change in control to the process. For the multiple model adaptive controller, an incremental or "velocity" form of the control law is computed from the "position" form as in equation (7.66).

$$\Delta u(k) = u(k) - u(k-1) \tag{7.66}$$

A common integrator transforms these incremental control calculations from either the model-based or rule-based controllers into the total control to apply to the process. The origin of the $\Delta u(k)$ in this strategy is transparent to this integrator and thus insures a smooth transition.

The model reference adaptive controller adjusts the set point of the compensated inner loop. Since the inner loop formed by the multiple model controller filters this signal, the model reference control adjustments will

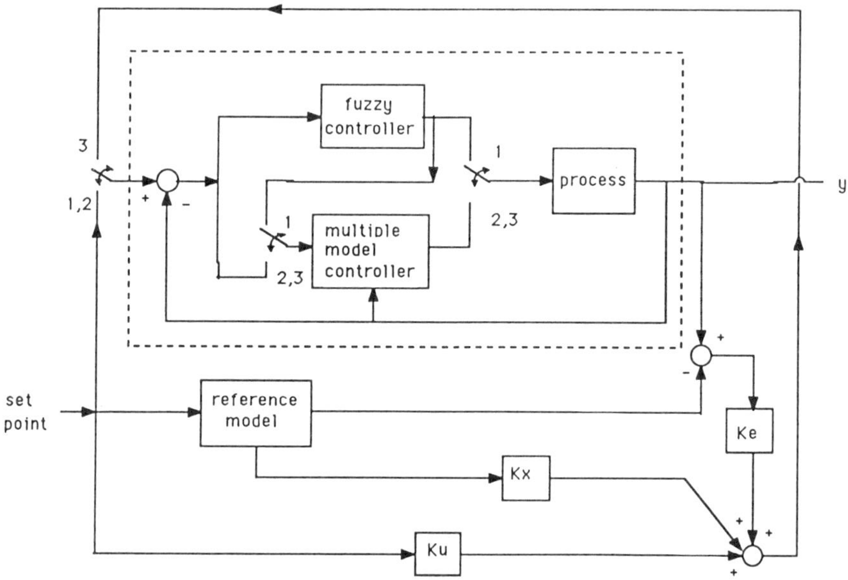

FIGURE 7.22. Block Diagram of Complete Hierarchical Controller.

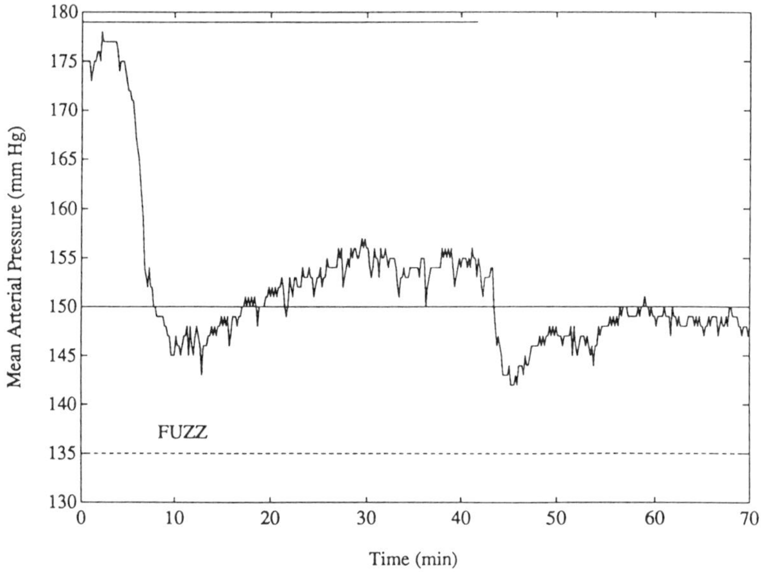

FIGURE 7.23. Plot of Mean Arterial Pressure (mm Hg) vs. Time (min). The Hierarchical Controller Contains only a Fuzzy Controller. (set point = 150 mm Hg)

not cause abrupt changes to the control value applied to the process. Thus, smooth controller transition is not an issue at this stage of the hybrid controller.

7.3.5 Experimental Results

The results presented in this section were obtained by applying the hierarchical controller to dogs in an actual animal laboratory. The hierarchical controller determined the necessary infusion rate of sodium nitroprusside in order to maintain the dog's mean arterial pressure at a desired level. Details on the experimental setup are given in [148]. Each subsection shows different aspects of the operation of the hierarchical controller by presenting time courses of physiological and/or controller parameters. When demonstrating controller switching, a curve at the base of the figure will indicate which of the three controllers is commanding the sodium nitroprusside infusion rate (fuzzy controller = FUZZ, multiple model adaptive controller = MMAC, model reference adaptive controller = MRAC). Phenylephrine, which has the opposite effect from that of sodium nitroprusside (causes an increase in mean arterial pressure), was infused periodically as a disturbance to the controller. A curve at the top of the relevant figures will indicate its infusion.

Hierarchical Controller Containing only the Fuzzy Controller

Figures 7.23 and 7.24 show how the fuzzy controller tackles the regulation of mean arterial pressure on its own. Fig. 7.23 demonstrates the ability of the fuzzy controller to move the pressure to the region of the set point quickly as a result of its highly nonlinear, coarse, control adjustments (Fig. 7.24). Note just after the transient, the controller makes no change in sodium nitroprusside despite pressure changes from approximately 5 $mmHg$ below the set point to 5 $mmHg$ above the set point. The coarse control nature of the fuzzy controller is again demonstrated by its response to the step off of phenylephrine which occurs at $t = 42$ min.

Hierarchical Controller Containing the Fuzzy Controller and the Multiple Model Controller

Figures 7.25 and 7.26 show how the hybrid controller regulates mean arterial pressure when the model bank contains a fuzzy controller and a multiple model adaptive controller. At start-up the fuzzy controller begins initial coarse control adjustments indicated in Fig. 7.26. When the multiple model identifier selects a model and the state $(e(k), \Delta e(k))$ of the system reaches the proper region, control switches to the multiple model adaptive controller (Fig. 7.26). In contrast to the fuzzy controller, the multiple model compensator makes adjustments in the control value whenever there is a

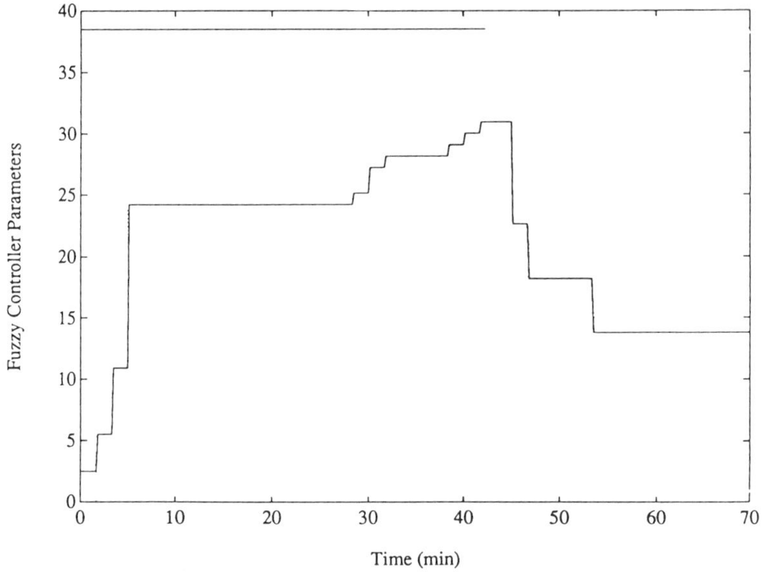

FIGURE 7.24. Plot of Sodium Nitroprusside Infusion Rate vs. Time (min).

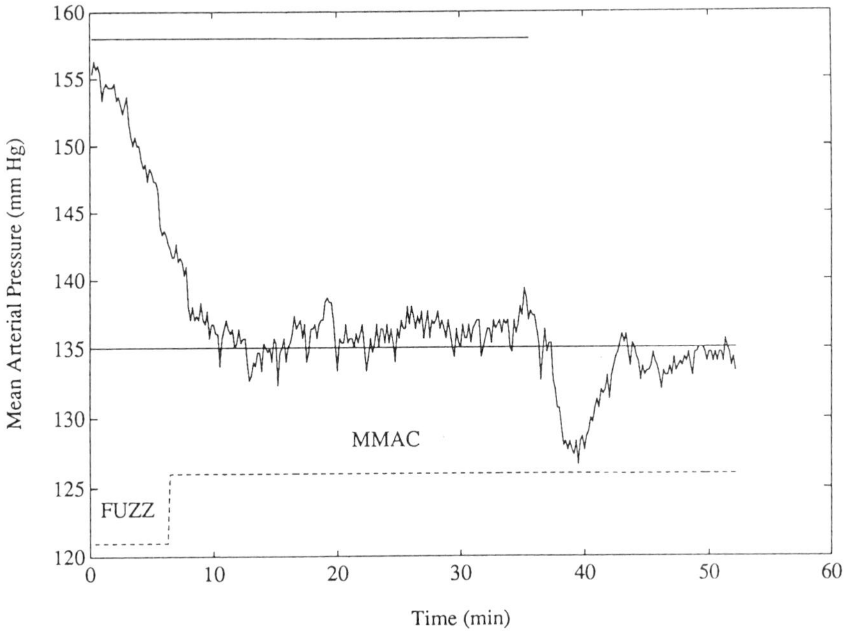

FIGURE 7.25. Plot of Mean Arterial Pressure (mm Hg) vs. Time (min). The Hierarchical Controller Contains a Fuzzy Controller and a Multiple Model Controller. (set point = 135 mm Hg)

non-zero error. This control scheme successfully maintains the pressure at the set point despite the changes in the dog's sensitivity to the drug. This is evident from the dramatic rise in control (Fig. 7.26) during the interval $t = 10$ to $t = 35$ minutes while the pressure remains in close proximity to the set point (Fig. 7.25). The step off of phenylephrine at $t = 36\ min$ demonstrates how the multiple model adaptive controller handles a sudden change in plant parameters.

Complementary Operation of the Hierarchical Controller's Model-Based Algorithms

Figures 7.27–7.30 present an example of the model reference adaptive controller operation. Compare the responsiveness of the model reference control action with that of the multiple model control adjustment (Figures 7.27 and 7.28). When each compensator is faced with approximately the same error (as indicated by the arrows in the two controller regions in Fig. 7.27), Fig. 7.28 shows that the model reference controller (adaptive outer loop) makes a more aggressive control adjustment as compared with the multiple model compensator (inner loop). To avoid excessive drug infusion levels, the disturbance (phenylephrine) was removed at $t = 38\ min$. The model reference adaptive controller responded with the same aggressive action as on the rising edge (Fig. 7.28) which results from its set point adjustment above and then below the multiple model controller set point level respectively (Fig. 7.29). Fig. 7.30 demonstrates an example of the model reference adaptive controller gain reinitialization process once the mean arterial pressure returns to the region of the set point. Note that $K_x(t)$ and $K_u(t)$ return to their approximate command generator tracker values ($K_x = 0$, $K_u = 1$) while $K_e(t)$ grows during the period of model reference controller operation. Once the hierarchical controller reverts to the multiple model controller, the model reference gains are reinitialized to their start-up values ($K_e = 0$, $K_x = 0$, $K_u = 1$). The plant exhibits highly nonlinear behavior during this operating time (marked by dramatic increases in drug infusion rate with small pressure changes). The multiple model adaptive controller addresses this behavior by classifying the plant into the lowest plant gain sub-space (range 1 in table 1), in which the inner loop compensator contains the highest gain value ($k_c = .696$). Despite this nonlinear response and the selection of $D_S = 0.4$ (which possibly violates a sufficiency condition based on a linear plant), the model reference adaptive controller successfully returns the pressure to the region of the set point at which time the multiple model controller resumes command of the drug infusion rate.

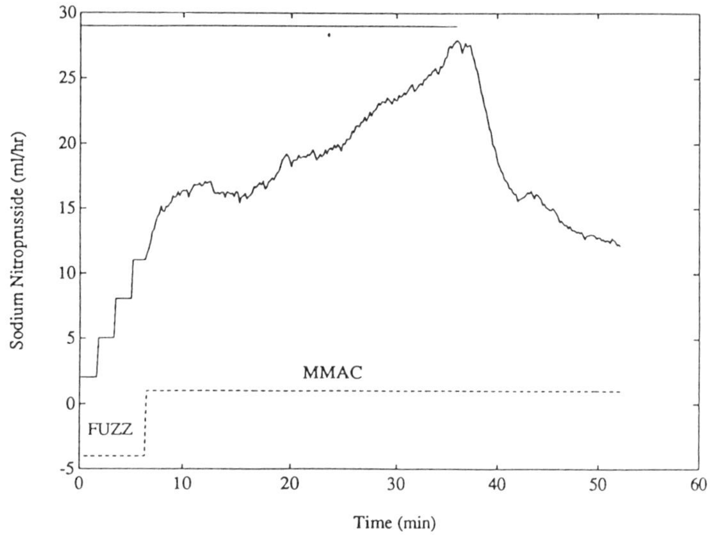

FIGURE 7.26. Plot of Sodium Nitroprusside (ml/hr) vs. Time (min). The Hierarchical Controller Contains a Fuzzy Controller and a Multiple Model Controller.

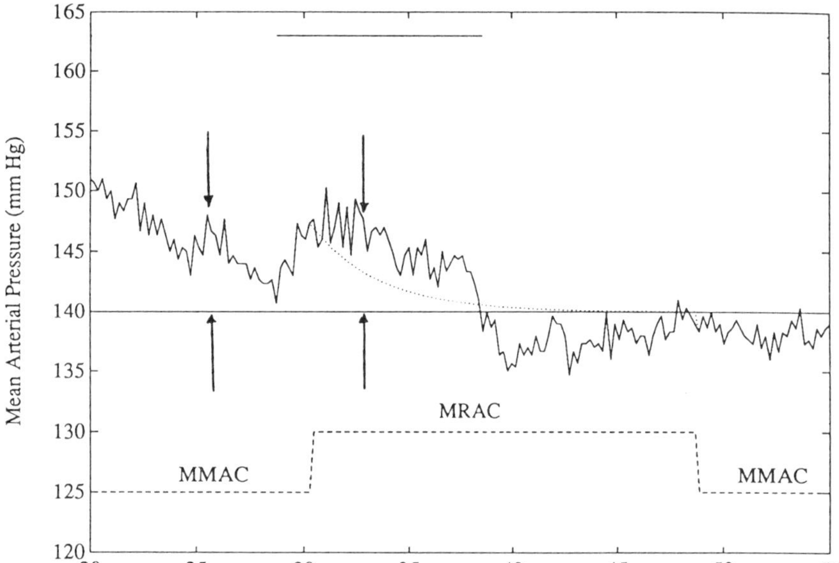

FIGURE 7.27. Plot of Mean Arterial Pressure (mm Hg) vs. Time (min) During Steady-State Operation of the Complete Hierarchical Controller Containing the Fuzzy, Multiple Model, and Model Reference Controller ($D_s = 0.4$), (set point = 140 mm Hg). Dotted Line Represents Reference Model Trajectory.

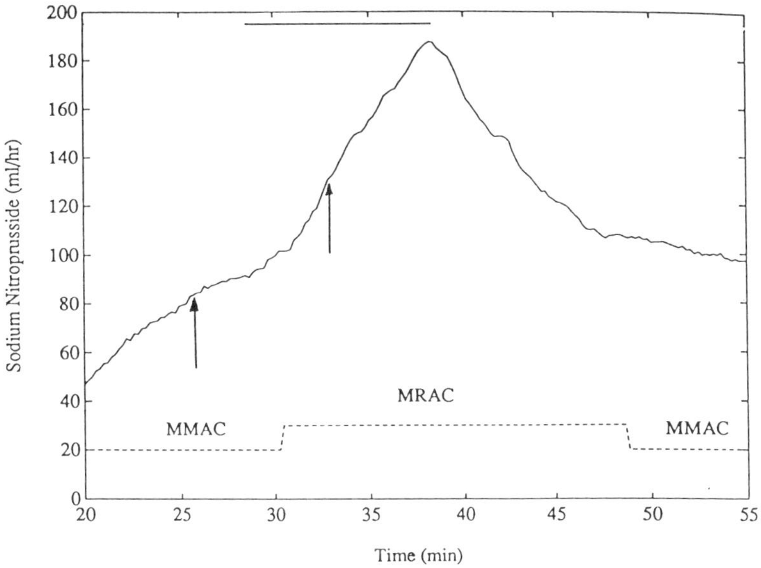

FIGURE 7.28. Plot of Sodium Nitroprusside (ml/hr) vs. Time (min) During Steady-State Operation of the Complete Hierarchical Controller Containing the Fuzzy, Multiple Mode, and Model Reference Controller (Ds =0.4).

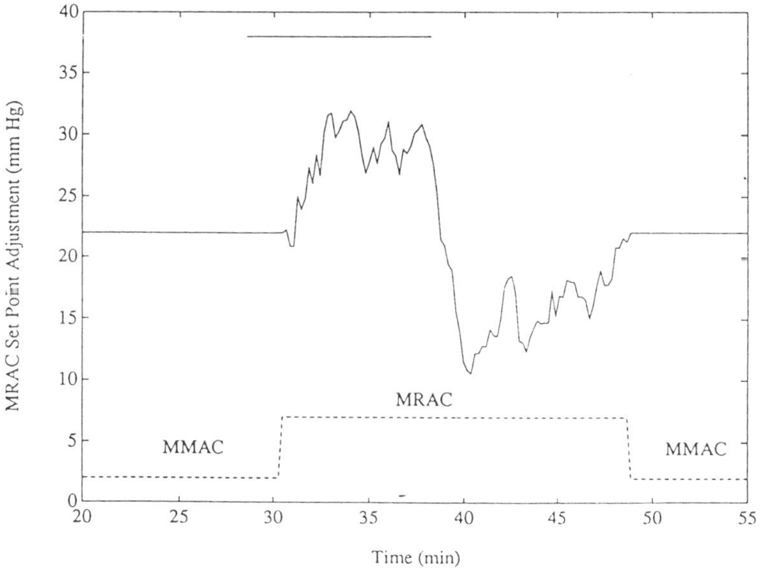

FIGURE 7.29. Plot of Adjustment to the Inner Loop by the Model Reference Adaptive Controller (mm Hg) vs. Time (min) During Steady-State Operation of the Complete Hierarchical Controller.

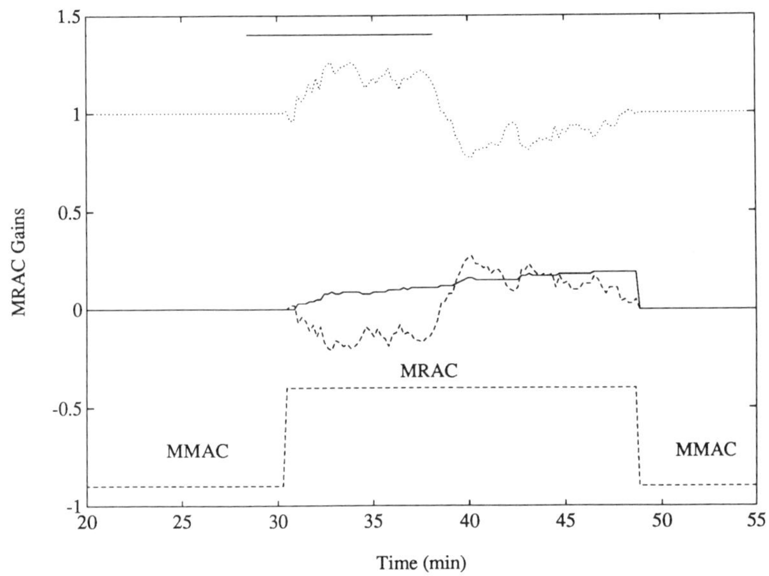

FIGURE 7.30. Plot of Model Reference Adaptive Controller Gains vs. Time (min) During Operation of the Complete Hierarchical Controller. K_e (solid), K_x (dashed) and K_u (dotted).

7.3.6 Conclusions

This control strategy combines rule-based and model-based control algorithms in order to provide the "best" control for a broad range of plant conditions. In the worst case with little knowledge of the process, this method has the potential to perform as well as a human operator; at best, the state(s) of the process will be driven to the desired set point in a manner prescribed by the chosen reference model. Combining these model-based and rule-based methods in the complementary manner prescribed yields a control structure with the following advantages:

- the controller contains a control strategy for start-up based on the human operator's experience,
- the controller contains a strategy in case of failure of the model-based methods, and
- the controller adaptively regulates (with zero steady state error) processes that satisfy the indicated assumptions without requiring the identification of all the plant parameters.

The inner loop structure resulting from the complementary operation of the two model-based algorithms results in a control structure with attractive characteristics. The model reference adaptive controller requires knowledge of D_S which the multiple model controller provides. This could be considered an indirect control method in this configuration, in which the multiple model controller "identifies" D_S for the model reference controller. However, in contrast to the traditional indirect adaptive control methods, which require explicit identification of plant parameters, this approach only requires knowledge of one parameter (D_S). In addition, since the hierarchical controller reinitializes the model reference controller gains to the command generator tracker values after its use, the gains are free to grow as required, to meet the desired performance specifications. This is in contrast to the σ modification suggested by Ioannou and Kokotovic [149] and applied to the command generator tracker model reference controller in Section 4.2. This modification to the control law increases the robustness of the algorithm while compromising performance. The hierarchical configuration thus allows the nonlinear control laws described by equations (7.62-7.64) to calculate the controller gains without restriction. A future area of research would be to generalize this inner loop concept to include other adaptive control algorithms.

7.4 Adaptive Control for a Relaxed Static Stability Aircraft

7.4.1 Introduction

Typical design of conventional aircraft stability augmentation systems uses classical techniques to compute a fixed feedback gain matrix for each linearized flight condition. The final design requires a form of adaptation known as gain scheduling, where gains are typically varied between the linearized flight conditions as a function of dynamic pressure. Such a design requires air data sensors and an air data computer to implement the gain schedule.

An alternative design is proposed here which may not require air data sensors and an air data computer for implementation. The proposed design utilizes the model reference adaptive control algorithms which were presented in Sections 2.4 and 3.4. The reference model has the aircraft design specifications incorporated within it. For example, a step input to the reference model would cause its outputs to respond with the specified rise time, overshoot, and damping ratio.

The aircraft and reference model work together to incorporate these characteristics into the aircraft. The pilot commands are fed into the reference model which responds with the design specifications which have been built into it. If the control system is designed properly, the inputs to the aircraft (which are generated from the pilot commands, the reference model states, and the error between aircraft and reference model outputs) drive the outputs of the aircraft to equal the outputs of the reference model.

For this study, the algorithms of Sections 2.4 and 3.4 are utilized for the design of a stability augmentation system for the longitudinal dynamics of a relaxed static stability version of a transport aircraft. This aircraft is one in which, due to fuel placement, the center of gravity (CG) is moved aft. The aft CG position allows the aircraft to fly with reduced control surface trim conditions and thus results in an aerodynamically cleaner configuration.

The performance of the two different MRAC systems are evaluated for a scenario which includes takeoff, climb and cruise conditions. The adaptive law is compared with a fixed gain controller designed using a pole placement algorithm. The MRAC system is shown to be superior to the fixed gain controller for obtaining the desired short period behavior.

7.4.2 Model Development

Aircraft Model

The aircraft is a wide transport with its center of gravity moved aft. The CG position varies between 0.322 and 0.331 mean aerodynamic chord (MAC) as compared with 0.25 MAC for the conventional vehicle.

mented with actuator dynamics is described by:

$$\frac{d}{dt}\begin{bmatrix}\alpha\\ q\\ \delta_H\\ \dot{\delta}_H\end{bmatrix} = \begin{bmatrix} Z_\alpha & Z_q & Z_{\delta_H} & 0\\ M_\alpha & M_q & M_{\delta_H} & 0\\ 0 & 0 & 0 & 1\\ 0 & 0 & -125 & -26.25\end{bmatrix}\begin{bmatrix}\alpha\\ q\\ \delta_H\\ \dot{\delta}_H\end{bmatrix} + \begin{bmatrix}0\\ 0\\ 0\\ 125\end{bmatrix}\delta_{HC} \tag{7.67}$$

where
α = angle of attack (rad)
q = pitch rate (rad/sec)
δ_H = horizontal stabilizer deflection (rad)
δ_{HC} = horizontal stabilizer command (rad)

The aircraft is assumed to vary linearly between the takeoff, climb and cruise flight conditions. These linearized operating points correspond to flight at sea level and Mach 0.22, 10,000 feet and Mach 0.5, and 37,000 feet and Mach 0.8, respectively. The aircraft stability and control derivatives for the three flight conditions are shown in Table 7.9.

Reference Model

The reference model development starts with a cruise model for the conventional aircraft at 37,000 feet and 0.83 Mach with the CG located at 0.25 MAC. This model is described by

$$\frac{d}{dt}\begin{bmatrix}\alpha\\ q\\ \delta_H\\ \dot{\delta}_H\end{bmatrix} = \begin{bmatrix} -.573 & .984 & -.125 & 0\\ -1.27 & -.616 & -4.88 & 0\\ 0 & 0 & 0 & 1\\ 0 & 0 & -125 & -26.25\end{bmatrix}\begin{bmatrix}\alpha\\ q\\ \delta_H\\ \dot{\delta}_H\end{bmatrix} + \begin{bmatrix}0\\ 0\\ 0\\ 125\end{bmatrix}\delta_{HC} \tag{7.68}$$

Next, a constant output feedback control law is designed for the preliminary model described by (7.68) so that the closed loop system has a damping factor ζ =.707 and natural frequency ω_n = 1.6 rad/sec. This feedback gain matrix is given by

$$K = [K_\alpha, K_q] = [-0.00109, 0.158] \tag{7.69}$$

	Takeoff	Climb	Cruise
Z_α	-.451	-.685	-.529
Z_q	.962	.969	.984
M_α	-.401	-.963	-.285
M_q	-.456	-.829	-.598
$Z_{\delta H}$	-.134	-.178	-.118
$M_{\delta H}$	-1.65	-4.83	-4.63

TABLE 7.9. Aircraft Stability and Control Derivatives

Combining the preliminary model described by (7.68) with the feedback law given by (7.69) yields the reference model shown below:

$$\frac{d}{dt}\begin{bmatrix} \alpha \\ q \\ \delta_H \\ \dot{\delta}_H \end{bmatrix} = \begin{bmatrix} -.573 & .984 & -0.125 & 0 \\ -1.27 & -0.616 & -4.88 & 0 \\ 0 & 0 & 0 & 1 \\ -0.137 & 19.7 & -125 & -26.25 \end{bmatrix} \begin{bmatrix} \alpha \\ q \\ \delta_H \\ \dot{\delta}_H \end{bmatrix} + \begin{bmatrix} 0 \\ 0 \\ 0 \\ 125 \end{bmatrix} \delta_{HC} \tag{7.70}$$

A block diagram of the reference model is shown in Figure 7.31.

7.4.3 Control Law Development

Fixed Gain Benchmark Controller

A fixed gain feedback control law was designed as a benchmark to be used in evaluating the performance of the adaptive control algorithm. The gain was computed to obtain a short period mode with $\zeta = 0.707$ and $\omega_n = 1.6$ at the takeoff flight condition. The takeoff flight condition was chosen as the design point because it is the most critical of the three operating points. A diagram of the fixed gain benchmark controller is shown in Figure 7.32.

A computer simulation of the response of the time-varying aircraft with the fixed gain controller is shown in Figure 7.33 with the reference model overplotted for comparison. The aircraft pitch rate does not follow the reference model pitch rate in steady state. However, more significant is that the aircraft pitch rate response is overdamped for $t \geq 40$ sec whereas the reference model pitch rate response is underdamped. Therefore, a fixed gain controller is unacceptable for the time varying aircraft.

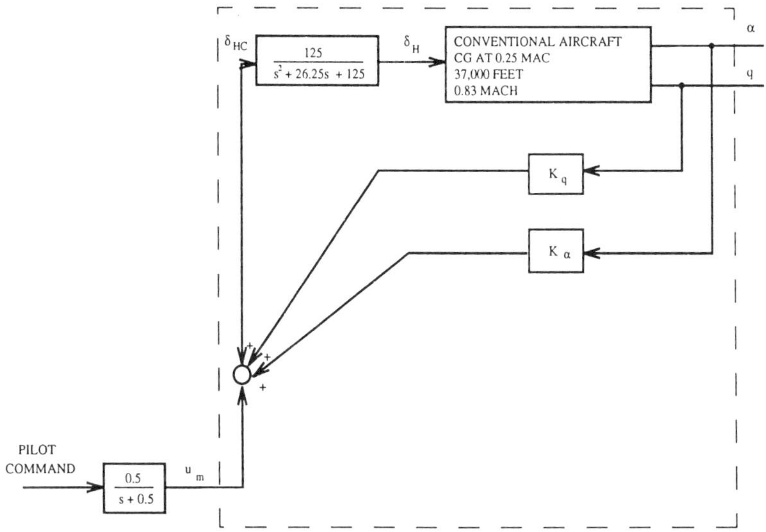

FIGURE 7.31. Reference Model

Adaptive Controller of Section 2.4

The algorithm described in Section 2.4, which requires that the plant be ASPR, is utilized to design an adaptive controller for the time varying aircraft. No attempt is made to demonstrate that the transfer function of the relaxed static stability aircraft is almost strictly positive real. Thus, in some sense, the design using the Section 2.4 algorithm is ad hoc. However, the adaptive system will be validated using a computer simulation.

Since the adaptive algorithm assumes that the aircraft's transfer function matrix is square, it was decided to match the aircraft pitch rate to the reference model pitch rate. The weighting matrices were tuned by trial and error, which resulted in the choice $T = \bar{T} = 500I$. The adaptive feedback gain K_e was initialized at the value in the fixed gain benchmark controller which was computed for the takeoff flight condition. This is valid because at the initial time when the aircraft leaves the runway, the aircraft is operating at the takeoff condition.

The simulated pitch rate using the adaptive algorithm is shown in Figure 7.34, where it can be observed that the pitch rate model following is good after an initial learning phase. The small steady state error results because the stability analysis is for a time invariant plant, whereas the aircraft is time varying.

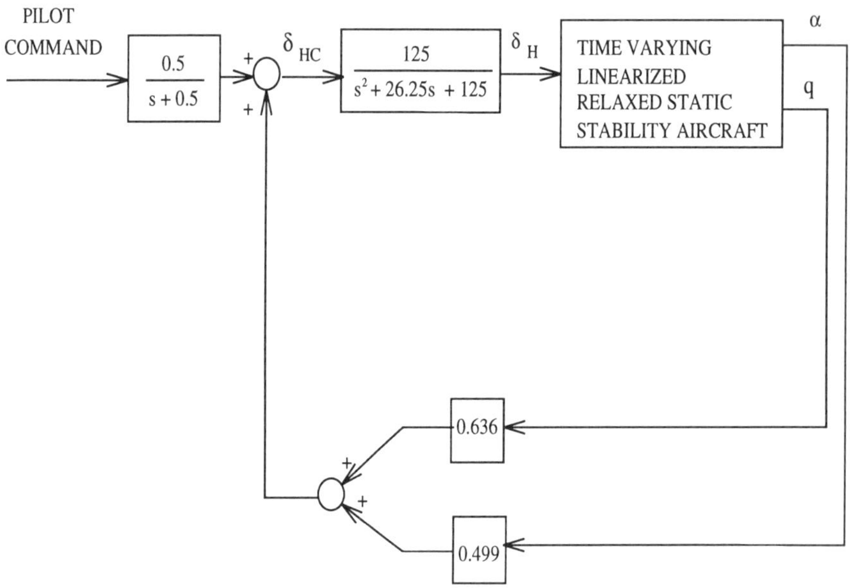

FIGURE 7.32. Fixed Gain Benchmark Controller

Adaptive Controller of Section 3.4

The algorithm described in Section 3.4 and denoted as Algorithm 2 is now utilized to design an adaptive controller for the time varying aircraft. This algorithm is defined by (3.136) with design method described by (3.260)–(3.265). The design method which is used here was illustrated for the Rohrs' Example in Section 3.4.3. First, the dynamic compensator is selected as:

$$G_c(s) = \frac{-10(s+1)}{s} \tag{7.71}$$

which can be shown to stabilize the plant when the parameters are frozen at either the takeoff, climb, or cruise flight conditions. No attempt will be made to determine whether or not this compensator stabilizes the time varying plant because the stability analysis in Section 3.4 is only valid for a linear time invariant plant.

In a manner similar to that which was used for the Rohrs' example in Section 3.4.3, (3.190)–(3.193) are used to obtain the parallel supplementary dynamics shown below:

$$\dot{x}_f(t) = x_f(t) - u_f(t) \tag{7.72}$$

$$y_f(t) = x_f(t) \tag{7.73}$$

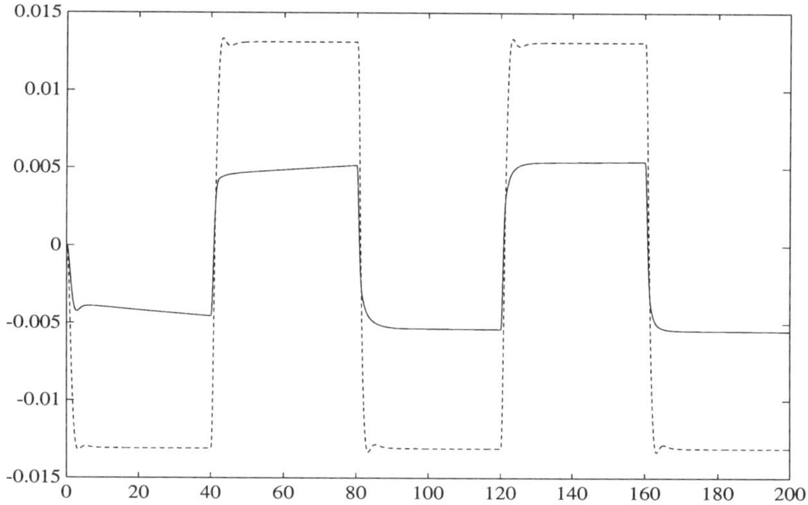

FIGURE 7.33. Plant and Model Pitch Rate, Fixed Gain Benchmark Controller

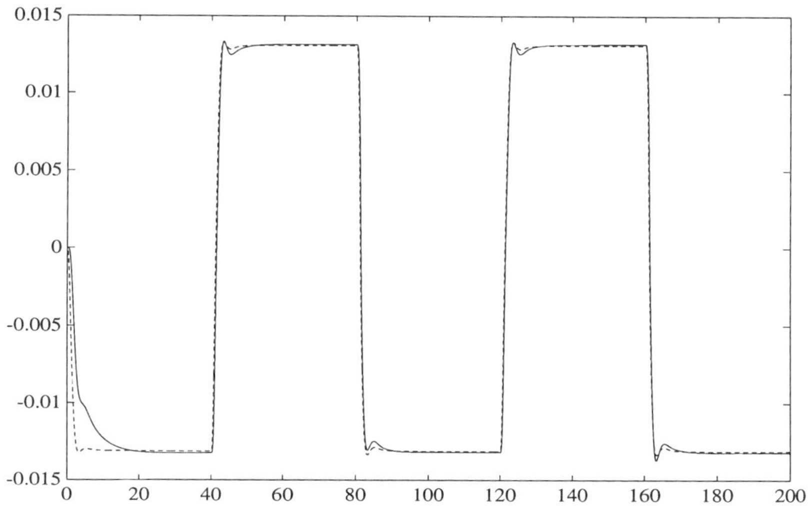

FIGURE 7.34. Plan and Model Pitch Rate, Adaptive Controller of Section 2.4

Since A_f is non-singular, Q_f can be arbitrarily selected as

$$Q_f = 20 \tag{7.74}$$

and then G and Q_p are chosen to satisfy

$$G = -Q_f A_f^{-1} B_f = -5.71428 \tag{7.75}$$

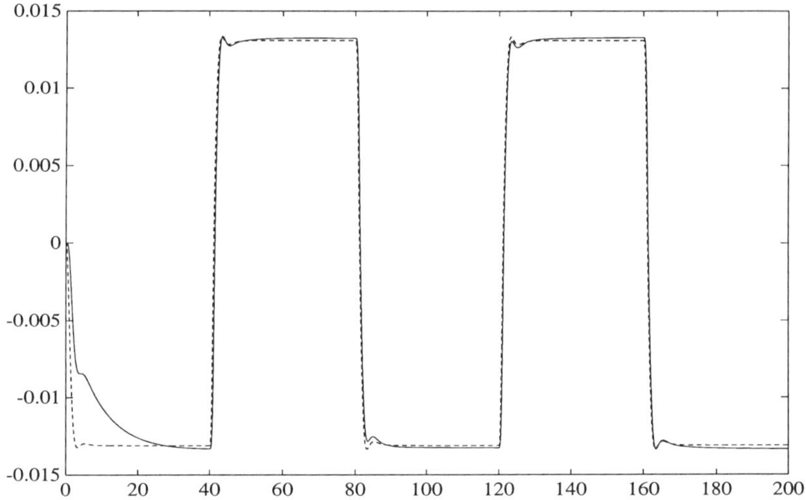

FIGURE 7.35. Plant and Model Pitch Rate, Adaptive Controller of Section 3.4

$$Q_p = GD_c = 57.1428 \tag{7.76}$$

Once again it was decided to match the aircraft pitch rate to the reference model pitch rate. The adaptive feedback gain K_e was again initialized at the value in the fixed gain benchmark controller which was computed for the takeoff flight condition. The weighting matrices were tuned by trial and error which resulted in the choice $T = \bar{T} = I$.

The simulated pitch rate using the adaptive algorithm is shown in Figure 7.35 where it can be observed that the pitch rate model following is similar to the response which was obtained with the algorithm of Section 2.4.

7.4.4 Conclusions

Two direct model reference adaptive control algorithms were utilized to design a stability augmentation system for the longitudinal dynamics of a relaxed static stability aircraft. The adaptive system's performance was evaluated for a scenario which includes takeoff, climb, and cruise flight conditions. The adaptive controller yields good performance after an initial learning period without the need for gain scheduling. Future work should consider combining the adaptive controller with a crude gain schedule for the purpose of eliminating the requirement for external air data probes.

References

[1] P. Wellstead and M. Zarrop, *Self–Tuning Systems.* Chichester, U.K.: Wiley, 1991.

[2] S. Sastry and M. Bodson, *Adaptive Control: Stability, Convergence, and Robustness.* Englewood Cliffs, NJ: Prentice-Hall, 1989.

[3] K. Narendra and A. Annaswamy, *Stable Adaptive Systems.* Englewood Cliffs, NJ: Prentice-Hall, 1989.

[4] G. Goodwin and K. Sin, *Adaptive Filtering Prediction and Control.* Englewood Cliffs, NJ: Prentice-Hall, 1984.

[5] K. Astrom and B. Wittenmark, *Adaptive Control.* Reading, MA: Addison Wesley, 1989.

[6] J. Maciejowski, *Multivariable Feedback Design.* Reading, MA: Addison Wesley, 1989.

[7] H. Bode, *Network Analysis and Feedback Amplifier Design.* Princeton, NJ: D. Van Nostrand, 1945.

[8] I. Horowitz, *Synthesis of Feedback Systems.* New York: Academic Press, 1963.

[9] L. Zadeh and C. Desoer, *Linear System Theory - The State Space Approach.* New York: McGraw–Hill, 1963.

[10] B. D. O. Anderson and J. B. Moore, *Linear Optimal Control.* Englewood Cliffs, NJ: Prentice Hall, 1971.

[11] H. Kwakernaak and R. Sivan, *Linear Optimal Control Systems.* New York: Wiley–Interscience, 1972.

[12] W. Hahn, *Stability of Motion.* New York: Springer–Verlag, 1967.

[13] M. Vidyasagar, *Nonlinear System Analysis.* Englewood Cliffs, NJ: Prentice Hall, 1992.

[14] J.-J. Slotine and W. Li, *Applied Nonlinear Control.* Englewood Cliffs, NJ: Prentice Hall, 1991.

[15] R. Mohler, *Nonlinear Systems.* Englewood Cliffs, NJ: Prentice Hall, 1991.

[16] H. Khalil, *Nonlinear Systems.* New York: MacMillan, 1992.

[17] I. Landau, *Adaptive Control - The Model Reference Approach.* New York: Marcel Decker, 1979.

[18] K. Astrom, "Theory and applications of adaptive control - a survey," *Automatica*, vol. 19, pp. 471–481, 1983.

[19] K. Sobel, H. Kaufman, and L. Mabius, "Model reference output adaptive control systems without parameter identification," *18th IEEE Conference on Decision and Control*, pp. 349–351, December 1979.

[20] I. Bar-Kana, "Adaptive control - a simplified approach," in *Control and Dynamic Systems - Advances in Theory and Applications* (C. Leondes, ed.), vol. XXV, pp. 187–235, 1987.

[21] I. Bar-Kana, "Almost passivity and simple adaptive control of nonstationary continuous linear systems," *International Journal of Systems Science*, vol. 21, pp. 835–862, 1990.

[22] I. Bar-Kana and A. Guez, "Simple adaptive control for a class of nonlinear continuous linear systems with application to robotics," *International Journal of Control*, vol. 52, pp. 77–99, 1990.

[23] I. Bar-Kana and A. Guez, "Simplified techniques for adaptive control of robots," in *Control and Dynamic Systems - Advances in Theory and Applications* (C. Leondes, ed.), vol. 40, pp. 147–203, 1991.

[24] I. Bar-Kana and H. Kaufman, "Simple adaptive control of uncertain systems," *International Journal on Adaptive Control and Signal Processing*, vol. 2, pp. 133–143, 1988.

[25] M. Aizerman and F. Gantmacher, *Absolute Stability of Regulator Systems.* San Francisco: Holden Day, 1964.

[26] A. Dewey and E. Jury, "A note on Aizerman's conjecture," *IEEE Transactions on Automatic Control*, vol. AC-19, pp. 482–483, 1963.

[27] A. Bergen, *Power Systems Analysis.* Englewood Cliffs, NJ: Prentice-Hall, 1986.

[28] I. Bar-Kana and R. Fischl, "A simple adaptive enhancer of voltage stability for generator excitation control," *Proceedings of American Control Conference*, 1992.

[29] I. Landau, "A survey of model reference adaptive techniques: Theory and applications," *Automatica*, vol. 10, pp. 353–379, 1974.

[30] R. Monopoli, "Model reference adaptive control with an augmented error signal," *IEEE Transactions on Automatic Control*, vol. AC-19, no. 5, pp. 474–484, 1974.

[31] K. Sobel, H. Kaufman, and L. Mabius, "Implicit adaptive control for a class of MIMO systems," *IEEE Transactions on Aerospace and Electronic Systems*, vol. AES-18, pp. 675–690, September 1982.

[32] P. Osborn, H. Whitaker, and A. Keezer, "New developments in the design of adaptive control systems," *Institute of Aeronautical Science*, 1961.

[33] C. Price, "An accelerated gradient method for adaptive control," in *Proceedings of the 9th IEEE Symposium on Adaptive Processes, Decision, and Control Conference*, pp. IV 4.1–IV 4.10, December 1970.

[34] R. Monopoli, "Model reference adaptive control with an augmented error signal," *IEEE Transactions on Automatic Control*, vol. AC-19, October 1976.

[35] A. Morse, "Global stability of parameter adaptive control systems," *IEEE Transactions on Automatic Control*, vol. AC-25, pp. 433–439, June 1980.

[36] A. Feuer and A. Morse, "Adaptive control of single–input, single–output linear systems," *IEEE Trans. Automatic Control*, vol. AC-23, pp. 557–569, August 1978.

[37] K. Narendra and L. Valavani, "Stable adaptive controller design–direct control," *IEEE Transactions on Automatic Control*, vol. AC-23, pp. 570–583, August 1978.

[38] K. Narendra and L. Valavani, "Direct and indirect model reference adaptive control," *Automatica*, vol. 15, pp. 653–664, November 1979.

[39] K. Narendra, Y.-H. Lin, and L. Valavani, "Stable adaptive controller design part II: Proof of stability," *IEEE Transactions on Automatic Control*, pp. 440–448, June 1980.

[40] C. Rohrs, L. Valavani, M. Athans, and G. Stein, "Robustness of continuous time adaptive control algorithms in the presence of unmodelled dynamics," *IEEE Transactions on Automatic Control*, vol. AC-30, pp. 881–889, September 1985.

[41] K. Astrom, "A commentary on the C.E. Rohrs et. al. paper, robustness of adaptive control algorithms in the presence of unmodelled dynamics," *IEEE Transactions on Automatic Control*, vol. AC-30, p. 889, September 1985.

[42] B. Peterson and K. Narendra, "Bounded error adaptive control," *IEEE Transactions on Automatic Control*, vol. AC-27, no. 2, pp. 1161–1168, 1982.

[43] G. Kreisselmeier and B. Anderson, "Robust model reference adaptive control," *IEEE Transactions on Automatic Control*, vol. AC-31, no. 2, pp. 127–133, 1986.

[44] P. Ioannou and P. Kokotovic, *Adaptive Systems with Reduced Models.* Berlin: Springer–Verlag, 1983.

[45] P. Ioannou, "Robust adaptive controller with zero residual tracking errors," *IEEE Transactions on Automatic Control*, vol. AC-31, pp. 273–276, August 1986.

[46] P. Ioannou and K. Tsakalis, "A robust direct adaptive controller," *IEEE Transactions on Automatic Control*, vol. AC-31, pp. 1033–1043, November 1986.

[47] D. Lindorff and R. Carroll, "Survey of adaptive control using liapunov design," *Int. J. Control*, vol. 18, no. 5, pp. 897–914, 1973.

[48] H. Erzberger, "On the use of algebraic methods in the analysis and design of model following control systems," Tech. Rep. D-4663, NASA, 1963.

[49] I. Bar-Kana, *Direct Multivariable Model Reference Adaptive Control with Applications to Large Structural Systems.* Ph.D thesis, Rensselaer Polytechnic Institute, 1983.

[50] I. Bar-Kana and H. Kaufman, "Some applications of direct adaptive control to large structural systems," *Journal of Guidance, Control, and Dynamics*, vol. 7, no. 6, pp. 717–724, 1984.

[51] I. Bar-Kana, H. Kaufman, and M. Balas, "Model reference adaptive control of large structural systems," *Journal of Guidance and Control*, vol. 6, no. 2, pp. 112–118, 1983.

[52] D. Bayard, C. Ih, and S. Wang, "Adaptive control for flexible space structures with measurement noise," *Proceedings of the 1987 American Control Conference*, pp. 368–379, 1987.

[53] C.-H. Ih, S. Wang, and C. Leondes, "An investigation of adaptive control techniques for space stations," *Proceedings of American Control Conference*, pp. 81–94, 1985.

[54] C.-H. Ih, S. Wang, and C. Leondes, "Application of adaptive control to space stations," *Proceedings of AIAA Guidance and Control Conference*, pp. 709–724, 1985.

[55] C.-H. Ih, S. Wang, and C. Leondes, "Adaptive control for the space station," *IEEE Control Systems Magazine*, pp. 29–34, 1987.

[56] F. Lee, I. Fong, and Y. Lin, "Decentralized model reference adaptive control ofr large flexible structures," *Proceedings of American Control Conference*, pp. 1538–1544, 1988.

[57] Y. Yan, S. Wang, and D. Boussalis, "Large flexible structure control technology experiment and facility design," *IFAC Proceedings Series*, no. Control of Distribtued Paramter Systems, 1987.

[58] I. Bar-Kana, "Parallel feedforward and simplified adaptive control," *International Journal of Adaptive Control and Signal Processing*, vol. 1, no. 2, pp. 95–109, 1987.

[59] D. Meldrum and M. Balas, "Application of model reference adaptive control to a flexible remote manipulator arm," *Proceedings of 1986 American Control Conference*, pp. 825–832, 1986.

[60] J. van Amerongen and A. U. Ten Cate, "Model reference adaptive-controller for ships," *Automatica*, vol. 11, pp. 441–449, 1975.

[61] I. Bar-Kana, "Comments on a paper by Kidd," *International Journal of Control*, vol. 48, no. 3, 1988.

[62] F. Fortunato and H. Kaufman, "Model reference adaptive control of thyristor driven DC motor systems with resonant loading," *Proceedings of 23rd IEEE Conference on Decision and Control*, pp. 375–380, 1984.

[63] D. Platzer and H. Kaufman, "Model reference adaptive control of thyristor driven DC motor systems subject to current limitations," *Proceedings of the IFAC 9th Triennial World Congress*, pp. 231–235, 1984.

[64] F. D. Busto, E. Sanchez, R. Urbietta, and H. Kaufman, "Application of a model reference adaptive algorithm to combustion control of a power unit boiler," *Proceedings of American Control Conference*, pp. 618–619, 1985.

[65] G. Neat and H. Kaufman, "Hierarchical adaptive control system design and application," *9th IFAC/IFORS Symposium on Identification and System Parameter Estimation*, pp. 573–578, July 1991.

[66] K. Sobel and H. Kaufman, *Direct Model Reference Adaptive Control for a Class of MIMO Systems*, vol. XXIV, ch. Advances in Control and Dynamic Systems, pp. 246–314. Academic Press, 1986.

[67] E. Sanchez, "Adaptive control robustness in flexible aircraft application," *Proceedings of the 1986 American Control Conference*, pp. 494–496, June 1986.

[68] W. Morse and K. Ossman, "Model following reconfigurable flight control system for the AFTI/F-16," *AIAA J. Guidance*, vol. 13, pp. 969–976, November-December 1990.

[69] I. Bar-Kana and H. Kaufman, "Robust simplified adaptive control for a class of multivariable continuous systems," *24th IEEE Conference on Decision and Control*, pp. 141–146, 1985.

[70] I. Bar-Kana and H. Kaufman, "Global stability and performance of a simplified adaptive algorithm," *Int. J. Control*, vol. 42, no. 6, pp. 1491–1505, 1985.

[71] I. Bar-Kana, "Positive realness in multivariable continuous–time systems," *Journal of the Franklin Institute*, vol. 328, no. 4, pp. 403–418, 1991.

[72] H. Kaufman, G. Neat, and R. Steinvorth, "Asymptotically stable MIMO direct MRAC for processes not measuring satisfying a positive real constraint," *Proceedings of First European Control Conference*, pp. 1813–1822, July 1991.

[73] R. Steinvorth, G. Neat, and H. Kaufman, "Direct model reference adaptive control with application to flexible robots," *IASTED Conference on Adaptive Control and Signal Processing*, pp. 9–12, October 1990.

[74] W. Su and K. M. Sobel, "A unified theory for the CGT approach to adaptive control," *International Journal of Control*, 19–.

[75] K. Narendra, R. Ortega, and P. Dorato, *Advances in Adaptive Control.* New York: Institute of Electrical and Electronics Engineers, 1991.

[76] M. O'Brien and J. Broussard, "Feed–forward control to track the output of a forced model," in *Proceedings of the 17th IEEE Conference on Decision and Control*, (San Diego), pp. 1149–1155, January 1979.

[77] R. E. Skelton, *Dynamic Systems Control.* New York: John Wiley, 1988.

[78] E. Davison and S. Wang, "Properties of linear time–invariant multivariable systems subject to arbitrary output and state feedback," *IEEE Transactions on Automatic Control*, vol. AC-18, p. 24, 1974.

[79] A. Barraud, "A numerical algorithm to solve $A^TXA - X = Q$," *IEEE Transactions on Automatic Control*, vol. AC-22, no. 5, p. 883, 1977.

[80] C. Rohrs, L. Valavani, M. Athans, and G. Stein, "Robustness of adaptive control algorithms in the presence of unmodeled dynamics," in *Proceedings of 21st IEEE Conference on Decision and Control*, (Orlando, FL), pp. 3–11, 1982.

[81] I. Bar Kana, "Rohrs' examples and robust stability of simple adaptive control," *Transactions of ASME, Journal Dynamic Systems, Measurements and Control*, pp. 721–728, 1991.

[82] I. Bar Kana, "On the Lur'e problem and stability on nonlinear controllers," *The Journal of the Franklin Institute*, vol. 325, no. 6, pp. 687–693, 1988.

[83] J. LaSalle, "Stability of nonautonomous systems," *Nonlinear Analysis Theory Methods and Applications*, vol. 1, no. 1, pp. 83–88, 1981.

[84] B. D. O. Anderson and S. Vongpanitlerd, *Network Analysis and Synthesis: A Modern Systems Theory Approach.* Englewood Cliffs, NJ: Prentice Hall, 1973.

[85] J. Wen, "Time domain and frequency domain conditions for strict positive realness," *IEEE Transactions on Automatic Control*, vol. 33, pp. 988–992, October 1988.

[86] J. Willems, "Dissipative dynamical systems," *Archive for Rational Mechanics and Analysis*, vol. 45, no. 5, pp. 321–393, 1972.

[87] P. Ioannou and G. Tao, "Frequency domain conditions for strictly positive real functions," *IEEE Transactions on Automatic Control*, vol. AC-32, pp. 53–54, 1987.

[88] M. C. M. Texeira, *Real Positive Systems and Adaptive Control.* PhD thesis, Pontificia Universidade Catolica, Rio de Janeiro, Brazil, May 1989.

[89] G. Gu, "Stabilizability conditions of multivariable uncertain systems via output feedback control," *IEEE Transactions on Automatic Control*, vol. 35, no. 8, pp. 925–927, 1990.

[90] L. Mabius, *An Implicitly Adaptive Model Reference Control Design for Linear Multi-Input Multi-Output Systems.* PhD thesis, Rensselaer Polytechnic Institute, Troy, NY, June 1976.

[91] D. Siljak, "New algebraic criteria for positive realness," in *Proceedings Annu. Princeton Conference on Information Science Systems*, pp. 329–335, March 1970.

[92] A. Vicino and A. Tesi, "Robust strict positive realness: New results for interval plant plus controller families," in *Proceedings of the 30th Conference on Decision and Control*, (Brighton England), pp. 421–426, 1991.

[93] H. Chapellat, M. Dahleh, and S. Bhattacharyya, "On robust nonlinear stability of interval control systems," *IEEE Transactions on Automatic Control*, vol. 36, pp. 59–67, January 1991.

[94] L. Hitz and B. D. O. Anderson, "Discrete positive real functions and their application to system stability," in *Proceedings of the IEE*, pp. 153–155, January 1969.

[95] T. Kailath, *Linear Systems*. Englewood Cliffs, NJ: Prentice Hall, 1980.

[96] I. Bar-Kana, "Parallel feedforward and simplified adaptive control," *International Journal of Adaptive Control and Signal Processing*, vol. 1, no. 2, pp. 95–109, 1987.

[97] L. Abida and H. Kaufman, "Model reference adaptive control for linear time varying and nonlinear systems," in *21st IEEE Conference on Decision and Control*, (Orlando), pp. 287–288, 1982.

[98] W. Cauer, *Synthesis of Linear Communication Networks*. New York, NY: McGraw–Hill, 1958.

[99] W. L. Brogan, *Modern Control Theory*. Englewood Cliffs, NJ: Prentice Hall, 1991.

[100] J. LaSalle and S. Lefshetz, *Stability by Lyapunov's Direct Method with Applications*. New York: Academic Press, 1961.

[101] J. LaSalle, "Stability theory for ordinary differential equations," *Journal of Differential Equations*, vol. 4, pp. 57–65, 1968.

[102] I. Bar Kana and H. Kaufman, "Direct adaptive control with bounded tracking errors," in *Proceedings of the 22nd IEEE Conference on Decision and Control*, (San Antonio, TX), pp. 181–182, December 1983.

[103] R. R. Fortesque, L. S. Kershenbaum, and B. D. E. Ydstie, "Implementation of self-tuning regulators with variable forgetting factors," *Automatica*, vol. 17, pp. 831–835, 1981.

[104] L. Hsu and R. R. Costa, "Bursting phenomena in continuous - time adaptive systems with σ - modification," *IEEE Transactions on Automatic Control*, vol. AC-32, no. 2, pp. 84–85, 1987.

[105] B. D. O. Anderson, "Adaptive systems, lack of persistency of excitation and bursting phenomena," *Automatica*, vol. 21, pp. 247–258, 1985.

[106] I. M. Y. Mareels and R. R. Bitmead, "Nonlinear dynamics in adaptive control - chaotic and periodic stabilization," *Automatica*, vol. 22, pp. 641–655, 1987.

[107] I. Bar-Kana, "Absolute stability and robust discrete adaptive control of multivariable systems," vol. XXXI, ch. Control and Dynamic Systems - Advances in Theory and Applications, pp. 157–183, Academic Press, 1989.

[108] I. Bar-Kana, "Bursting phenomena without external excitation," in *Proceedings 1900 Conference on Information Sciences and Systems*, (Princeton, New Jersey), pp. 259–264, March 1990.

[109] G. Rey, R. R. Bitmead, and C. R. Johnson, "On bursting in adaptive systems," in *Proceedings of 1989 Conference on Information Sciences and Systems*, (Baltimore, Maryland), pp. 78–83, 1989.

[110] B. D. O. Anderson, "Quadratic minimization, positive-real matrices and spectral factorization," Technical Report EE-6812, University of Newcastle, New South Wales, Australia, 1968.

[111] D. J. Hill and P. Moylan, "Dissipative dynamical systems basic input-output properties," *J. of the Franklin Institute*, vol. 309, no. 5, pp. 327–357, 1980.

[112] C. A. Desoer and M. Vidyasagar, *Feedback Systems: Input - Output Properties.* Orlando, Florida: Academic Press, 1975.

[113] I. Bar-Kana, "Positive realness in discrete-time adaptive control systems," *Int. J. Systems Sciences*, vol. 17, no. 7, pp. 1001–1006, 1986.

[114] I. Bar-Kana, "Positive realness in multivariable stationary linear systems," *J. Franklin Institute*, vol. 328, pp. 403–417, 1991.

[115] J. M. Schumacher, "Almost stabilizability subspaces and high gain feedback," *IEEE Trans. Automatic Control*, pp. 620–628, 1984.

[116] A. MacFarlane, *Design of Modern Control Systems*, ch. Complex Variable Methods in Feedback Systems Analysis and Design. London: Peter Peregrinus, 1982.

[117] I. Bar-Kana and A. Guez, "Neuromorphic adaptive control," in *Proceedings of the 28th IEEE Conference on Decision and Control*, (Tampa, Florida).

[118] M. Fu, A. Olbrot, and M. Polis, "Introduction to the parametric approach to robust stability," *IEEE Control Systems Magazine*, pp. 7–11, August 1989.

[119] B. Barmish, "Extreme point results for robust stabilization of interval plants with first order compensators," *IEEE Transactions on Automatic Control*, vol. 37, pp. 707–714, June 1992.

[120] C. Chieh-Li, T.-C. Yang, and N. Munro, "Output feedback pole assignment procedure," *International Journal of Control*, vol. 48, pp. 1503–1518, 1988.

[121] B. D. O. Anderson and J. B. Moore, *Optimal Control.* Englewood Cliffs, NJ: Prentice Hall, 1990.

[122] R. Steinvorth and H. Kaufman, "Direct model reference adaptive control of robots," *CISS*, pp. 667–672, March 1991.

[123] B. Armstrong, O. Khatib, and J. Bardick, "The explicit dynamic model and inertial parameters of the puma 560 arm," in *Proceedings of the 1986 IEEE Robotics and Automation Conference*, (San Francisco, CA), pp. 510–518, 1986.

[124] C.-H. Ih, D. S. Bayard, S. J. Wang, and D. B. Eldred, "Adaptive control experiment with a large flexible structure," *AIAA Guidance, Navigation, and Control Conference*, August 15-17 1988.

[125] C.-H. Ih, D. S. Bayard, A. Ahmed, and S. J. Wang, "Experiments in multivariable adaptive control of a large flexible structure," *AIAA J. of Guidance, Control and Dynamics*, vol. 16, pp. 14–20, January-February 1993.

[126] C.-H. Ih, D. S. Bayard, A. Ahmed, and S. J. Wang, "Experimental study of robustness in adaptive control for large flexible structures," *AIAA J. of Guidance, Control and Dynamics*, vol. 16, pp. 9–13, January-February 1993.

[127] D. Boussalis, D. S. Bayard, C. Ih, S. J. Wang, and A. Ahmed, "Experimental study of adaptive pointing and tracking for large flexible space structures," in *AIAA Guidance and Control*, vol. AIAA-91-2691-CP, (New Orleans, Louisiana), pp. 769–778, AIAA Guidance and Control Conference, August 1991.

[128] I. Bar-Kana, "Robust simplified adaptive control of large flexible space structures," in *Proc. Fourth IFAC Symposium on Control of Distributed Parameter Systems*, (Los Angeles, CA), UCLA, 1986.

[129] R. Livneh, *Direct Multivariable Adaptive Controller with Application to Wing Flutter.* PhD Dissertation, University of Cincinnati, Cincinnati, OH, 1989.

[130] D. S. Bayard, C.-H. Ih, and S. J. Wang, "Adaptive control for flexible structures with measurement noise," in *Proceedings of the American Control Conference*, (Minneapolis, Minnesota), pp. 368–379, June 10-12, 1987.

[131] C. Ih, D. S. Bayard, and S. J. Wang, "Space station adaptive payload articulation control," in *Fourth IFAC Symposium on Control of Distributed Parameter Systems*, (Los Angeles, CA), July 1986.

[132] I. Bar-Kana, H. Kaufman, and M. J. Balas, "Model reference adaptive control of large space structures," *J. of Guidance, Control and Dynamics*, vol. 6, no. 2, pp. 112–118, 1983.

[133] I. H. Mufti, "Model reference adaptive control for large structural systems," *AIAA J. Guidance, Control and Dynamics*, pp. 507–509, September-October 1987.

[134] D. S. Bayard, "Adaptive stabilization and control of large space structures with rigid body dynamics," Internal Document EM 347-87-210, Jet Propulsion Laboratory, 1987.

[135] D. S. Bayard, "Transient analysis of an adaptive system for optimization of the design parameters," *IEEE Transactions on Automatic Control*, vol. 37, pp. 842–848, June 1992.

[136] D. S. Bayard, "An averaging approach to optimal adaptive control of large space structures," in *Advances in Adaptive Control* (K. S. Narendra, R. Ortega, and P. Dorato, eds.), (New York), IEEE Press, 1991.

[137] R. Clough and J. Penzien, *Dynamics of Structures.* New York: McGraw-Hill, 1975.

[138] L. Meirovitch, *Analytical Methods in Vibrations.* New York: Macmillan, 1967.

[139] G. Strang and G. Fix, *An Analysis of the Finite Element Method.* Englewood Cliffs, NJ: Prentice Hall, 1973.

[140] J. Wen, *Direct Adaptive Control in Hilbert Space.* PhD Thesis, Rensselaer Polytechnic Institute, Troy, NY, June 1985.

[141] M. J. Balas, H. Kaufman, and J. Wen, "Stable direct adaptive control of linear infinite-dimensional systems using a command generator tracker approach," in *Proceedings of the Workshop on Identification and Control of Flexible space Structures*, no. 85-29 in JPL Publication, pp. 127–144, JPL, April 1985.

[142] R. J. Benhabib, R. P. Iwens, and R. Jackson, "Stability of large flexible structure control systems using positivity concepts," *J. Guidance and Control*, no. 5, pp. 487–494, 1981.

[143] H. C. Vivian, "Flexible structure control laboratory development and technology demonstration," Final Report Prepared for USAF-AFAL and NASA-OAST JPL Publication 88-29, Jet Propulsion Laboratory, October 1987.

[144] J. B. Slate, *Model-Based Design of a Controller for Infusing Sodium Nitroprusside During Postsurgical Hypertension.* Phd Thesis, University of Wisconsin-Madison, 1980.

[145] L. A. Zadeh, "Fuzzy sets," *Information and Control*, vol. 8, pp. 338–353, 1965.

[146] L. A. Zadeh, "Outline of a new approach to analysis of complex systems and decision processes," *IEEE Transactions on Systems, Man and Cybernetics*, vol. SMC-3, pp. 28–44, 1973.

[147] W. G. He, H. Kaufman, and R. Roy, "Multiple model adaptive control procedure for blood pressure control," *IEEE Transactions on Biomedical Engineering*, vol. BME-33, no. 1, pp. 10–19, 1986.

[148] G. W. Neat, *Expert Adaptive Control: Method and Medical Application.* Phd Thesis, Rensselaer Polytechnic Institute, Troy, NY, 1990.

[149] P. A. Ioannou and P. V. Kokotovic, *Adaptive Systems with Reduced Models.* Springer Verlag, 1983.

Index